Corporate Occupiers' Handbook

Simon Woodhead and Howard Cooke

2007

A division of Reed Business Information

Estates Gazette
1 Procter Street, London WC1V 6EU

ISBN 978-0-7282-0511-6

Cover design by Rebecca Caro/Photo courtesy of Rex Images
Typeset in Palatino 10/12 by Amy Boyle
Printed by Short Run Press

Contents

Table of Cases

Table of Statutes

Page numbers in **bold** refer to the appendices

Preface

It may be asked why we have written a book aimed at the corporate occupier. Indeed who or what is a corporate occupier?

Corporate occupier is a term which describes an occupational tenant, or an occupier of freehold premises, where the occupation is for the purposes of a business. While many corporate occupiers are tenants, and one of the principal purposes of this book is to address common issues which arise from occupational leases, the term is much broader. A corporate occupier may be a freeholder that occupies premises for its business. It may be a tenant which occupies the premises for the purposes of it operations. Throughout the book we refer to this context as the operational context. However there is a second perspective which a corporate occupier can face, where the freehold or leasehold premises become surplus to the corporate occupiers needs. In this scenario, and in particular where the premises are held on a lease with many years to run, the corporate occupier may find its roles reversed and become a 'reluctant landlord'. If the lease is overrented making assignment difficult, unless the lease contains a break clause the corporate occupier may have no option but to sub-let the premises to generate an income stream to offset against head-rent. The corporate occupier will then have to manage out the premises until lease expiry unless a surrender deal can be negotiated with the superior landlord. If the corporate occupier has a number of surplus properties held on leases, this is often referred to as a 'legacy portfolio'.

The corporate occupier's approach to property is to treat it as a point of production or the place where it provides services. It can also be an important statement about the nature of the business and is where a business invests in its people as well as in the building and plant. With the exception of property companies, property is not the end in itself but it is an environment where the business operates from, although with increasing home working, the business environment it broadening. A corporate occupier is a term which relates to the business and is not restricted to a company. It can be a partnership or a sole trader. It can range from a part-time sole trader at one end of the spectrum, to multinational companies or governments at the other. While there will be differences between corporate occupiers, all share some similar characteristics particularly in relation to the issues which arise out of their need to occupy property.

Property remains for most corporate occupiers the second largest cost which a business has to carry. The highest cost is people. However, in the UK the cost of labour is a more flexible cost base than property. Generally speaking, it is still a fairly straightforward process to hire and fire people. The costs of such a process can be determined within very narrow parameters. Property is much more inflexible whether taking space or looking to dispose of surplus space. Therefore, when this is taken into account, property can out rank labour as the greatest cost a business has to bear.

Property journals and books on commercial property tend to be dominated by articles and books written from an investment landlord's perspective. A glance at any property journal in a typical week will reveal numerous articles on lettings by landlords, investment acquisitions/disposals or development projects. The corporate occupier gets very little coverage. A corporate occupier will want to make its mark in relevant industry and trade journals and not in property journals. In our experience that lower profile has generally resulted in a poorer understanding by many management surveyors and lawyers about the property perspective of corporate occupiers than might otherwise be the case. In particular, there is often little awareness of the factors which influence a corporate occupier's decisions on property issues, and the strategies (both long, medium and short term) which will best suit its business objectives. This can result in a corporate occupier's overall property costs being considerably higher than necessary, particularly in relation to surplus properties held on long overented leases. In our experience, there is an opportunity for management surveyors and lawyers to make a very real difference to a corporate occupier's property costs.

This book is principally aimed at corporate occupiers, their management surveyors and legal advisors. It deals for the most part with common issues which arise out of occupational leases and the particularly difficult problems thrown up by surplus leases. The message is a very simple one. A corporate occupier can in many cases reduce its overall property costs significantly by taking care when negotiating its lease, and thereafter devising strategies in good time at key points in the lease journey ie when carrying out alterations; assigning or sub-letting; operating break clauses; at rent reviews and lease renewals and last but by no means least, when dealing with dilapidation liabilities. It is increasingly important that management surveyors who act for corporate occupiers have a broad view of their task. They need to draw on the skills of a management surveyor, investment surveyor, rent review expert, accountant and lawyer. Most importantly, the management surveyor must understand the corporate occupier's business and its needs.

This book is not, however, the preserve of the corporate occupier. Often a lack of understanding by the landlord and its advisors can lead to frustration and a break down in its relationship with the tenant. In the transactional context this can mean that a deal unravels after many months of painstaking negotiations. A key purpose of this book is therefore to offer some insight to the landlord and its advisors about the corporate occupier's perspective.

The book is broadly structured around the life cycle of a lease starting with a consideration of the business decision to take new property, and the factors which should be taking into account when reaching a decision on this important step. We have endeavoured to provide a broad and comprehensive discussion of each topic, with reference to specific issues which may affect a corporate occupier. Inevitably, given the breadth of the topics discussed and the constraints on space, we have not been able to cover every issue. We have attempted where possible to keep the text practical including suggested precedents and prescribed forms at the end of each chapter. For completeness and ease of reference, copies of the relevant sections from the principal statutes are contained at the end of the book.

A book of this nature is inevitably the result of a long gestation period and the consequence of considerable (sometimes bitter) experience. To that end we would like to thank all those with whom we have worked over the years and who have helped in shaping our views. A particular vote of thanks must go to Guy Fetherstonhaugh QC who has on many occasions guided our clients expertly through some very difficult and often stormy waters. We would also like to thank Simon's colleagues, Marcos Toffanello and John Scannell for their support, and a special debt of gratitude to Alison Flood who has provided substantial help throughout, and in proof reading every word, has saved us from making countless mistakes. Finally, and most importantly, our thanks must go to our clients without which there would be no book.

Any views expressed in the book are entirely ours (and not necessarily those of our respective firms), and any mistakes are also ours. We have set out to summarise the current state of the law as at 31 July 2007.

There is one further group of people who deserve a last word. Our families probably thought that we had gone quite mad in undertaking this project. To Sarah, William and James, and Gillian, Joshua, Olivia and Jacob — our love and thanks — it was only temporary insanity.

Simon Woodhead
Birketts LLP

Howard Cooke
GVA Grimley LLP

31 July 2007

1 The Business Decision

1.1 Introduction

Within a company very few business decisions are taken for property reasons alone. This is because the property decision will follow on from the business decision. It is the business that drives a company forward not property. Ideally the property strategy will follow the business strategy. Property is only the purpose of a business and its *raison d'etre* when the company is a property company (ie a company whose sole business is in owning and managing property) or alternatively is a property developer, whose business is developing new property and/or refurbishing old property. For other companies property is part of the fabric of the business and it is the point of service delivery or point of production, whether that is, for example, the production of widgets or providing insurance polices. It therefore ranks alongside plant, machinery and desks as an element that facilitates the production process or service delivery. For the most part few corporate occupiers will spend time contemplating their property portfolio other than to look at reducing its costs, realising value from it or seeking to determine how they can improve their business by making property changes.

It follows that when making property decisions the corporate occupier will be looking first and foremost at what is needed for the business. If a new manufacturing line is going to be developed, should it be housed in an existing facility, in a new building on an existing site or on a brand new site? The key focus will be the delivery of the finished products to the customers and not on the property *per se*. For example, will the new location result in improved transportation of the goods compared with the existing site? What will the lead-time and total cost impact be for a new building compared to stripping out a line from an existing one? If an existing office can be refurbished what will be lost in output while the work is being carried out compared to moving to a new building? After considering business issues first, a corporate occupier may consider the property consequences. There are times, however, when the decision is all about the property. This may be because property is used to generate value through a sale-and-leaseback or possibly because of the appearance of the property. This tends to be the case when the decision to take more space is driven by the status it affords to the company.

Businesses tend to be in a constant state of change. They will be growing or contracting; acquiring or disposing of other businesses; innovating and developing or static and deteriorating. Where a company is static, in reality it is contracting. The market place for all businesses is now a global one. No one has the luxury of standing still anymore; otherwise they start to fail.

To cope with change, all the factors of production or service delivery need to be flexing and responding to changing business needs and ideally need to be capable of relatively short term adjustment. This is the case with the flow of materials and in most markets it is the way with labour costs. However, it is not always the case. A number of countries have employment protection policies which inhibit short-term changes. The UK does afford protection for staff but it is comparatively limited when compared with countries such as France and Germany. Consequently, in the UK, corporate occupiers can continually change their people and the sole issue is one of cost (either recruitment and training costs or redundancy costs). In contrast, property in the UK is not a flexible resource and it is considerably less flexible than in other countries. The decision-making process which a corporate occupier undertakes before acquiring new space must, therefore, take this factor into account. The process of acquiring and disposing of property in the UK is relatively slow; both for freehold and leasehold property, and the ability to terminate a lease early is very limited without the existence of a break clause in the lease. Consequently, and in general terms, the UK has flexible employment and inflexible property; Europe has inflexible employment and flexible property and the USA has flexible employment and flexible property.

1.2 Property

When considering what role property fulfils in a business and the reasons why the role changes there are a number of factors to take in to account.

1.2.1 Property functions

Property fulfils a number of functions:

- a location of manufacturing — where the manufacturing of goods takes place
- a point of service delivery — eg the repair of a product, the answering of customer enquiries
- an investment focus — where money, in all its forms, is invested in people and machinery
- it provides a statement about the business — eg a functional business, a high quality business, an innovative business
- a matrix within which social interaction takes place — increasingly people are the rarest resource of a business and so they need to be treated as such. Property will provide a place in which people can interact with each other to develop the common knowledge for the business (a function that is becoming progressively important).
- a marketing medium — eg posting adverts on a building.

While there have been some changes over the last decade in the way businesses use property, there has not been a fundamental change across the board. Advancements in technology have allowed a more fluid workforce, most prominently in the office sector. Improvements in computers and the spread of broadband have allowed people to work from alternative locations, in particular from home. As wireless networks in public locations have increased it has offered fluidity to the working environment with people able to work in coffee shops and the like.

This change dynamic is going to increase over time. It will be driven in the short-term less by improvements in technology but more by failures in the transport system and peoples' reluctance to spend time travelling. The explosion in travelling (primarily by car) over the last 30 years to and from work, as well as for work, is likely to be reversed as the transport systems fail. Increasingly traffic jams are grid locking the road system and the railways are not a viable alternative because of poor quality, lack of capacity, and overcrowding. The lead in time for major infrastructure projects will mean that it is 10 to 15 years before any major changes take place. These transport issues will be the push to change as employers start to recognise the need to change working practices so that they can maximise their resources. Technology will continue to evolve at a rapid pace, but it is likely to act as an empowerer, not as the driving force for change.

All of the functions that property fulfils require investment by a business, either directly eg fitting the premises out, or indirectly with costs such as installing plant and machinery or removal costs. Consequently, a decision by a corporate occupier whether to lease property has to be thought through very carefully. However, in practice, property decisions are seldom properly thought through. Often the decision maker is not someone who understands fully the ramifications of the decision on the business.

1.2.2 The need to change

The decision to acquire property will be the result of one or more of three drivers:

- replacement — this involves finding another property to replace an existing one, possibly because of a lease expiry with the current property
- enhancement — this stems from changing an old building for a new one, or moving to a better location or a larger property
- obligation — this arises when the corporate occupier is forced into a change, for example when a property no longer complies with health and safety requirements, or when the lease has expired and the landlord has decided to redevelop the site.

There are times when all three apply. A good example is a corporate occupier that is moving to new premises to improve the space it occupies because the current premises do not fully comply with current legislation, (such as the Disability Discrimination Act 1995) and the change is being made on lease expiry.

1.3 Decision making

No two businesses are alike even in the same sector. Their journey to a particular point will have followed different paths and as a consequence they will have different 'baggage' from the past. Importantly, there will be different personalities involved in both the past and the current decision-making process. Nevertheless, there are some common issues that arise.

1.3.1 Investment considerations

When a corporate occupier is considering an investment in premises it should have regard to some or all of the following issues:

- What is the decision in response to?
- What is the 'no action' choice — what will happen if no action is taken and what impact will it have on the business?

- What are the corporate occupier's competitors doing and what will be their response to the decision made?
- What are the alternatives to taking additional space?
- Will this be the best choice of expenditure for the business unit or the business overall?
- How does this decision sit with a corporate occupier's overall property strategy?

1.3.2 Decision making criteria

The decision making process needs to be based on a quantitative and qualitative analysis of the project. It is important to have an analysis of the hard facts and figures which can be incorporated into a calculation that determines the worth of the proposed action. Typically this analysis will be carried out by using the techniques of Payback Period or Discounted Cashflow (DCF), both of which are explained in detail below. However, the modelling of costs and projected income provides only one dimension. There are other factors that are less well suited to analysis by means of a DCF and are more subjective and qualitative in their output. These factors also need to be taken in to account.

The decision to take on new space usually involves the following criteria.

- *Financial*: generally the common currency for all businesses, even those organisations that are non profit making bodies, is money. It provides a simple approach to determining the before and after position, as well as looking at the cost of investment in the proposal. This could be measured by changes to the asset base or the amount of time committed to a project.
- *Human resources*: what resources will be needed? How will those involved in the project be affected? For example, how many staff will be lost with a move of five miles? Will the new production process require a re-training of staff? Can the new office be used to improve motivation and reassure the staff as to the future?
- *Production*: what is the availability of the new plant and what are the lead times?
- *Operational*: what will the down time be with the change of buildings? As one line closes and a new one opens what will be the crunch points in productivity until the new line is in place and the scheme completed?
- *Sales and marketing*: what will be the impact of the decision? Does it provide a competitive advantage for the business?
- *IT*: how much software needs to be written to allow the project to start? What is the obsolescence profile for the cabling?
- *Business sector*: how will competitors react? Does it provide a competitive advantage for sufficient time to justify the cost and the risk?
- *Economic*: what is the prognosis for the country of domicile and what of the countries where the goods are sold? What impact will possible interest rate changes have on the investment decision?
- *Political*: what potential hurdles exist from closing down the production unit in a particular country or area? What political capital can be gained from the decision?

In making a decision on any proposal a business should have a set of criteria to judge the proposal against. These should be the same fundamental criteria for any business related decision so as to provide consistency and ensure that projects of totally different types are susceptible to a common assessment.

Such decision making should not be too rigid or narrow otherwise the business will not be able to innovate. The criteria need to be capable of taking into account subjective as well as objective factors.

Having set payback criteria is one approach but that needs to be allied to less rational considerations such as 'gut feel'. This can be the case when deciding to go forward with a new product or in taking a new property. The fundamentals need to be right but the decision making process involves risks which may also offer opportunities for businesses.

1.4 Appraisal techniques

It is not the intention here to consider the various forms of capital valuation for property, but to look at techniques that can be considered in the context of a business case which may involve property. There are a number of options to consider but in each case it is a matter of looking at the payback period for the investment.

1.4.1 Payback period

The initial assessment involves looking at how soon the original investment is repaid from profits generated from the activity. Consequently, if the cost of the investment is £100,000 and the profit is £40,000 pa the payback is two and a half years. It is a very simple and quick technique, but it does not take into account factors such as risk and the effect of the time value of money. Therefore most appraisal techniques are driven around DCFs, which are dealt with at 1.4.2 below.

A simple comparison between DCF and Payback demonstrates how they work. A manufacturer invests £40,000 in a piece of machinery that has a life span of six years with the following cashflow profile:

Time	*Cashflow*
Year 1	5,000
Year 2	5,000
Year 3	7,000
Year 4	8,000
Year 5	10,000
Year 6	10,000

Assuming there is no residual value to the machinery at the end of the sixth year, we can derive the following Payback Period model.

The first period of a cashflow is always regarded as zero because at that initial point there has not been any opportunity to earn a return.

Time	*Cashflow*	*Cumulative*
Year 0	–40,000	–40,000
Year 1	5,000	–35,000
Year 2	5,000	–30,000
Year 3	7,000	–23,000
Year 4	8,000	–15,000
Year 5	10,000	–5,000
Year 6	10,000	+5,000
Total	5,000	

The payback takes place in year 6 and notionally a £5,000 profit is made (a 12.5% return) ie

$$\frac{£5{,}000}{£40{,}000} \times 100\%$$

1.4.2 Discounted cashflow

There are two elements to a DCF.

- *The cashflow*: over the life of a project there will be expenditure and income, and the costs need to be netted off against each other for each period under consideration eg each year. The total of the cumulative cashflow is the net benefit of undertaking the project. This is the same approach as calculating a Payback Period; indeed there is no difference between the two at this point.
- *Discounting*: the difference between a DCF and the Payback model is when the discount element is taken into account which shows the time value of money. In other words, £1 today is worth more than £1 in a year's time because if £1 is placed in a building society today it will receive interest over the next 12 months. Consequently, if an investor puts £100 in a building society today and receives interest at 10% pa, in one years time the investor will have the original £100 plus 10% of £100 ie £10 making a total of £110. After two years the investor will have £110 plus 10% of £110, ie £11 making a total of £121. This same calculation will continue year on year.

  The alternative view, if the investor wants to receive £100 in one year's time, is to identify what will have to be put into a building society now to generate that sum in twelve months time. This is know as the present value of a future sum and is the reverse process of receiving interest. Consequently, where the calculation of the interest is £100*(1+i) we divide by it to produce £100/(1+i), which is

Future Sum	£100
Divide by (1+i)	1.10
Present Value	£90.91

  The discounting process works on the basis of the following formula:

$$\frac{1}{(1 + i)^n}$$

  Where i is the interest rate and n is number of years, which it accumulates. Consequently, the further away the return is on an investment the less worth it has in comparison to something that provides a return in the short term.

On the basis of the example given above, adopting a DCF approach with a discount rate of 10% produces different figures.

Time	*Cashflow*	*Discount Rate @10%*	*Discounted Cashflow*
Year 0	–40,000	1.00	–40,000
Year 1	5,000	0.91	4,545
Year 2	5,000	0.83	4,132
Year 3	7,000	0.75	5,259
Year 4	8,000	0.68	5,464
Year 5	10,000	0.62	6,209
Year 6	10,000	0.56	5,645
Total	5,000		–8,745

Consequently the investment makes a loss of £8,745, which is a return of minus 22%, when the true value of money is taken in to account:

$$\frac{-£8,745}{£40,000} \times 100\%$$

This is caused by the income stream being concentrated largely towards the end of the cashflow period. If the income was more concentrated in the early years there would be a positive cashflow total.

Aside from the cash payments and receipt, another major element is the discount rate that is adopted. If we look at an alternative discount rate of 5% we can see the effect:

Time	*Cashflow*	*Discount Rate @5%*	*Discounted Cashflow*
Year 0	–40,000	1.00	–40,000
Year 1	5,000	0.95	4,762
Year 2	5,000	0.91	4,535
Year 3	7,000	0.86	6,047
Year 4	8,000	0.82	6,582
Year 5	10,000	0.78	7,835
Year 6	10,000	0.75	7,462
Total	5,000		–2,777

While this project still makes a loss, it is much smaller (–7%). It follows that with a discount rate of circa 4% the project will break even.

There are two approaches to DCF's that need to be considered; Internal Rate of Return (IRR) and the Net Present Value (NPV) of the scheme. With an IRR the calculation looks at what the total return is over the life of the project, where as the NPV will look at the returns against a given discount rate. Using the same example, if the IRR is 8.0% then with a discount rate of 8% the NPV should be zero, although there may be minor mathematical variances. They are slightly different ways of looking at the same thing, and the choice between them depends on what suits the business best. The question is this: does the investor want to compare the return it gets on this compared to another project with a basic building society interest rate, or does the investor know what return it wants to achieve and so see what additional cash return it could get over and above that return.

In deciding which method to use there are issues which arise in the application of the techniques, especially IRR. To use DCFs the basic assumptions are that:

- future cashflows can be estimated with reasonable accuracy so that risk and uncertainty need not be taken into account
- the opportunity cost of capital is known
- the project involves an initial outflow of capital followed by inflows of money
- projects are independent of each other.

Clearly these assumptions are not always correct. At times of inflation, risk and uncertainty need to be considered and can make NPVs difficult to use because the value produced is an absolute one, whereas the IRR is a rate of return. Indeed the rate of return is helpful in business as it allows easy comparison. In contrast the benefit of the NPV is that it does produce an absolute number. When the IRRs are fairly close (say 8.0% and 8.2%), understanding the real difference can be difficult, whereas under an NPV of 8% the absolute difference can be seen in £s.

The techniques do not automatically produce the same results because there are cases were cashflows are not simple investments. There can be negative cashflows during the course of the project rather than just at the beginning. These negative cashflows (for example caused by refurbishment works in year 4) can give misleading results under the IRR method. Finally, it is rare for investment projects to be mutually exclusive; often the technique is used to choose between the two projects. A worked example of a DCF can be found at 1.7. This has been calculated as both a NPV and an IRR to show the difference.

The techniques do not provide an answer in themselves. They provide an output that has to be considered. For example it is possible to get two projects to come out with an IRR of 10%, but the reality is that they are two dissimilar projects and carry different risk profiles. It would be wrong therefore not to consider other factors.

One of the great benefits of the DCF model is that all the parameters can be expressly identified so that carrying out a sensitivity analysis is relatively easy. If the key components are changed by plus and minus 5% and 10% from the assumed figures the effect on the feasibility can be readily seen. Looking at two projects returning 10%, if one has a low degree of sensitivity to changes in the key parameters while the other sees major changes then the probability of those variances needs to be considered. How those variances affect the project will be important, in particular if it is a negative impact. An increase of 5% in production costs reduces the return by 20%; where as a reduction of 5% makes only a marginal improvement in the return. Such a project would seem to have considerable risk. All too often this is the type of profile that results from property decisions where the structuring of the business case has not been adequately thought out.

A good example is where a corporate occupier takes a lease of a property for 10 years to service a five-year contract, but with a customer only option to extend it for a further five years. Assuming the contract makes money over the five-year period (including the cost of the property) then it may be seen as a viable decision. However, what happens for the second five-year period? If the customer does not renew the contract the business will be left with a surplus property for five years. A prudent consideration will be to look at what will be the total profitability if the property remains as a liability for the full five years. The critical question therefore is whether the profit for the first five years covers the downside on holding the property for a further five years. Ideally the corporate occupier should secure a break clause in the lease to protect against the customer not requesting an extension of the contract. However, break clauses are still an exception rather than rule, they are often restrictive in how they operate and more often than not the timing of the break and the business change are not linked. Consequently, while the customer may extend the contract for one year, the landlord is unlikely to be as flexible.

Despite these thoughts there are countless examples of companies that have totally ignored the risk and have signed up to leases that run well beyond their business needs.

The choice of which investment appraisal technique a corporate occupier should use will depend upon which technique that best suits its business needs across all investments made. NPV will usually be the preferred option because of the ability to use it with negative cashflows during the period of the project.

1.5 Property investment decisions

Cashflow models are ideal techniques for assessing property decisions because they allow all of the relevant parameters to be set out in particular:

- lease length
- rent
- service charge
- rates
- capital costs
- inflation in costs
- rental growth
- capital benefit on works.

When considering the acquisition of a freehold or long leasehold property, a corporate occupier or its advisors should have regard to a valuation text book such as the Red Book (*RICS Appraisal and Valuation Standards*, RICS Books) to provide detailed advice on valuations. Whether to acquire a freehold or take a short lease needs more than the Red Book valuation albeit a Red Book valuation should be part of the appraisal, both in looking at the costs of purchase and, second, in the value that might be achieved on disposal.

While a formal valuation might be required for funding purposes, such a valuation will not provide the worth of the property to the business. It is important for a corporate occupier to consider the difference between value and worth when contemplating property investment. The value of the property is generally regarded as the price the property would achieve in the open market if it were exposed to the market for a reasonable time. Worth is a narrower concept in that it represents the actual value to the owner or occupier. For example, the value of an old factory site might only be £1.5 m, but it could be worth £3m to the adjoining factory owner if the two sites were amalgamated. A major influence on worth to a Corporate Occupier is often the costs to the business if a facility did not exist or it had to be moved elsewhere. Consequently, costs of moving plant and machinery should be factored in to the equation if they have to be replicated elsewhere. Achieving consistency of approach is very difficult but important to securing a good quality decision.

It is rare that a decision, for example, to acquire one property can be made in isolation, yet frequently a corporate occupier will look at the business case solely in terms of the unit which is the subject matter of the decision and not consider the entire property portfolio. However, in the same decision making process consideration will be given to the impact of the changes on HR across the entire business. A full appraisal of the property portfolio is not likely to be needed for each property decision, but the impact of the new property on the portfolio should be considered, and in particular the opportunity to use an existing property for the project should not be overlooked. Such considerations become even more important when the decision concerns relocating from one property to another. Crucially, what will happen to the surplus property? Will the surplus property be disposed of at the same time as the new one is taken? If not, what will be the impact on the business of that additional cost (in particular if it is held on a lease) and how long will it take for the property to be disposed of? These are very important factors which are often overlooked and can result in hidden long term costs.

1.6 Accountancy issues

1.6.1 Introduction

It may seem strange that a section on accountancy matters is included in a book on corporate real estate, especially in a chapter on decision making. However, the reality is that with increases in the regulatory environment in the UK, USA and Europe, financial requirements now have a major influence on property decisions. This influence may be at the point of acquisition or vacation or how the effects of alterations are treated. A corporate occupier needs to have an appreciation of the main financial standards. It needs to be alert to potential issues and to be aware of how to deal with the consequences proactively. It also needs to be aware of the effect of portfolio transactions such as sale-and-leaseback on the accounts of the company. While a deal may make sense from a purely property perspective, that does not necessarily mean that it makes sense when considered against the total impact on the business. A corporate occupier when making property decisions should take on board the effect on the business of issues such as accounting treatment and tax ramifications. Tax implications are not covered in this book. Each company has its own unique tax structure and tax rules change much more frequently than do accounting regulations.

The UK accounting regime is aimed at providing a 'level playing field' so that anyone looking at two sets of accounts, for example, to decide which one of two companies to invest in, will be safe in the knowledge that the accounting approach is common. In the US this is achieved by prescriptive rules, where as in the UK and most of the other main western economies it is done by means of a set of principles. Under the UK model therefore, there is scope for taking slightly different approaches to the calculation of a particular item; it is not an absolute set of rules.

Increasingly property has formed a specific part of the International Accounting Standards (IAS) because of events in the past. The calculation of the capital value of a property portfolio has developed into a standard know as IAS16, with the nature of the calculation being well established and providing clear guidelines so that the key variables are determined. For less straight forward properties other techniques have been established. For example, a simple office block with one tenant may have its capital value calculated by reference to the rent passing, the current open market rent and the appropriate yield, where as a very specialised building such as a hospital may be assessed on a replacement cost model.

Part of the reason for change and the introduction of accountancy principles was as a consequence of corporate activity that exploited property valuations. There were two types of activity that highlighted the need for control. First, when a company had a large freehold property portfolio, which had not been re-valued for a long time the property assets were set out in the balance sheet at a figure well below their true value. On occasions this led to companies being taken over for those property assets. The purchasers then immediately re-valued the properties and were able to cover the cost of the acquisitions from the sale of one or two of the properties.

The second area which highlighted the need for control was in the provision of liabilities for surplus leases. This was required when a company no longer occupied or used a property that it held under a short lease, generally regarded as one of less than 50 years when granted, and a provision would have to be made for the net cost of the leased property to the company until the end of the lease. One approach was that in the year after the takeover of a company the purchaser would make a very large provision for every conceivable liability — a 'big-bath' provision. Often this pessimistic provision would provide for all possible costs of a property and ignore any subletting income. However, when profits were under pressure in subsequent years (possibly because the effects of the merger that were touted at the time of the takeover had not materialised) the provision would be revised with a more aggressive and optimistic view taken on outcomes and monies then released from the provision. This manipulation of the provisions and hence profits led to a masking of the true picture of the health of a company.

Accounting for property within a corporate occupier is an extensive area and there are many different regulations that apply depending on where the properties are, the corporate structure of the business and the domicile of the parent. There is an argument for also considering UK GAAP, US GAAP and the International Standards. There has been a general move to harmonise all of the standards and there has been a considerable alignment of standards in recent years.

1.6.2 Surplus property — IAS37

For a number of years surplus property has been synonymous with FRS12 (Financial Reporting Standard 12). However, recently the emphasis has switched to IAS 37. What are the differences between FRS12 and IAS37? The International Accounting Standards Committee (IASC) and the Accounting Standards Board (ASB) developed these two standards in parallel; published them on the same day and the text is virtually identical. FRS12 gives slightly more guidance on the discount rate to be used and on the recognition of an asset. In short, the differences are minimal. They are to all intents and purposes one and the same. Changes may result from efforts to achieve convergence of the IASB standards and those of the US Financial Accounting Standards Board. If this is achieved it will ultimately simplify matters in relation to US GAAP reconciliation for those operating in the US.

Both IAS37 and FRS12 were introduced in 1999 and in the UK the focus was on FRS12 until 2006 when there was a move towards IAS37. This was triggered by larger companies starting to comply with international financial reporting standards. The two standards continue to apply and the standard adopted is determined by whether the company follows ASB or IAS standards. Throughout the chapter reference is to IAS37, but unless stated to the contrary, the same points apply to FRS12.

The subject of lease liabilities has now come to the attention of large corporate occupiers together with a realisation that it is not something that can be hidden away any longer. From the point of view of a corporate occupier, having a standard reference point helps in dealing with the conflicts that occur between the real estate needs and the demands of the finance department. From the finance department's viewpoint the pressure is usually to save money and therefore it seeks to avoid increasing provisions, whereas the corporate occupier needs to ensure that the provision for surplus property is sufficient to cover the liability and ideally at a level that will allow deals to happen. In particular, there needs to be adequate provision to allow the corporate occupier to undertake works to bring the property up to a lettable standard and ideally to agree surrenders with landlords at a sensible level.

IAS37 is not just aimed at surplus properties: it applies to all onerous contracts and includes, for example, agreements for photocopier leases and fleet cars. However, surplus leasehold liabilities are a prime example of onerous contracts and there are a number of considerations that need to be addressed in determining a provision.

(a) The basic considerations

The rationale behind IAS37 was to avoid what became known as 'big-bath' provisions, where a corporate would provide for anything and everything it could think of as a means of tidying up the company for the future. This was seen on several occasions, often post take-over of a company, when large provisions were made and would often turn the company from profit to loss. Over time the provision would be released to deal with the specific costs incurred, but in certain cases it has been used to boost profits and as a technique to manipulate the company results and hence its share price.

FRS12 came into effect for accounting periods on or after 23 March 1999, whereas IAS37 came in on 1 July 1999. The objective of IAS37 is to ensure the proper recognition of the item and that correct measurement techniques are applied to provisions to provide appropriate disclosure.

A provision should be made when:

- an entity has a present obligation as a result of a past event
- it is probable that an outflow of economic benefit will be required to settle the obligation and
- a reliable estimate can be made of the amount of the outflow.

A contingent liability is not covered as this refers to a potential liability and measurement cannot take place until that liability has crystallised. In simple terms a provision will be needed for a leasehold property that is surplus and is either vacant or has been sublet. A contingent liability will arise if the property has been assigned. In the case of an assignment, while there is a risk of default by the assignee, until it happens, the liability for the company cannot be measured other than in the circumstances where there is some subsidy to pay, for example, a rent subsidy for a period of time. In property terms this is less logical than for other contracts. When a lease on a property is assigned there remains a real risk that the liability can revert back to the company and there is a legal framework for this to take place (see Chapter 10).

(b) Fundamental points

The fundamental points for the corporate occupier when considering making a provision are set out below.

- *The obligation is current*
  This does not mean that the lease must still be in existence. A dilapidations liability will require a provision where a lease has expired but where there are ongoing settlement discussions. What may be relevant is whether there is activity on the case or not. If the lease expired three years earlier, nothing has been heard from the landlord and it has not initiated a dilapidations claim then it is arguable that the corporate occupier should consider this a contingent liability.

- *There is a cash liability*
  What is required is a commitment (usually contractual) to pay money. Rent is a prime example of this.

- *The cashflow is to be measured*
  What is envisaged is a detailed calculation of the cashflow that will withstand the scrutiny of auditors. For example, making a simple assumption that two years rent will be sufficient is unlikely to pass muster if it is just a guess.

- *Best estimate*
  The cashflow should represent the best estimate that can be made of settling the obligation. This is not necessarily the sum that a landlord will accept for the surrender. It is the difference between the likely expenditure and the projected income stream from a subletting. In short, this will not be the gross costs to expiry. It will be the net costs. It could also be the premium figure that is likely to be paid to an assignee to take the lease on, or the premium payable to the landlord for taking a surrender, or the gross costs to expiry if the property is never likely to be sub-let. Which one is taken depends on which is the most likely. The corporate occupier should not simply look at one solution for a portfolio, but should take a rational view of each property in the portfolio.

- *Estimates should be based on the experience*
  Estimates should be based on the experience of the management team and if appropriate on evidence from independent experts. This requires the management team to understand the requirements of IAS37 at first hand. They also need to have market evidence of the property to substantiate the assumptions on reletting income. Having a default assumption that the headlease rent will be achieved throughout the remainder of the contractual term is not realistic, particularly for an office portfolio where rents were set at the market peak at the end of the 1980s. Consequently the approach should be conservative and if anything result in a slight over provision.

- *Risk*
  The corporate occupier must address the issue of risk in any assessment. This should be done by considering the nature of the properties in a portfolio and market conditions. If the letting market for a particular property is difficult with a considerable over supply then the corporate occupier needs to take that into account. It should not make the letting assumptions that it would do in a buoyant letting market. Reassessing the IAS37 figure annually should help, in part at least, to deal with this element.

- *Range of answers*
  Calculation of the provision could result in a range of answers. A good example is an office suite that has been sublet on terms that include a break clause in the subtenants favour in three years time and where there is then a further five years before the headlease expires. Does the corporate occupier assume the sub-tenant will break or not? If there is an assumption that the break will be exercised will the property re-let or not? Potentially there are three outcomes to calculation without even considering when the property would re-let. While the wording of the standard suggests that the mid-point be taken, this is not always appropriate and there are issues of probability and hence risk. The guiding principle has to be one of prudence rather than out-right optimism.

- *Time value of money*
  The standard refers to the time value of money and states that it is material in the calculation of a provision. The choice of discount rate needs to be a pre-tax rate reflecting the markets' view of the

combination of money, time value and risk. The discount rate should avoid double counting, so if the cashflow has already dealt with specific risks the discount rate does not need to take that element in to account. The choice of rate is not specified so it is down to the individual company to determine its own. Clearly the higher the rate adopted the smaller the actual provision, so from prudence point of view a lower rate should be applied. One way of looking at the issue is to ask this question. What rate of interest would a bank guarantee over the period of the lease so that by depositing the provision today (and if everything assumed came true) the last payment would take the deposit down to £0? The best rate that banks seem willing to guarantee at times of economic stability and over what could be a reasonable length of time is the Base Rate. It follows that investing the money today will see the last penny available to cover the last penny of monies due at the expiry of the portfolio after interest has been added.

- *Reassessment*
  The provision will change over time as events unfold. This will be driven by a number of factors, both property specific (such as a subletting of the unit) or at a macro level, eg significant changes in interest rates. Another factor that has to be carefully monitored is changes in the law. A prime example of this is the increasingly onerous health and safety and environmental legislation. These changes should be picked up when the provision is reassessed annually.

(c) Calculation of a provision

To create a cashflow a number of elements are required:

- A cashflow model — whether a macro, simple spreadsheet or a piece of paper, the corporate occupier needs to decide on the methodology to be adopted, and then remain consistent with it (see 1.4.2 above).
- Accurate data on the property for the normal costs of rent, rates, service charge, insurance and maintenance costs.
- Lease details that have been verified including restrictive lease terms, especially on alienation and rent review provisions. This has to include all of the lease documents, because, by way of example, an agreement for lease entered into 10 years ago might include an obligation to construct a building in the car park by the 15th year of the term, and if that has not been done the obligation will still have to be met and factored into the provision.
- Estimates of other costs such as dilapidations and potential one off service charge costs eg replacement of lifts.
- Knowledge of the market for each property supported by letting advice from local letting agents on vacant properties. The largest swing factor which will affect a provision is the ability or otherwise to sublet a property or the likelihood of a property becoming vacant. While making a standard set of assumptions, and using them as the default basis is a good starting point, the corporate occupier should look at each property and aim to make property specific assumptions. These assumptions will include when the property is likely to let, for how long, at what rent and with what incentives (all of which will provide an income stream to offset the gross cost).
- In appropriate circumstances the solution is to sublet a property in several parts. A typical example is a multi-storey office building, with individual floors being sublet. This creates other costs as the corporate occupier will need to run the building and in particular, the common parts, with a service charge and all the costs will need to flow through to the provision.

- A grey area can arise where a property remains part occupied; does that property fall to be provided for or not? It would appear not because it is not an onerous contract — benefit is still being obtained from the lease and consequently it does not need to be provided for. However, the corporate occupier should consider the benefits against the costs for the property.
- For a sublease, knowledge of the sublease terms is essential to cover issues such as service charge caps and schedules of condition which will restrict dilapidations liabilities.
- Second-guessing what a tenant is going to do on a lease expiry or break is impossible and the appropriate way to deal with this is for the corporate occupier to run different scenarios and take a figure in the middle.
- Part of the assessment required is the likely increase in rents at reviews, whether they are fixed uplifts or to open market rents.

There are two areas that sit outside the property remit , tax and VAT. Companies have different tax regimes and their ability to recover VAT will differ. How tax is dealt with is a matter for the finance department at the macro level rather than being calculated as part of the IAS37 model.

(d) Key considerations

It is important to remember the Pareto principle, which is also known as the 80–20 rule. The principle states that for many phenomena, 80% of the consequences stem from 20% of the causes. In the case of an IAS37 calculation 80% of the sensitivity comes from 20% of the variables. These key factors need to be the focus of accuracy and understanding.

(i) *Data*

As with all considerations of a portfolio the essential starting point is the data and having up to date information. Ideally the information is the best feasible but there is a diminishing return from trying to improve the data, so getting the most recent insurance demand from the landlord is unlikely to make a material change to the outcome.

(ii) *Lease documentation*

Arguably this forms part of the data for the property, but increasingly there is a separation between the two. While a corporate occupier might have a spreadsheet of data often it will not contain all of the leases. Ideally in undertaking an assessment of the IAS37 liability the lease documentation needs to be reviewed in depth. In particular focus needs to be given to the key areas of the demise; lease term; rent review terms; user; break clause provisions; alterations; alienation provisions; licences — for alterations and subletting, together with any deeds of variation.

While spreadsheets will provide useful summaries, over time there is a risk of the data being compromised and errors creeping in. This is true of all databases and undertaking a regular review of the source information is important. While this should be an exercise that is within the capability of a corporate occupier, there may be certain areas that need to be referred to a property litigation solicitor, such as the operation of a break clause.

There can often be assumptions made in respect of the documents that are erroneous because while something has been agreed, for whatever reason it has never been implemented.

(iii) *Demise*

It may seem obvious but checking the extent of the demise is a key element. Has it remained the same throughout the term and what exactly is included? Obvious examples are variations which take out or put in an area. This could include corrections of an original error or the tenant taking an extra area, such as a basement storage unit. This will impact on the rental value for the whole and the obligations on maintenance. On the other side of the coin, there may be an assumption that car spaces are demised with the offices whereas in fact only some of them are, the rest being held under licences that can be terminated on three months notice. For a city centre location that could be crucial factor affecting the corporate occupier's ability to sublet the space.

(iv) *Lease term*

When does the lease expire? Are there rights of renewal under the Landlord and Tenant Act 1954? This will affect how a potential assignee will view the space if there is a short term remaining on the lease.

Where a sublease expires well before the headlease, it cannot automatically be assumed that the subtenant will renew. An assessment will need to consider the options of no renewal, no renewal but subsequent letting and of renewal on open market terms. Which one is adopted will depend on the probability assigned to each of the three scenarios.

(v) *Rent review terms*

In general terms, the wording of rent review clauses has become fairly standard in recent years. Leases from the 1980s may have quirks that have been removed from subsequent leases because of case law highlighting these flaws. There may be some residual matters that have not been picked up in the past but could crop up in the future. The biggest problem is something which will increase the rent passing to a figure in excess of the current market value.

There are factors which will depress rents. A good example is an assumption contained in the rent review clause specifying that the lease term will be the longer of 15 years or the remainder of the term. Occasionally a clause will provide an assumption that the hypothetical lease is to be the original term granted which could be 25 years. In recent years there has been a move to much shorter leases with a norm of around 10 to 15 years. Consequently a long lease term with no break clause will reduce the rent that a tenant is prepared to pay. Often such leases were drafted and the rent set in the boom market at the end of the 1980s and the leases have been over-rented ever since. What these restrictions do is reduce the chance of any rental uplift in the headlease. Therefore in assessing the IAS37 liability the CREM needs to be aware of these influences on the rental value and not automatically assume an uplift in the head rent.

One assumption that is often made is that at the next review the sublease rent will match the headlease rent. That is rarely the case in practice unless the rent review clause in the sublease provides a matching clause ie a clause that explicitly states in the sublease that at the next rent review the rent in the sublease will match the rent in the headlease. The reason for the mismatch is usually because the terms of the two leases will tend to vary on issues such as demise; term and repairing obligations (especially the effects of a schedule of condition on rental value). There is a possibility that the sublease rental value may exceed that of the headlease.

(vi) *User*

Restricted user clauses still exist, although they have become less frequent in recent years. This is because on renewal under the Landlord and Tenant Act 1954 a court is likely to fix the new rent at substantially below the market rent to reflect the restriction. Nonetheless there are clauses that are not deemed to be particularly onerous but will affect the ability to dispose of the property.

(vii) *Alienation*
Leases tend to fall into the following three categories:

- those which permit the assignment of the whole of the premises and prohibit subletting or
- those which permit the assignment or underletting of the whole of the premises only or
- those which permit the assignment of the whole of the premises or permit subletting either whole or in part.

Leases that provide for the assignment of the whole but prohibit (or significantly restrict) subletting tend to be short-term leases. Where the length of term remaining exceeds three years this can be a problem, and is compounded when the property is over-rented. That leaves a corporate occupier with one option only — assign and pay a premium to the assignee (see 5.6.3 (c)).This will expose the assignor to the risk that if the assignee becomes insolvent during the term of the lease, a proportion of the premium will be lost.

The ability to sublet the whole does add to a corporate occupier's flexibility in that it can then control the property and where it is over-rented subsidise the shortfall each quarter.

Subletting of the whole or in parts can be restricted by reference to specific parts and number of occupiers. This is to prevent too excessive fragmentation of the building. A head landlord will look for additional protection in such circumstances by seeking to have the subleases excluded from the security of tenure provisions of the Landlord and Tenant Act 1954. Indeed the trend is for all sublettings to have this provision unless it involves a large building where any subtenant will be investing a considerable amount of money, for example, in fitting it out.

One problematic area is where a tenant is prohibited from subletting except at a rent, which is the higher of the market rent or the rent passing under the headlease. This clause can have a major impact on the ability of a corporate to dispose of excess space, in particular when the headlease rent is at or above market value. This problem is explored in detail at 5.6, and in particular, the leading case of *Allied Dunbar Assurance plc* v *Homebase Ltd* [2002] 2 EGLR 23. In *Homebase* the Court of Appeal held that a side deed used to circumvent the rental restriction did not prevent the tenant from being in breach of the alienation provision, and therefore did not work.

Since that judgment the side deed route to circumvent the restrictive alienation provisions has been closed. The extent of this problem varies between types of property, but the problem occurs mainly with offices below 10,000 sq ft let on leases granted in the early to mid 1990s. One of the areas particularly affected by the problem is the financial services sector, such as insurance companies; where in some cases up to 30% of their leased properties have such a restriction.

Dealing formally with such a clause can be difficult as there is as yet no obligation on a landlord to agree to relax or vary such a restriction (however a number of investment landlords have signed up to the Commercial Lease Code 2007 which addresses this problem). Therefore, if it chooses, a landlord can charge whatever premium it wants calculated on whatever basis it pleases. Often the landlord will seek to take a proportion of the rent a corporate occupier secures from the proposed subletting. An IAS37 calculation dealing with the impact of a *Homebase* problem should address at least two possibilities. The first should be on the basis that no subletting is possible until the market rent exceeds the headlease rent. The second approach should be on the basis that it is necessary to pay a capital sum to the landlord to vary the alienation clause.

(viii) *Alterations*

Tenants generally carry out works to a property to make the premises suitable for their own occupation eg installing demountable partitions or loading doors. Most leases or licences for alterations contain reinstatement clauses requiring a tenant to take out the alterations and put the property back into the condition it was at the commencement of the lease. Alterations should be capable of being tracked by means of licences for alterations granted from the landlord to the tenant. Problems arise when tenants carry out works without obtaining consent and thus making it difficult to assess the impact of the alterations on any terminal dilapidations claim for the property. In addition, alterations can be a barrier to letting space and a unit that has a very specific fit out may struggle to be let in the market. This is discussed in more detail in Chapter 6.

(ix) *Break clause provisions*

The specific wording of a break clause is very important. It will determine how easily a lease can be broken, be it a headlease or a sublease. The detailed legal rules governing this importance topic are dealt with in Chapter 12.

Break clauses tend to fall into three general categories:

- no compliance required — aside from providing vacant possession the tenant has no obligation to do any works or pay monies
- compliance is relatively simple to achieve — eg all that is required is that the tenant serve six months prior written notice and pay all sums due under the lease up to the determination date
- rigorous compliance is required — often a mixture of paying all sums due under the lease and complying with onerous repairing, redecoration and reinstatement covenants.

The latter group causes the most problems in assessing an IAS37 calculation. Can the break be operated in reality? If it is unlikely that the break can be operated successfully, then where the break is contained in a headlease, there should be a default assumption that the break will not take place. However, if a sublease also contains a break right, the corporate occupier must assume that any subtenant will successfully operate its break right. This is, of course, the worst case scenario for a corporate occupier. It is not prudent to assume that a corporate occupier with the benefit of a break right will automatically be successful in exercising that right. There are many cases where corporate occupiers have failed to comply with the strict conditions precedent imposed by the lease.

(x) *Current market conditions*

In the same way that it cannot be assumed that the contractual term of a subletting will be co-terminus with a headlease, it also cannot be assumed that it will be possible to achieve, for example, a subletting in one year, granting a six month rent free and achieving the headlease rent.

The specific market conditions should be looked at to determine:

- the rental value
- the incentives that will need to be granted
- the term of the sublease
- an estimate as to how long it will take to achieve a letting.

One of the key factors in determining the net liability is how soon a property is likely to let and the timing of when the rent payments start. The longer the void period the higher the liability. Once the premises have been sublet, costs such as rates and service charges are likely to be covered as they tend to be picked up from the start of the letting even if there is a rent free period. Letting agents tend to have an in built optimism for when a unit is going to let, and it is essential that the corporate occupier investigates the true position.

An awareness of what works are required to the unit will impact upon the speed of letting as well as the terms achieved. The marketing works are important and the relationship with speed of letting is difficult to prove in absolute terms. Finance directors of corporate occupiers will want to be assured that by undertaking capital expenditure on a surplus property that they will get an immediate return and will require some form of empirical evidence to support it. That does not exist. However, experience shows that if a property is in need of extensive works, carrying out those works will provide a much better chance of a deal happening; leaving the space dilapidated will virtually guarantee no deal.

The structure of a likely deal with rent frees and breaks needs to be carefully considered as part of this overall assessment of the market.

If the solution is to divide a building in to smaller units, for example letting individual floors of an office building, the cashflow needs to reflect the additional costs of creating lettable units. This will include the costs of splitting services and possibly the provision of a receptionist.

(xi) *Building condition*

Putting the property into a condition where it can compete with other available space in the area is very important and should minimise the length of time the property is vacant. In the majority of cases, undertaking works will improve the lettability of a property. For a small number of properties no amount of refurbishment will improve their chances of being let. This may be due to restrictions on subletting; restricted user clauses or because there is simply no demand for that type of property in that location. Consequently, for each property there needs to be consideration of what can be done and what should be done.

A further consideration is potential dilapidation claims. Virtually all leases contain an obligation to keep the premises in repair and to undertake decoration on a regular basis. When the premises are in disrepair the tenant is in breach of those obligations and at risk from action by the landlord. The landlord may serve an interim schedule of dilapidations or may be entitled to enter the premises and carry out the works and then recharge the cost to the tenant (see 14.4.2).

When considering what marketing works to carry out the corporate occupier should consider the risk from these events and also take a long term view (ie to the end of the lease) and take into account what works will need to be carried out at lease expiry. A common dilemma which arises with office suites is whether to strip out partitions as part of the marketing works or whether to leave them in and merely redecorate them. The lower cost option is generally to leave the partitions in and redecorate. A key consideration is then whether a potential subtenant would want to see the premises partitioned or not. If the answer is yes, it makes good sense to leave the partitioning in the premises for use by the subtenant.

However, it is often better to strip the partitions out and let any incoming tenant deal with their partitioning needs themselves. The rationale for this is four fold. First, it is unlikely that the partitions will be in the right place for the incoming tenant. Second, many tenants are unable to get a clear view of the amount of space when a unit is heavily partitioned and it will be more attracted to the open plan space of the same size. Third, if the partitioning is not stripped out prior to subletting, the definition of the demised premises is likely to differ between the headlease and sublease in that the subtenants will have the partitions as part of their demise and will not have an obligation to remove them at the end of their lease, whereas the headlessor is likely to be under an obligation to strip out. Finally, by stripping out and taking the unit back to the original condition the corporate occupier is dealing with a significant part of any dilapidations claim at that stage, and consequently if the space does not let, significantly less works will be needed at the end of the lease. This is in contrast to redecoration and carpeting of partitioned space, which a corporate occupier could end up paying for twice if the partitioning remains and these works are carried out on re-letting, but then need to be carried out again once the partitioning has been stripped out as part of dilapidations works on lease expiry.

Assessing a potential dilapidations liability can be difficult (see Chapters 14 and 15). First there is the need to determine what works are likely to be required, and then to estimate the probable cost of carrying out those works. Overriding this is a statutory cap on damages contained in section 18(1) of the Landlord and Tenant Act 1927 which limits the damages recoverable in terminal dilapidations claims to the lesser of the costs of the works or the diminution in the value of the landlord's reversion, and also provides that no damages are recoverable if the landlord intends to redevelop the site shortly after lease expiry. For older properties it could be said that the allowance for dilapidations contained in the IAS37 calculation should be zero because in view of the age of the premises it is most likely the site will be redeveloped. However, the landlord cannot be forced into redevelopment and if it decides to carry out repairs with a view to re-letting the premises, the tenant may be caught out by having under provided. Therefore the prudent approach is to provide for the full reinstatement cost and seek to mitigate the liability at the time.

With some properties there will be a risk of environmental problems, which can include asbestos issues and oil contamination. Generally this should be considered outside of the IAS37 provision and the appropriate allowances made. Ideally the remediation works should be carried out promptly to ensure that the premises comply with repairing obligations and statute.

(xii) *Discount rate*

The time value of money needs to be taken into account and accordingly the cashflow should be discounted. At what rate, should the cashflow be discounted? Is it at a risk free rate or does the rate express some risk in itself? In part the question is one of approach. The more explicit the cashflow is in looking at the variables and the more focussed the organisation is on considering the sensitivity of the cashflow to key variables, the less the discount rate has to take into account. It is arguable that the best 'risk free' rate should have no discounting so that the provision is the cash provision. Generally organisations will have rates they use for projects, which could be applicable. However, the key difference is that unlike projects that have to produce a rate of return, the purpose of the discount rate is to ensure that if you put the provision into a bank account, it will pay all the bills in the future should that scenario be achieved. What interest rate will a bank pay over the longer term? They will be reluctant to make long term commitments because of the volatility of the market, but from research it is probable that current base rate could be achieved for the life of a liability of 10 to 15 years.

Does the discount rate need to be any different? Should the Weighted Average Cost of Capital (WACC rate) provide a better figure? This is unlikely because the WACC rate is generally 4 to 5% above Base Rate and therefore there will be a possible shortfall between the figures invested and what are needed. The other factor to bear in mind is what premium a landlord will be looking for when discussing a surrender. One view the landlord will have of any potential deal is its income stream and it may look at this on a cashflow basis against Base Rate. Alternatively the landlord will be looking at it by applying the yield for the property, which generally will be a low figure in the current market, and consequently the corporate occupier should not be too far apart from that figure.

(xiii) *Miscellaneous*

There are a number of other points that need to be considered:

- *Risk of tenant default*

  There is a possibility that a sub-tenant will default and go into liquidation or bankruptcy. How should that be dealt with? One option is to create a model that takes into account the chances of the income being received and assumes for a particular tenant, say, that only 80% of the income will be received. This will work when there is only one surplus property. In a portfolio situation, arguments can be put forward to the effect that the risk of all or a substantial number of the subtenants defaulting is either low or high. It can be low because of the mitigation from the diversity of the portfolio, and high because if the properties are of a similar type, the market for those properties will be the dominant factor. Consequently, if the secondary retail market is having tough trading conditions it will affect all such retailers and if a portfolio is solely in that sector, there will be a disproportionate rise in risk of tenant default. Care needs to be taken in looking at tenant default to ensure that there is no double counting with the assumptions made in the individual cashflows.

  A sound approach to this issue is to consider the portfolio holistically once the individual calculations have been done and take a broad-brush approach. The calculations do not provide hard numbers only indicative ones because there are too many variables. Consequently, a broad allowance should cover the default risk.

- *Tax*

  Should allowances be made for tax, in particular VAT? It is impossible to make generalisations on this because the tax position differs between companies. The simplest approach is to carry out the calculations without tax and then the specific tax position of the company can be taken in to account by the accountants in the appropriate part of the annual accounts.

  VAT is difficult in that financial services companies, for example, do not fully recover VAT, so the question of VAT may have an impact on the total cost. In particular the impact can be felt on service charge costs. All occupiers will be affected by non recoverable VAT where there is a substantial service charge cost and the building has not been elected, because the landlord will then be recharging the costs gross without the VAT shown separately. In those cases the gross cost needs to be taken in to account. However, if the corporate occupier is disposing of a portfolio of surplus properties the probable purchaser will be unlikely to be affected and will be able to recover VAT. It is therefore better, again, to deal with this issue post calculation.

- *Part occupation*

  There can be cases where a tenant only occupies part of the property. Does a provision have to be made in such a case? The key is the lease structure; if the vacant part is governed by one lease, and the occupied part by another, then a provision needs to be made in respect of the lease of the vacant part. Where, for example, it is one lease of two floors in an office building, and one floor is vacant then that is more difficult. It is a matter of whether the lease has become an onerous contract and how prudent the company wants to be. The benefits need to be considered against the liability. IAS37 does not deal with these specific issues and there is no right answer. It is something that is likely to be determined by the financial health of the company. If it is in poor health, it will seek to avoid making the provision.

- *Timing of the release of the provision*
  When certainty has been reached then the balance of monies can be released. The point of certainty can vary and is determined by the nature of the event that has triggered the release. In the case of a surrender, it should be when the deed of surrender is completed provided there are no residual matters to resolve. On assignment, the risk then becomes a contingent one, and the timing of the release will depend on how contingent liabilities are dealt with in the company's accounts. If the trigger event is lease expiry, the question of dilapidations must be resolved. If no immediate claim is brought by the landlord, the tenant may nevertheless have to retain a proportion of the provision up to the expiry of the six year limitation period. However, as the likelihood of a claim diminishes over time in line with the landlord's ability to prove its case, it may be appropriate to release a proportion of the provision.

  A trickier decision is whether to reduce the provision if a sub-tenant is found at a much earlier date than anticipated. Should the IAS37 calculation be re-run and the excess released? Prudence is important because a sub-tenant might go in to liquidation or exercise a break clause, and this would result in the need to make an increase in the provision.

(e) Application of principles

When applying the principles set out above and calculating IAS37 figures for a portfolio, it is important to remember what the figures will be used for and the contradictory pressures that apply. On the one hand there will be a desire to minimise the provisions to avoid a negative effect on the balance sheet in a particular year. However, at the same time there is a need to set the provision at a level that will allow surrenders with landlords, and too low a provision will prevent that.

1.6.3 Valuation of freehold and long leasehold operational properties — IAS16

The International Accounting Standards Board (IASB) has sought to reduce the number of alternative treatments for specific areas but within IAS16 two alternatives remain, cost less depreciation or a fair value basis. The first approach reflects the historic accounting convention that a freehold property is capitalised when acquired and recorded in the balance sheet as an asset. In subsequent years a proportion is charged to the profit and loss account for as long as there is a benefit from the asset, such that at the end of the useful life of the asset it will have been written down to the disposal value or zero if it has no value. The intention of the IASB is to move to the fair value principle for the future.

If the revaluation model is adopted then the fair value is determined by market value evidence for the property, in the event that the building is specialised and such evidence does not exist, the appropriate measure may be calculated by the income approach or the depreciated replacement value method. The frequency of revaluation is not laid down in IAS16, but it stipulates that it should be sufficiently regular to ensure that the amount carried in the balance sheet does not materially differ from the fair value. This places the onus on the directors to be aware of the way that values are changing — up or down.

The RICS publish guidance notes and practice statements on property valuation in the Red Book. For the corporate occupier the key issues are set out below.

- Understanding the principles of the valuation techniques that are applicable to its portfolio. If it comprises entirely freehold office buildings then a detailed knowledge of the income approach is unlikely to be needed.
- When was the portfolio last professionally valued? Use of out dated valuations can leave a company vulnerable to takeover, because the asset base of the company does not reflect its true worth. There have been cases in the past where the entire cost of a purchase of a company has been funded by a valuable freehold not being included in the asset base of the company in the balance sheet.
- Are the right valuers being used for the portfolio in question and do they have the right information on which to properly carry out their valuation?
- There should not be an automatic acceptance of what the valuers say. The corporate occupier should examine the approach and the details provided to ensure that the figures that are adopted are the fair value of the property.
- It should be remembered that fair value is not a reflection of what the property might be worth to the company or indeed to another company. Nor does it take into account special purchasers of the property. If this is significantly different from the fair value and there is a prospect of a third party buying the property for a higher figure, this needs to be reflected in the valuation report.

1.6.4 Short leasehold properties — IAS17

(a) General

IAS17 has been in place for many years and the principles have existed in the UK under SSAP21. The requirements have not changed for the last two decades. IAS17 was the first standard to deal with substance over form and to incorporate the present value basis of measurement.

The division of leases into finance leases and operating leases is based on risk and reward. A finance lease is a 'lease that transfers substantially all of the risks and rewards incident to ownership of an asset'. Whereas, an operating lease is defined as 'a lease other than a finance lease'.

There is not a definitive point at which the change happens nor is there a numerical guideline. Under SSAP21 an operating lease changed into a finance lease when the present value of the minimum lease payments exceeded 90% of the fair value of the asset. There is now a more principle based approach that focuses on the substance of the transaction rather than the form. Examples of issues that would point towards a finance lease are:

- the lease transfers ownership at the end of the term to the lessee
- the lessee has an option to purchase at a price significantly below the fair value
- at the commencement of the lease the present value of the minimum lease payments amount to substantially the fair value of the asset.

The classification occurs at the commencement of the lease unless there is a material change during the term of the lease, which had it existed at the commencement, would have resulted in it being defined as the alternative at that time.

(b) Finance leases

In UK property law all leases of buildings are leases of land and there is not the separation that exists in other countries which results in clear separate values between land and buildings. The revision of this standard in 2004 saw the need to show the values separately. One effect is that it is possible to have an operating lease for the land but a finance lease for the buildings.

Finance leases are shown as assets or liabilities on the balance sheet of the lessees. They are shown either as the fair value of the asset, or if lower, the present value of the minimum payments. The liability decreases by the rent paid, but there is then a finance charge added to the balance to provide the carried forward balance. In effect, the rent repays the principal and the finance costs of the deal.

(c) Operating leases

IAS17 treats operating leases as an expense to the profit and loss account on a straight line basis. The matter of incentives is somewhat ambiguous in that they are regarded as part of the total consideration and should be treated on a straight line basis over the lease term. By way of example, a one year rent free on a 10 year term should be spread over the 10 year period and consequently the £10,000 pa rent becomes, in effect, £9,000 pa.

A complication arises where the lease contains a break clause after five years. Does the same principle apply or is the net rent for the first five years £8,000 pa? Further, what is the effect of an upwards only review clause at year five? Does this lead to the straight line being just five years or is it 10 years with a deduction of £1,000 pa from the rent? There is no absolute answer and the issue is open to interpretation. This is one of the advantages (and also disadvantages) of the UK accountancy system of setting principles rather than absolute rules.

The differential treatment of the two types of leases is important to the corporate occupier. If the transaction results in a finance lease then there is an immediate impact on the balance sheet compared to the same deal where it results in an operating lease. Such issues rarely arise in the 'normal letting market'. It is generally the result of sale-and-leaseback structures.

There is a move to remove the distinction between operating and finance leases so that both will be recognised at fair value in the balance sheet of lessees, such a change would mean that the conceptual approach for the two would be the same and align it to the theoretical approach for surplus leases under IAS37. This would then lead to the identification of the total commitment for each lease in the balance sheet and the economic benefit that will flow from it.

For both financial and operating leases there is a requirement to disclose future minimum rent payments for the leases split into leases of less than one year; those of more than one but less than five, and for those of more than five years. However, this does not provide a total profile of the liability of a company for leases, especially those that have portfolios with long leases. A number of retailers have undertaken sale-and-leaseback programmes over the years some with terms of 30 years, which will provide a long obligation for the company and therefore a substantial commitment. Focussing on rent alone only provides part of the picture because other costs can be more important than the rent. There may be long leases in the portfolio which are on a low rent, but this might be because of a future obligation of the tenant to spend a considerable amount of money on the property.

1.7 An example of a discounted cashflow

A business buys a machine that costs £100,000 and receives £30,000 at the end of the first year, and then £40,000, £50,000 and £60,000 at the end of the subsequent years, at which point the machine is worthless and unable to produce anything else. The business has a cost of capital of 15%, so what is the true benefit to the business?

Year	Cashflow (£s)	PV of £1 at 15%	Net Present Value (£)
0 (today)	(100,000)	1	(100,000)
1	30,000	0.870	26,100
2	40,000	0.756	30,200
3	50,000	0.658	32,900
4	60,000	0.572	34,300
Cash Profit =	80,000	Net Present Value =	23,500

So the investment is worthwhile, and indeed the purchaser could pay more for the investment, up to £123,500 and still maintain the 15% return.

This is the Net Present Value approach; if we look at the same figures for seeking the Internal Rate of Return, the approach is as follows:

Year	Cashflow (£s)	PV of £1 at 24%	Net Present Value (£)	PV of £1 at 25%	Net Present Value (£)
0 (today)	(100,000)	1	(100,000)	1	(100,000)
1	30,000	0.806	24,200	0.800	24,000
2	40,000	0.650	26,000	0.640	25,600
3	50,000	0.524	26,200	0.512	25,600
4	60,000	0.423	25,400	0.410	24,600
		Net Present Value =	18		(2)

The technique of iteration is then used to calculate the IRR. Trial rates are used until two rates very close to zero are found and then linear interpolation is used to find the rate. The formula is

$$\text{Lower trial + rate} \quad \frac{\text{NPV at lower rate}}{\text{NPV at lower rate + NPV at higher rate}} \times \text{difference of trail rates*}$$

* Ignoring +ve and –ve signs

For our example this provides

$$24\% + \frac{18}{18 + 2} \times 1 = 24.9\%$$

The calculation can be represented on a graph to undertake the calculation but spreadsheets on computers can calculate this very quickly.

Portfolio Issues

2.1 Introduction

Property portfolios fall in to three categories — operational, surplus and mixed, the latter comprising both operational and surplus properties. It is the latter category that is most prevalent in the UK. The commercial lease market in the UK has until recently favoured longer leases of 20 to 25 years, and for many years there was no option but for a corporate occupier to take leases of this length whether that fitted with its business model or not. Until the 1960s long leases were regarded as anything over 50 years, most being 99 or 125 year leases. That definition has changed and now long leases are considered to be leases of 20–25 years. The 125 year lease is now regarded as a ground lease or building lease. Most organisations, whether in the public or private sector, have evolved over time and as they have changed as too have their property needs. As a consequence of such changes and the length of leases they were historically forced to accept, organisations have been left with legacy properties. This has led to corporate occupiers holding a mixture of operational and surplus properties.

With some corporate occupiers, the majority of their property portfolio is made up of surplus properties because their business model has changed so dramatically in recent years that they no longer have an operational need for property. A good example of this is television rentals. At one stage there were two businesses providing national cover, Radio Rentals and Granada, as well as a number of smaller regional or local operators. The dramatic fall in the cost of ownership of televisions as a proportion of wages, the changes to the availability of finance (not least seen with the boom in credit availability and credit cards) and the marked improvement in reliability of the product all contributed to the demise of the business. Consequently, the sector has gone from two competing companies having a shop in every town to a merger of the two businesses in one location. However, even the merged business has faced difficulties over the last five years because in the UK the business model is not as effective as it used to be. This has led to the closure of a large number of shops. While the business continues, it is as a virtual company on the Internet rather than through a retail network. Consequently, while the demand for space in that particular business has disappeared, the burden of the properties has remained resulting in substantial portfolios of surplus properties.

The two distinct parts of the property portfolio, operational and surplus, place different demands on the corporate occupier. It is not uncommon for a corporate occupier to employ an in house surveyor entrusted with the task of overseeing and managing its property portfolio, often known as a corporate real estate manager (the CREM). While there are common issues in the day-to-day management of operational and surplus portfolios, the approach is fundamentally different and defined by the impact each type of portfolio has on the business.

2.2 Operational property

There are occasionally exceptional circumstances which result in the operational portfolio not being the CREM's principal concern. A good example is where a corporate occupier gets into serious financial difficulty or where it has recently been through an extensive restructuring that has significantly increased the surplus property portfolio. The latter can occur where there has been a disposal of a large part of the business but the leased properties have remained or where there has been a merger of two previously competing businesses leading to the duplication of locations.

However, absent these exceptional circumstances, it is the operational portfolio that should command the highest priority for the CREM. On a day-to-day basis these properties provide the fabric for the business units to operate in. Without functioning operational properties there can be no operation. If the factory is uninhabitable because of fire damage then the production of widgets cannot take place. Equally, water pouring through an office building caused by failure of a sprinkler system will prevent an insurance company processing claim forms, and a damaged shop front will prevent a retailer trading. How crucial such events are to a corporate occupier depends on what alternatives are available to it. If the unit is one of several, then while there will be an interruption to production, it may have a relatively small impact compared to a business that has just one production facility. If there is only one facility, the key issues become how quickly repairs can be carried out or an alternative found. The exposure is greater with more specialised buildings and where there are substantial amounts of fixed plant and equipment in the unit. Consequently, replacing an office building and installing desks will be far easier and hence quicker than finding a factory and adapting it to take particular plant and equipment.

These are the principal concerns of the CREM, with avoidance of business interruption being a high priority. As such, the CREM and his team will tend to be part of the support team to the main business and not a profit centre.

2.2.1 Leasehold v freehold

There has been much debate in the UK as to whether corporate occupiers should own or lease their property portfolio. For some, ownership is essential because there is little chance of finding suitable leasehold sites (eg power stations), whereas for others, the fragmented nature of the space needed, such as many small office suites, makes leasing the only option. Aside from the two extremes, is there a perfect balance that can be achieved? Theoretically the answer is yes. Research on this subject identified certain attributes that determine the optimum proportion of freehold and leasehold properties. However, the number of variables that exist make it very difficult to construct a model that works well. Some of the factors that need to be taken into account include:

- the kind of industry that the business operates in — if a business operates in several then that will make the model more complex
- the size of the company
- the structure of the company
- the financial aspects of the company, including debt ratio and profitability

- innovation in the sector and the rate of change.

With a number of factors coming together, any modelling of what should be the right balance is difficult and is exacerbated by human elements. The current and previous decision-makers' influence on a business can be very significant to a corporate occupier's property ownership profile. Whether a corporate occupier has in the past tended to own the buildings it operates from or alternatively lease them is often more the product of human foibles than anything else.

That being said, part of the role of the CREM should be to provide guidance on the subject and to encourage the debate. The guiding principles are set out below.

(a) Certainty of occupation

How crucial is it for a business to remain in a certain location? It can be very important for some businesses. A good example is some industrial activities where the costs of relocation of plant can be very high either because the building has to be constructed around the plant or the bases for the plant have to be specifically constructed, and the plant is then built in situ. In such cases continuity of occupation is very important because relocation costs and the additional costs associated with business interruption are very high.

Is the unit regarded as a core property? The core and flex approach to space has become a model for businesses that use large amounts of office premises. The core buildings tending to be freehold with a commitment to develop the business around the core sites, using flex leasehold space to satisfy short term additional space requirements.

It is important for corporate occupiers to be aware that even when a lease falls within the provisions of the Landlord and Tenant Act 1954 this does not provide certainty of permanent occupation. While a long lease might be an option, they have tended to fall out of favour in recent years. Long leases were used by local government and government bodies, such as new town development corporations, as a means of encouraging investors to develop sites. In exchange for a 125 year lease at either a peppercorn rent or a low geared rent, the lessee would construct a property on the site and then sublet it on conventional lease terms. This method of funding development continues to be used by charities and the like for redevelopment of properties. However, it is rare for a long lease to be granted to a corporate occupier to allow it to erect a property to its own specification because there are a limited number of landlords who are interested in long-term low-income properties, and most occupiers recognise that they are not necessarily the best developers of space. Equally, most large developers take a particular approach to designing buildings which results in space that does not always reflect contemporary ways of working.

(b) Flexible space

A key factor in the lease versus freehold debate is how much change is likely within the fabric of the building and how difficult will it be for those changes to be implemented under normal lease arrangements if the landlord's consent is required each time? While the warehouse and industrial sectors provide the obvious examples of buildings evolving to allow the business to adjust to its changing business needs, the same can also be the case with the office sector if alterations beyond the addition or removal of demountable partitions are likely. It is usually possible to negotiate a lease that allows partitions to be changed without the landlord's consent on each occasion, particularly if the building is of substantial size. But does that provide a corporate occupier with sufficient flexibility? If the freehold of a building is bought because it offers scope for development as the business grows and is likely to be required to accommodate more staff, then having the freehold and carrying out works that add value will be of great benefit. By the same token, is a corporate occupier likely to carry out works that will benefit the business, for example extensive canteen facilities, but which will not necessarily add value to the building? If so, when considering the leasehold option, the corporate occupier should in addition to the costs of the works, factor in the likely costs of reinstatement (as the landlord is likely to require reinstatement at lease expiry) as well as the costs of obtaining the landlord's consent.

Flexible space means flexible within the fabric of the building or site. It also means the flexibility within the portfolio to change the nature of the buildings occupied to reflect growth or contraction of the business. The latter requires leases to be flexible. This is achieved by negotiating short-term leases or better still, leases which contain regular tenant break options that are unconditional and simple to operate. Corporate occupiers often crave flexibility but less than 30% of all break clauses are ever exercised, so is that craving a real need or a perceived one? Part of the problem is that currently break clauses in leases tend to be at fixed periods, at best every five years, and are often subject to compliance requirements which make exercising the break option very difficult and on occasions virtually impossible (see Chapter 11). This works against the corporate occupier because of the uncertainty of being able to exercise the break, but more importantly because the business drivers that would precipitate the change rarely coincide with the break date. Whether a rolling break would help remains to be seen. On the one hand, it would provide total flexibility, but on the other this could on occasions cause problems in itself, as corporate occupiers can be susceptible to change their minds frequently on such issues. It is not unusual for a break clause to be exercised and then within a short period the operational part of the business changes its mind and wants to remain in the unit, or even extend the previous lease. This can be challenging and creates numerous problems for the CREM and therefore knowing from the outset what flexibility the business really wants is an important factor in determining whether a freehold or leasehold option is the best course to take.

(c) Financial issues

Freehold property forms part of the asset base for a company, whereas leaseholds are a liability. For some businesses there is great benefit in having leasehold property rather than freeholds. This is a function of the nature of the business; the gearing the business has; whether the business is driven by cash or assets; and how analysts perceive the business.

Certain sectors have tended to have a particular approach to ownership. For example the retail sector tended to look to freehold ownership and this is why retailers such as Marks & Spencer created such large portfolios of freehold properties. This changed with the advent of shopping centres when owning the freehold was not an option, even for M&S. Some retailers moved across to be leaseholders selling off freeholds to boost returns for investors, whereas others retained a substantial proportion as freeholds. In recent years the ownership of freeholds has been a mixed blessing as investors have looked to take over businesses that have under valued the freehold portfolio.

2.2.2 Lease flexibility

Businesses evolve and as they do so the real estate needs of the corporate occupier change to reflect the evolution of the business. What a corporate occupier wants to avoid is having too much space or too little space. Too little space can be worse than too much in many respects because the lack of space can impinge on trading and hence will affect profitability.

The ideal solution is space that the business can expand into easily when growing, or vacate easily (and terminate the leases without penalty) when contracting. For retailers, this is virtually impossible, unless they live in the virtual world ie on-line retailers. For corporate occupiers that rely on short term accommodation to sell goods from, short term retail space is usually only available as poor quality units on the very fringes of the retail area. The problem facing the industrial corporate occupier lies in the costs of manufacturing plant and costs of installation, and again therefore, it is very difficult to build in flexibility. Warehousing tends to be simpler in that in traditional units there is not a great deal of fixed machinery, but the very modern high bay units often have specialist racking systems which are very expensive to fit. The flexibility can be obtained for the low-tech space, which may be adequate for dealing with certain aspects of the business.

The greatest flexibility exits in the office sector, where short-term solutions are available. While fit out costs for offices can be high and can constrain flexibility, it is possible to create a portfolio that has a reasonable degree of flexibility. Serviced office operators have led the way in providing short term space that is already fitted out with desks and with IT infrastructure, so that the corporate occupier only has to supply computers, people and files. This works very well for providing space for short term project teams (ie less than 12 months) where the business does not have free space or deliberately wants the team to be away from the rest of the business. In addition the office sector does generate more leases with break clauses in than the other sectors. This then allows a corporate occupier to create a portfolio of properties with a wide range of flexibility. The core properties, which are deemed essential for the business and are either owned freehold or have been taken on long-term leases, will comprise the major hubs for the business, including its headquarters. The second tier will be those properties that are deemed to be a medium term hold, for instance, regional properties where the business wants to achieve certainty for the five to 10 year time-frame, or perhaps some shorter leases or those containing break clauses. The final group will be the short term flex space that includes short-term leases with breaks or serviced offices depending on the length of the project term. Beyond a 12 month period the costs of serviced offices are expensive compared to conventional leases with break clauses. However, landlords are reluctant to provide rolling break clauses as opposed to fixed breaks, and which route to adopt will depend upon the certainty and timing of the break rights that can be negotiated.

Lease flexibility is often looked at in respect of the desire to avoid too much space and in particular having to make provision in a company's accounts for surplus space. This is very laudable, but lease flexibility comes at a cost. The addition of break clauses will generally result in smaller incentives being offered by the landlord to take the lease initially, and is also likely to result in higher rents on review. Serviced offices are more expensive than conventional offices both in terms of the rents charged and the costs of ancillary services such as telephone costs.

A corporate occupier should, therefore, consider carefully whether in reality it is likely to exercise the break clause, and what the additional cost of having the break cost will be. It should carry out a cost benefit on the inclusion of break clauses in a lease to determine the parameters of the decision-making. As indicated above, the number of corporate occupiers exercising break clauses tends to be less than 30% of those who have the negotiated break rights.

2.2.3 Metrics

Metrics are the various measures that can be used to assess the relative performance of a particular aspect of a property portfolio. It could involve measuring the cost of electricity between different buildings or it could be an analysis of the performance of the total portfolio. A key issue for a corporate occupier is the use of the space that it owns or holds under leases. As costs of occupation have continued to rise, the question of efficiency and effectiveness of occupation has become more important. Is the business getting the best from the space it has?

This is an extensive subject and has been covered in a number of academic papers. The need to assess the use of space is crucial to a business in understanding what value it is generating from the space it occupies. This is true of a business with one property or one hundred. The difference is that with a portfolio there is the ability and the need to compare units by the same measure.

How does a CREM decide which measures to use and how to apply them? No one method will deal with everything and consequently a number of metrics need to be used to allow various aspects to be evaluated. Recent research shows that the majority of people use at least two measures when trying to establish the effectiveness and efficiency of their property (see CBI/GVA Grimley *Property Trends* Winter 2006 Supplementary Report).

The use of metrics varies between building types. The office sector has spawned a number of different metrics including:

- total occupancy cost
- cost per m^2 (or sq ft)
- m^2 (or sq ft) per desk
- m^2 (or sq ft) per person.

The retail environment tends to be more driven with sales per m^2 (or sq ft) and profit per m^2 (or sq ft) as key drivers. Industrial buildings have output per m^2 (or sq ft). Overlying all sectors is turnover per m^2 (or sq ft) and profit before interest and tax (PBIT) per m^2 (or sq ft).

Clearly for a business operating in one sector the range of metrics can be consistent and relatively limited, for a business in a number of sectors the range of metrics will be broader and will need to be driven at the business level rather than group one. To make the most use of the metrics they should be from the smallest unit upwards with conglomeration of figures and examination of variance. For a retailer the matter is generally by branch upwards, for an office user it can be slightly more complicated. Is it at the office building level or per floor? The ideal must be per floor because that will then identify under utilised space in a property. An added complication may be that different division's share a building and again that needs to be broken back to the individual occupier per floor if possible.

The aim of carrying out such an exercise is not to simply produce a block of data, but to produce dynamic data that is capable of statistical analysis that allows the CREM to continually evaluate the properties in the portfolio so as to maximise and achieve the best usage of the space. At a simple level the availability of this data may cause a corporate occupier to avoid taking on additional space by taking advantage of excess capacity in one of its existing buildings which may otherwise have been overlooked. However, a key constituent in this is determining the benchmarks. What is the right amount of space per desk for example? This will be governed by the nature of the business. A data processing centre will have different needs to a firm of solicitors, and a US firm of lawyers will have slightly different needs to a UK firm.

A further variable is the amount of time a desk is being occupied, something that has rather inelegantly been termed the velocity of occupation. It identifies how long someone is actually sat at a particular desk in any 24 hour period. For a number of businesses this figure can be fairly low, especially in an environment where a large part of the workforce spends time away from the office, for instance, sales or audit teams. Even with a business that has people in the office full time the amount of desk time can be lower than expected. For example, with a workforce employed on an eight hour day, their desk is only going to be occupied for one third of the day. When bank holidays, annual holidays, sick leave, maternity or paternity leave and lunch breaks are taken into account the actual time the desk is occupied can be surprisingly low. This may seem like an HR issue, but it is where HR and property meet, and may offer scope for the CREM to add value to the business. Understanding the dynamic of desk occupancy can lead to changes in the way the business operates to reduce the costs of space, which in turn, will feed through to increased profitability of the business. It is not a simple matter of getting people to share desks, but creating the right environment for a business to migrate to flexible working. This is an area that a CREM needs to understand and apply to the corporate occupier's business.

So what are the elements that need to be measured? Looking at the office environment, the following are key essentials:

- space — gross and net floor areas; space usage; the allocation of space between different uses — meeting rooms, cellular offices, open plan offices, canteen facilities, break out space, reception areas and the like
- space use — people employed; desks per department unit and the total for the property; people with fixed offices; people with fixed desks; flexible space users; home workers
- property costs — rent; rates; building insurance and service charge
- facilities costs — energy; cleaning; repairs and decoration; furniture costs and the like.

The objective is to provide an analysis at various levels and to benchmark a corporate occupier's buildings for efficiency and effectiveness. This needs to be undertaken within the portfolio and also compared with other businesses in the same sector and outside of the corporate occupier's group. As with most comparative exercises, the bigger the sample of comparative portfolios, the less opportunity there will be for portfolios at either end of the spectrum to skew the results.

Before starting the exercise, the CREM needs to determine suitable standards for the business. Is the average space per desk to be 120 sq ft or 150 sq ft? Adopting a standard at the outset enables the current performance of the portfolio to be measured. This is only the first stage of the exercise because the results are not an end in themselves. They mark the start of a process of improvement. Consequently, if the benchmark has been set from other portfolios at 120 sq ft per desk and the outcome is 130 sq ft accross the portfolio, the corporate occupier will want to move towards the 120 sq ft over time to reduce costs and increase profitability. However, the benchmark may need to be adapted to reflect ongoing changes. By way of example, a benchmark of 130 sq ft per desk may also represent 130 sq ft per person, that is one desk per person. However, by moving to more flexible working it may be possible to change to eight desks per 10 people in which case savings start to flow, with a 20% reduction in desks, resulting in additional savings.

Clearly, savings which result from efficiency of use are quite independent from any additional savings which might be achieved by evaluating the base costs of the space. While some of the savings may be relatively small, others may be more significant and these savings flow through to the bottom line of the business, without any other issues being taken into account.

2.3 Surplus property

2.3.1 Development of legacy portfolios

The surplus property portfolio is also known as the legacy portfolio, it being the legacy from previous business decisions. A legacy portfolio may be the consequence of a bad decision (or series of bad decisions), or because of a change in direction of that business. A good example of legacy portfolios created by business change is those that have been created in the insurance industry. The traditional property model led to a local branch from which the sales representatives would sally forth to find new customers and collect premiums from existing ones. The prime example of this was the 'Man from the Pru'. The salesman would head out on his pushbike from a local office to see his customers weekly or monthly and while collecting premiums would be looking for new clients or to write new policies. The branch network had to be fairly dense to reflect the multiple short journeys, but as the use of the car took hold the network became more flexible, and fewer branches were needed. It is interesting to note that all of the general insurance companies followed a similar pattern with their town centre offices to the extent that it was often possible to name a city and guess which building an insurance company was occupying.

This property model worked well for many years but by the early 1990s the nature of the business started to change. The costs of providing services from local branches started to be disproportionate to the premiums received. In addition the role of the collector and the salesman were separated, with the former done centrally by the business, leaving the branches to become sales offices. Over time the area that one individual was expected to cover became larger with improvements in the road network and wider use of cars so that fewer branches were needed. The birth of the legacy portfolios ironically started with the death and closure of the small local branch office. This was however an incremental change in comparison to the next change, which was the move away from direct selling face-to-face to the use of telesales and then the internet.

The first brought in call centres, large semi-industrial buildings that housed large numbers of workers selling products on the telephone. As a consequence a large number of branches were closed. The situation was then compounded by the internet and on-line sales. There was a sudden growth in call centres in the UK but the boom was fairly short lived, before the costs of the centres and the problems associated with running them begin to impact on businesses. One problematic issue for the UK call centre providers was the global nature of the business and the move towards a 24-hour business day. To improve time coverage, reduce costs and improve quality, there was a move away from the UK to India. Employment and property costs were considerably lower in India so in a business that was seeing ever-tighter margins, a number of insurers moved their operations off-shore. Consequently, the original call centres joined the legacy portfolios of the insurers. The dramatically shorter business cycles when compared with lease durations (25 -30 years) mean that the original legacy properties are still in existence. The next phase may already be on the horizon in that the offshore call centres are now under pressure, in part from rising costs because so many others have followed the same route, but also in view of poor standards of service. We are now seeing a number of companies bringing these operations back to the UK, but then looking for a different type of building and different locations.

Business change is a fact of life and the CREM needs to be aware of that and plan the portfolio accordingly, Nevertheless, it is clear that because business and property will never be totally aligned, there will be a continual supply of surplus properties generated by ever changing business needs.

2.3.2 Traditional solutions

A detailed consideration of break rights and alienation (ie assignment and subletting) is contained in Chapters 12 and 5 respectively. The discussion in this section focuses upon the practical considerations that a CREM needs to take into account when evaluating the various options and strategies available for surplus property.

(a) Surrender

The ideal solution to a surplus property problem is the surrender of the unit(s) back to the landlord(s). Very few leases provide the tenant with a unilateral right to surrender. It follows that a surrender can only take place with the agreement of the landlord. It cannot be forced upon the landlord. If the landlord does not want to take a surrender of the unit there is nothing the tenant can do to force the issue.

In almost all cases (except perhaps where a landlord has its own agenda for recovering possession such as redevelopment), a landlord will be looking to secure a premium as compensation for interrupting its income stream. Surrender negotiations tend to be difficult as a result of the different ways in which a landlord and a tenant approach the assessment of the value of the reverse premium.

The tenant's valuation will tend to be influenced by the parameters that determined the provision it has made for that lease in its accounts and use that as the benchmark against which to negotiate. For example, if the tenant has been advised it is likely to find a subtenant within two years subject to giving a six months rent free incentive, is the appropriate surrender premium 2.5 years rent? Unfortunately (from the tenant's perspective) this is too simplistic and will be challenged by a landlord for (at least) the following reasons.

- The value of the premium must take into account the total costs of the lease. The value of the rates, service charge and insurance will all need to be added.
- In many cases there will be a difference in the dilapidations liability between the headlease and the sublease. This may be because the definition of the demise is different, for example partitions are included in the sublease but not the headlease, or because the obligations in the sublease are reduced by a schedule of condition.
- It is not uncommon for a subtenant to insist on the sublease containing a break clause. Clearly this will impact on the certainty of the income stream from the sub-rent.

Even with these additional points factored in the valuation is unlikely to be the same between the landlord and the tenant, except perhaps if the lease is coming to an end. This is because the landlord is likely to approach valuation from a very different perspective. The landlord will take into account the impact a surrender has on the capital value of the property, and not just cashflow issues. So while it may agree on the length of time it will take to re-let the premises and the likely rent, the landlord will then input that data into a capital valuation model that will tend to result in a bigger net impact than the tenant's cashflow model. There are other factors which may influence the landlord's approach.

- *Covenant strength*: if there is not a like for like replacement of the covenant, factoring in the loss, for example, of a triple A covenant will make a bigger difference because this will result in a yield movement.
- *Mortgagee*: most landlords typically borrow 60% of the cost of purchase from a bank. There are two consequences from this. First, from a cashflow perspective the landlord will be acutely aware

of having to continue to pay its loan instalments when cash is not being generated from the property. Second, and more importantly, the bank is likely to be risk adverse and it will be difficult for the landlord to persuade the bank to agree to a surrender. The bank will be keen to retain the covenant and a well let property so as to ensure that it can easily sell if the landlord defaults to recoup all or part of the loan.

- *Human element*: why bother? A landlord may ask itself why it should bother making work for itself in having to find a new tenant to replace the good covenant that it already has? With this point in mind, the premium must contain an additional amount over and above the other sums to give the landlord an incentive to do the deal otherwise the most convenient thing to do is nothing.

In view of the above points, in most cases it will be difficult for the landlord and the tenant to agree a reverse premium. A surrender can become more viable as the lease gets closer to expiry, and in particular in the last two to three years when the capital value of the property is starting to see the imminent lease expiry impinge upon the yield and possibly the rent as well. However, a surrender of a lease towards the end of its contractual term does very little to help a corporate occupier's surplus portfolio, other than to get rid of the tail end of a lease. The other situation where a surrender can be more viable is where the corporate occupier has found a replacement tenant thereby removing any risk of the landlord suffering a rental void. However the dilemma which the tenant then faces is whether the level of surrender premium which the landlord seeks justifies the finality (and avoidance of contingent liability) which the surrender will bring. Avoiding the payment of a premium by merely assigning or sub-letting the premises even with the risk of future contingent liability may be more attractive.

(b) Break options

An important way of mitigating the surplus property portfolio liability is to ensure that break clauses are exercised. Missed breaks dates or break rights that are not properly exercised (resulting in the loss of the break right) are wasted opportunities. Failure is often caused by the break details being incorrectly recorded in the corporate occupier's database or missed out altogether. Another common cause of break failure is an omission by the tenant to comply with essential conditions which must be complied with before the break right can be exercised. Often, for example, the tenant will fail to comply with its repairing, redecoration or reinstatement covenants before the break date, which depending on the wording of the condition precedent, may result in the tenant losing its break rights (see Chapter 12).

A break clause in a surplus property can be an aid to agreeing a surrender deal with a landlord, even if the break is several years away. For the tenant, agreeing a surrender of the lease may be far easier than trying to comply with the terms of the break clause. In addition, while a break if properly exercised will bring the total liability to an end, it can be detrimental in trying to create short term income from a subletting, because there is usually the perverse turn of nature that results in the potential tenant of a property with a break clause wanting a long lease, whereas the lease with no break will see a potential tenant wanting just a short term.

(c) Assignment

A further solution to a surplus leased property is to assign or sublet it to a new tenant. However, is it better to assign or to sublet the surplus lease? The answer to that depends on the CREM and the business. At present, accountancy rules do not require a corporate occupier to make a provision for an assigned lease in the same way that it is required to do where a property has been sublet. However, where a property has been sublet, the corporate occupier remains in control of the property and is able to deal with issues that arise directly, rather than run the risk of the assignee defaulting at some point in the future and then having to become involved in a long and often costly process of regaining control (see Chapter 10 on contingent liability and in particular obtaining an overriding lease or vesting order).

Often there will be a desire within a company to get surplus property off the books and proceed by way of assignment. However, the problems arising from contingent liability should not be underestimated and in many instances the commercially prudent approach is to retain control through subletting rather than assigning, so as to deal with issues as they arise. Importantly, by collecting the rent from the subtenant, the corporate occupier will be best placed to pick up at an early stage if the subtenant is getting into financial difficulties and take decisive and prompt action to collect arrears or forfeit the sublease if the need arises. The same is not the case where an assignee defaults in the payment of rent. It can take a number of months before the landlord pursues a former tenant for the arrears, and then a further number of months for the former tenant to obtain an overriding lease and then forfeit and relet the premises. During the whole of this period the former tenant will be liable to pay the rent (plus interest and cost) and have no opportunity of mitigating its loss by reletting.

An assignment of a surplus lease stands a better prospect of succeeding where the lease is at a rack-rent and is assigned to a covenant that is as good if not better than the assignors.

(d) Subletting

As indicated above, the great benefit of subletting is that it leaves the corporate occupier in control of the space, so that it can then manage out any of the issues that might arise during the course of the remainder of the lease. In many cases, subletting can be the only viable option because the remainder of the unexpired lease exceeds the length of term that potential tenants want. The length of lease required by occupiers has fallen dramatically in the last 15 to 20 years with the occasional blip, at times of shortage of space, when landlords have been able to force tenants into taking longer leases. Often a corporate occupier will decide to vacate space and seek to dispose of its surplus leases at times of economic down turn or expansion when there is a large amount of available property on the market. As a consequence, potential sub-tenants are usually in a stronger bargaining position and able to force the landlord to accept favourable terms (and build in flexibility) in the sublease.

The mismatch between what the corporate occupier is seeking to dispose of and what potential sub-tenants want to take, results in either a subletting for a term considerably shorter than the headlease, or an assignment of the lease coupled with an option for the assignee to require the assignor to take the lease back after a certain period or defined periods (ie akin to a break option). The assignment option can be too complex for most and will often prejudice the ability of the corporate occupier to dispose of the lease by way of assignment.

2.3.3 Sublet portfolio

The need for a separate management system to run a sublease portfolio will be determined by the size, scale and complexity of the portfolio. With a small number of relatively straightforward properties they can be managed within the system established for the operational portfolio provided it has a fair degree of flexibility. However, generally speaking, the larger and more complex a portfolio is the more benefit will be gained by having a separate management system.

(a) Sub-tenant issues

Most modern leases will require that any sublease should mirror the headlease terms and so with the possible exception of the length of the sublease, there tends to be little flexibility to change the terms. If some flexibility is permitted by the headlease alienation clause this is likely to involve the repairing obligations and occasionally rent and rent free periods.

As far as the length of the sublease is concerned this must be shorter than the headlease. As a headlease gets closer to its expiry it will become more difficult to find a tenant which wants such a short term sub-lease (other than perhaps a tenant looking for a small office suite). This is down to the substantial costs associated with occupying new space such as:

- moving costs
- IT fit out costs
- telephone and internet connections
- stationery costs.

If the corporate occupier has a surplus lease with a long unexpired term, usually a subtenant will insist upon a short lease or a longer lease with regular break clauses. The difficulty with this is that a corporate occupier is likely to be left with the fag end of a lease that will be difficult to assign or to sublet again. By the same token, where a sublease has been granted in the past, there can be significant value to a corporate occupier in negotiating the removal of the tenant break rights (eg by a deed of variation in return for a lower rent or a relaxation of a user restriction), but this is generally under-estimated by a corporate occupier.

The other common difference between the sub-lease and the headlease is the repairing clause. While the principal wording may be identical the obligation is often qualified by a schedule of condition so that the tenant is not required to hand the premises back in any better condition than evidenced by that schedule (see 14.2.5). The main impact of this occurs at the end of the leases. If the head and sub-lease are co-terminus then the dilapidations liability and negotiations will differ between the two leases. There is a significant risk for the corporate occupier of becoming the 'piggy-in-the-middle' of the tripartite negotiations which will take place between the corporate occupier and the head landlord on the one hand, and the corporate occupier and the sub-tenant on the other. The risk can be illustrated by looking at the scenario where a corporate occupier has leased offices and during its occupation added partitioning and air conditioning units. When the premises become surplus to requirements they are vacated and subsequently sublet with the partitions and air conditioning in place. The sub-tenant's repairing and re-instatement covenants are qualified so that it does not need to be put the premises into any better condition than that evidenced by a schedule of condition. On expiry of the leases, under the headlease the corporate occupier is required to strip out the partitions and the air conditioning, but under the sublease the subtenant is only required to redecorate and repair the premises in accordance with the schedule. In these circumstances the subtenant may well argue that any works that it was required to carry out would be wholly superceded by the much greater strip out works that were required under the headlease (see 14.2.5).

(b) Rent arrears

The issue for a corporate occupier who has sublet space is cashflow — positive and negative. Having a subtenant in place not only provides rental income it generally removes the obligation to pay rates and service charge. In many ways the removal of those costs is more important than the rent paid, as they will tend to start as soon as the lease term starts and in the case of rates means that the corporate occupier does not have to pay money out.

Where a subtenant falls into arrears the corporate occupier has to carry out a difficult balancing act when deciding whether or not to seek possession of a unit. It may be inclined to follow a landlord's approach and focus on the rent arrears, but for a surplus property it must also consider the increased costs which it will have to bear and there can be little upside from forfeiting the lease. While a freeholder would also have to bear the costs of rates and service charges it would have the benefit of its interest in perpetuity so that if it takes a year to relet the premises, the freeholder can still let the unit for 10 years if that is what the market is demanding. However, with surplus property if there is only 10 years left on the lease, waiting for a year means it can only grant a nine year lease and that year will never be recovered. Therefore weighing the pros and cons is much more important for a corporate occupier.

2.3.4 Vacant portfolio

The management of individual vacant properties and disposal issues are covered in Chapter 11. When the entire portfolio is vacant it is generally at the point of a major change having taken place. A good example of this is immediately after a decision has been made to close down a group of properties or an entire division. Such decisions tend to be made and implemented very quickly, for example, a telephone call made to all the managers requiring them to send everyone home as the division has been closed down. For the CREM this causes logistical problems. With one or two units it is not difficult to go to the premises to check/secure them and read the meters. With a number of premises that is considerably harder and a different approach is needed.

If, for example, a corporate occupier has 50 vacant properties, third party involvement will be needed unless the corporate real estate team is sufficiently large. One option would be to use local agents. However, this will be slow and cumbersome involving probably more than 50 phone calls. The overriding immediate priorities following closure of the business are security and safety. To that end a national security company is probably the best option. A good second choice would be a cleaning company, especially if there has been a national contract for all of corporate occupier's branches.

The immediate tasks which need to be carried out are to:

- collect all the keys from the key holders
- check all windows and doors and ensure they are secure
- ensure the premises are secure (changing the locks as appropriate)
- take photographs and/or video footage of the premises
- check gas, water and electric meters
- ensure there is no risk from heating or power items — all electrical appliances should be turned off and disconnected from the electrical sockets.

The above tasks should be evidenced by a checklist containing a comments section so that whoever is carrying out the inspection can tick off the items and add comments.

The provision of photographs will help with the preparation of an action plan for disposal of furniture and the like. While the initial focus is on security and health and safety, the CREM needs to have an eye on the disposal of the properties in the future, and therefore these initial steps will help formulate an action plan. Without understanding what is in the premises it is difficult to implement removal of equipment and furniture, and it is unlikely that the former employees will be too helpful in the process. It may also assist the CREM to obtained copies of relevant contracts for the property eg cleaning, alarm provision, water cooler and the like. Indeed central contracts do make coping with this event considerably easier. Clearly from the CREM's point of view, it would be beneficial to have prior warning of the board's likely decision to close down a large number of units. Once on notice, the CREM could then start to make preparations for the task that lies ahead.

The process of mothballing the premises and the pricing of the IAS37 liability should be carried out straight away. It is not uncommon for the closure programme to be carried out very swiftly, but for the CREM to fail to do anything other than to instruct local agents to dispose of the properties on a piecemeal basis. As a consequence, nine months later the offices are in exactly the same state as they were on the day the office had been closed, resplendent with papers and coffee cups on desks, and the heating still running!

When a unit first becomes vacant there are a number of matters that need to be addressed.

(a) Security

Security is paramount. What will be required will depend on the nature of the building, its location and configuration. Issues to address are likely to include:

- external locks — change them immediately
- install roller shutters to protect lobby areas
- padlock gates and use internal bars on doors so that there is just one point of entry
- windows — use existing window locks
- board up windows if necessary, especially for single glazed windows without locks on the windows
- letter boxes — seal them and redirect the post.

(b) Services

- electricity — consider leaving the power on for safety reasons
- water — switch off and drain down when possible with industrial units
- heating — for industrial units turn off. For offices ideally leave it on to prevent dampness.

(c) Exterior

It is important to ensure that fabric is maintained.

(d) Insurance

Insurance policies often require the tenant to notify the insurer of any change of occupation and this includes vacating the unit. Insurers may require certain steps to be taken such as the draining down of the services and importantly the carrying out of inspections on a regular basis. Such inspections can be very frequent and as such could involve the corporate occupier significant additional costs. The insurer's requirements need to be considered carefully. It may be necessary to enter into a dialogue with the landlord and its insurer to ensure that their requirements are sensible.

(e) Squatters

The risk of squatters is always a problem and great care should be exercised to avoid the risk.

2.4 Portfolio structure

One aspect of owning and managing a portfolio of properties that is seldom given sufficient consideration is which legal entity should take on the lease or acquire the freehold. There can be very significant benefits or disadvantages from a particular group company being granted a lease or purchasing a freehold and this is an aspect that warrants careful consideration.

2.4.1 Parental ownership

There are several advantages in placing all of the real estate assets and liabilities in the parent company:

- the parent company is usually the strongest covenant in a group of companies and therefore it should be easier to take leases
- the central management team will receive all demands and will be in control of expenditure
- if the scale of the portfolio is such that a CREM and/or team is needed that team will be at the heart of the business and should therefore have a clearer understanding of what the business is seeking to achieve
- compliance issues have a visibility at the top.

The disadvantages are:

- disposal of subsidiary companies can be difficult. Properties in which the subsidiary operates will need to be assigned across if leasehold, or sold at value if freehold. This can slow down the transaction and will mean that external parties will have to be involved in the transaction.
- the individual businesses do not have direct control of their property costs and this can lead to a lack of responsibility for property within the operating companies.
- the best quality covenant is given away at the start of the lease.

2.4.2 Operating company

If each individual company in the group takes on the running of their own portfolio the advantages are:

- the business sees its property costs and will tend to be more circumspect than if the costs are a remote central charge
- there is a direct link between the property decision maker and the business, avoiding third party involvement.

The disadvantages are:

- the operating companies will tend to seek to mitigate expenditure on the properties, especially in relation to maintenance and this can lead to issues around health and safety compliance as well as dilapidations.

- if there is generally a lack of expertise within the operating companies in respect of property issues that can lead to problems and costly mistakes.

2.4.3 Property company

One further option, which does not to arise that often in practice, is the creation of a property company within the group, which will take on and deal with all property portfolio issues. It tends to be more appropriate for larger companies which have large portfolios and a number of different operating companies. The advantages of having a property company are:

- property expertise can be focused in one company and available for dealing with the portfolio
- control of health and safety issues can be more easily undertaken
- accounting issues are dealt with centrally.

The disadvantages are:

- the accounts for the property company can look strange and make it difficult to get approval from landlords for new leases
- disposal of group companies can be difficult. Properties in which the sister company operates will need to be assigned across if leasehold or sold at value if freehold. This can slow down the transaction and will mean that external parties will have to be involved in the transaction
- there is separation between decision making and the business in occupation.

There is no right answer to ownership; it depends very much on the particular business in question. Nor is there just one solution. For example, in some circumstances a property company could be created to hold all the surplus properties and act as managing agent for operational properties, thus controlling all costs and ensuring a professional approach to the day-to-day issues. Costs are then charged back to the operating business on a monthly basis to ensure that the appropriate costs are met.

2.5 Disposal options

Chapters 5 and 11 consider in detail the three basic approaches to disposing of surplus property — surrender, assignment and subletting. Within a portfolio of legacy properties, these options may provide solutions on an individual property basis but they do not provide a comprehensive solution to the overall problem. It is important to remember that property is not a core function of a corporate occupier; surplus property is just a drain on finances and management time and ideally most would like to dispose of in one fell swoop. How is that achieved?

There are a number of possible options and these can be categorised as follows:

- manage out the portfolio to expiry of the last lease
- sale of the surplus leases to a third party
- sale of the companies who have the liability
- use of the covenant strength to unlock the negotiating impasse with a landlord.

2.5.1 Manage out

At the simple level, managing out involves a process of work through the portfolio and looking to sublet surplus space, surrender where possible and assign the remainder. This provides certainty for the business in that it allows it to control the portfolio through to the end of the last lease, and to seek to maximise income while mitigating the liabilities. This is the base option for a corporate occupier and one that alternative options should be judged against.

If the IAS37 assessment has been properly carried out and the outcomes are broadly in line with the assumptions made, the corporate occupier should see the out turn cost match the provision that was made. In other words, if the assumptions become reality, the provision will match the expenditure. There is clearly scope here for the corporate occupier to either under perform or out perform the IAS37 figure. This can be caused by a number of factors:

- an aggressive or conservative approach to the IAS37 assessment initially
- changes to the market place, so what was a good market at the time of assessment becomes a poor market with space difficult to dispose of and long rent frees having to be granted
- sub-tenants exercising break rights when it had been assumed they would not, or not exercising when the assumption was that they would
- luck.

If there is out performance of the IAS37 figure then the corporate occupier will have an excess provision to release back to profits. Should there be under performance then the corporate occupier will have to make good the shortfall at the end of the leases. This illustrates the importance of reviewing the IAS37 figure on a regular basis. Once a variation in an original assumption is identified the next assessment of the provision should be increased to reflect the revision to the projected outcome. If the position continues to worsen there will be a need to carry out further adjustments to reflect the deteriorating position. The effect is likely to be a drip feed of the extra costs at the end of the life of the portfolio. Companies dislike making revisions to provisions because of the impact it has on the year in which the further provision is made (in particular a reduction in the amount of money available for distribution to the shareholders).

If the corporate occupier manages out the portfolio, irrespective of whether the outcome is above or below provision, the business remains in control of the portfolio throughout its life. That offers both advantages and disadvantages.

The advantages include:

- the corporate occupier remains in control of the portfolio
- risks are kept at the property level ie the outcome is down to how a particular property performs and is not dependent on a third party's performance
- while a provision has been made in a corporate occupier's accounts the expenditure is over time and consequently the money remains in the company as long as possible.
- The do nothing option tends to be the easy option.

Disadvantages to retaining the properties are:

- it requires the corporate occupier to have a continuing involvement in the portfolio
- it ties up considerable management time

- there are continuing accounting issues
- it is necessary to ensure that compliance issues are dealt with (both financial and health and safety)
- it may remain visible as a sign of previous mistakes by the company.

How these advantages and disadvantages are viewed in comparison to the pros and cons of alternative options is something that will differ from company to company. There has been a move by some corporate occupiers to try and offload their surplus portfolios as they seek to focus on their core business. This has gone hand in hand with companies looking at the best way to gain competitive advantage and thereby improve profits. This usually involves the carrying out of an overall review of many functions and considering the possibility of outsourcing part of the business (eg telephone support or sales to call centres) or relocating offshore those function which are not seen as core to the business and which can be better performed by specialists and/or at a lower cost. As with all such concepts, it does not necessarily suit every company and a number have tried such changes to find that they are an expensive mistake.

2.5.2 Sale of leases

There is a market, albeit limited in size, for surplus leases. There are companies that will acquire surplus leases (and their liability) in exchange for payment from the corporate occupier. The business model is based upon an assumption that the purchaser can outperform the corporate occupier in managing out the liabilities (by achieving more favourable surrenders, sublettings or assignments than the corporate occupier would be able to achieve) leaving the purchaser with a surplus from the transaction which can be retained as profit. In other words, the premium paid to the purchaser must be sufficient to be applied to managing out the properties and also generate a profit (after all of the surplus leases have been disposed of, surrendered or their terms expired).

The sale of a surplus portfolio will usually involve the following steps.

(a) Decision to sell

The decision to sell is not always simple and can be fraught with difficulties. In arriving at a decision to 'sell the liabilities' (which in effect requires a payment to a third party to take on those liabilities) a corporate occupier must consider carefully the following factors:

- does it have money available to pay the purchaser for the liabilities?
- does it have the resources available to manage the process of disposal? Depending upon the size of the portfolio, such disposals can involve enormous amounts of management time. The corporate occupier will use external advisors for a number of the activities (eg surveyors and lawyers) but nevertheless there will be a considerable call on time for the company, especially the finance and legal teams and the CREM
- what risk profile does the corporate occupier want to bear in the future?

If the decision is taken to sell the portfolio without key factors having been considered fully, there is a real risk that the process may be aborted and resulting in a waste of time and costs.

(b) Revision of IAS37 provisions

While the accountancy rules require a corporate occupier to review and adjust a provision for a surplus property annually, many companies will not do this, but will make adjustments over time to the provision perhaps carrying out a full review every second or third year. Prior to embarking on a surplus portfolio sale a full review of the provision is needed. It is essential that the business knows what the likely costs of the transaction are going to be and the biggest element of the disposal cost will be the liabilities of the surplus properties. It is therefore important to have an up-to-date provision figure to enable the corporate occupier to evaluate whether the asking price (ie reverse premium) put forward by the potential purchaser is attractive or not.

(c) Data gathering

It is essential that all relevant data is collected, organised and made readily available for potential purchasers to view. Often such data can be stored and accessed by a potential purchaser in an e-data room. While the organisation of the data does not necessarily need to be done at the outset, it is a task that can take a considerable amount of time to complete, especially for a large portfolio that has been accumulated over a period of time and from a variety of sources. Often in complex sales, the corporate occupier will instruct its lawyers to review and organise the deeds.

(d) Tender exercise

Obtaining prices from interested parties can be carried out by means of a tender, formal or informal, or it can be by approaching one interested party and negotiating directly with them. The choice of approach depends on whether the business wishes to undertake the disposal process confidentially or is looking to be seen to have achieved best value. In most cases a degree of competitive bidding is essential to ensure that the pricing by a potential purchaser is market tested. The choice of purchaser should not simply be determined by price; it is important that the organisations can fit with one another. This may seem strange in that in one sense the two organisations are simply vendor and purchaser. However, the process of disposal of the leases is generally a fairly long one and they need to be able to work closely together. This cooperation will of necessity need to continue past the completion date of the disposal because of the very nature of property.

(e) Transfer of leases

There are basically three ways a sale may be structured:

1. a share sale of a property company that holds all the surplus portfolio leases
2. a legal assignment of each and every surplus lease to the purchaser
3. a virtual assignment (possibly).

The cleanest and quickest sale occurs where all the surplus properties are contained in a property company. The purchaser can purchase the shares in that property company and no protracted lease assignments are necessary. If the company has been the principal property holding company for many years, it will have the added advantage for the vendor of avoiding contingent liability altogether. The only exceptions to this are:

- in the context of an old tenancy, where the lease has previously been assigned by a group company (and which will therefore have contingent liability from its status as former tenant) or where a group company is a guarantor or
- in the context of a new tenancy, a group company has given an AGA or stands as a guarantor for the property company.

If the property company does not hold all the surplus leases which are intended to be sold, or a special purpose vehicle (SPV) is set up, leases will have to be assigned into the property company or the SPV before the share sale. This can give rise to two problems. First, the process of obtaining landlord's consent can be lengthy and it is important that a proper application is made as quickly as possible, and the landlord be reminded of its duties under the Landlord and Tenant Act 1988 (see Chapter 5). Second, if the property company simply comprises all of the group's onerous surplus properties, it will be a pretty unappealing covenant prospect with accounts likely to put off even the most relaxed of landlords. In such cases it may be necessary for a group parent company to offer a guarantee covenant to smooth over the assignment and enable the sale to proceed.

The second possible structure involves the laborious process of assigning each and every surplus lease to the purchaser. This can be a very lengthy process as it will require landlord's consent to most, if not all, of the proposed assignments. It may also raise difficult questions of covenant strength. In most cases however, a landlord will back down if it thinks the tenant is serious about litigating the issue. Precedent letters can be found in Chapter 5. One useful tactic can be to put a landlord on notice that the assignment is part of a much larger transaction and that if consent is not forthcoming within a reasonable time and the transaction flounders, the landlord could be liable for substantial damages.

In view of the delays caused by the assignment route, innovative lawyers (under pressure from clients and surveyors) have invented a device often referred to as a virtual assignment. A considerable amount has been written about virtual assignments and the benefits they offer. The concept was developed in the 1990s as part of the outsourcing of property portfolios predominantly by the government. The portfolios that were outsourced were generally of a mixed status including operational and surplus properties. The intention by the government was to bring in the right expertise to manage what had become rundown properties and also to inject much needed capital to improve its property stock. The government had neither the capital available nor the expertise in house to run such a process itself. It therefore entered into an arrangement with a supplier or number of suppliers whereby all the property interests were to be transferred and the government entered into a contract for 35 years.

However, a difficulty which faced the parties was how to transfer the lease liabilities across to the third party supplier. Straightforward assignments would not work easily because of the scale of the portfolio. There was also a concern that because the covenant strength of the government would be difficult to match, large-scale rejection of applications to assign were likely. The other major factor was that of speed. To undertake such an exercise would take a considerable length of time to complete and would delay the start of the contract. Virtual assignment was the solution put forward to deal with these problems. This involved splitting the legal interest in the lease and the beneficial interest. The legal interest would remain with the government, so that there was no need for the lease to be formally assigned, but the beneficial interest would be transferred to the supplier by means of one contract that covered all the properties in the portfolio. Consequently, the desired result was implemented within one document.

There have been a number of variations since in the way that the virtual assignment is legally documented. So far this device has worked well for this type of arrangement, where the purchaser of the leases continues to provide a service to the vendor. However, the surplus property was usually a very small element in the overall deal. The primary focus has been on the operational property side, in particular, the services provided by the purchaser under the facilities management part of the contract such as the refurbishment of existing space and the acquisition of new space.

It is very doubtful whether this device will provide any assistance in the context of a stand-alone disposal of surplus leases. Most well drafted alienation clauses will prohibit the tenant from parting with the beneficial interest in the lease and/or parting with possession. If a sub-lease is put in place, and the supplier is given a right to receive rent, this will constitute an assignment of interest in land and therefore breach the alienation clause. While virtual assignments have been tested in the VAT context (see *Abbey National plc* v *Commissioners of Revenue & Customs* [2006] EWCA Civ 886) they have yet to be tested in a case where a landlord alleges that there has been a breach of the alienation clause and seeks an injunction to reverse the unlawful alienation. It is suggested a corporate occupier's energies would be better rewarded by pursuing options one or two above, and not relying upon what on any view is a doubtful and untested device.

Even if the wording of the relevant alienation clause permits a virtual assignment it is not an ideal solution. It leaves the corporate occupier directly involved. Any legal matters concerning the lease will need to come through the company. It is also indirectly involved because as far as landlords are concerned, the purchaser is merely the managing agent for the corporate occupier. Consequently there are ongoing risks to the business eg from problems with health and safety issues or the like.

(f) Payment

The simplest means of payment is for the corporate occupier to transfer to the purchaser in cash the agreed price. For some businesses this may be a problem because of liquidity or banking covenants. One option is to transfer freehold assets in lieu of cash. However, this adds an extra dynamic because not only must there be agreement on the liability pricing, but there must also be agreement on the value of the freeholds. It may also slow the process down as it provides additional work for the parties' professional advisors. If the freeholds are being offered with vacant possession or with an existing lease in place the additional work may not be too significant, but if there is to be a sale-and-leaseback then the terms of the lease and in particular the rent will have to be agreed. Accounting issues will also need to be addressed with the rent and, in particular, whether the lease is to be an operating lease or a finance lease.

(g) Credit enhancement

One of the most important issues to consider is to what extent the corporate occupier is at risk of the liability bouncing back through contingent liability if the seller becomes insolvent. The complex rules relating to contingent liability are fully explored in Chapter 10. With complex portfolios, corporate occupiers will usually instruct lawyers to review all the deeds and provide contingency liability reports to identify any circumstances where the liability might bounce back.

If the sale is by way of a share sale of a property company the only possibility of contingent liability is if a group company was a former tenant or guarantor of an old tenancy, or if it stood as surety for the property company of a new tenancy.

However, if the sale is to be by way of either the purchase of a SPV or the assignment of all of the leases, the corporate occupier as former tenant is likely to have a significant contingent liability and this will be an important factor for the corporate occupier in negotiations. The corporate occupier can seek to mitigate these risks by seeking additional security from the purchaser (eg by way of bank guarantees or bonds).

(h) Completion

The precise nature of completion will depend on which structure has been used and whether the transfer of the leases has been:

- direct to the purchaser
- into a SPV that the purchaser has created
- into a property company within the corporate occupier's structure.

The most common structure usually involves transferring the surplus leases into a SPV. Whether the SPV is owned by the purchaser or the vendor depends on the particular circumstances of the transaction. If it is within the corporate occupier's group it has the advantage of simplifying the process because the transfer involves the sale of a company to a third party and the corporate occupier is then in control of the assignments into the SPV, and the funding of the company to match the obligations. Completion of the transaction is then a transfer of shares in the property company or SPV, which is a relatively simple and everyday corporate transaction and which has well-established protocols for dealing with apportionments and the like.

The process involved with a portfolio sale is therefore relatively cumbersome, and the benefits need to out weigh the costs (which can be very significant) in the broadest sense to justify going down such a route. If the premium and professional costs are likely to be substantially less than the provision that will be released to profits if the transfer goes ahead, and the corporate occupier has some measure of protection from contingent liability, this means of disposal may be a very effective way of dealing with a surplus portfolio. However, it is not for the faint hearted and the process does raise the question whether there are other less difficult and risky alternatives.

2.5.3 Disposal of dormant companies

If all lease liabilities are already held by one company, then as touched upon above, consideration should be given to whether it would be feasible to sell the company itself. That will depend upon what the company is currently doing. If it is the holding company or a main trading company then this is unlikely to be a feasible option. However, if it is a dormant company, that is one that has ceased trading, or a company that is undertaking little trading activity, then it is possible to create an SPV from an existing company. There are a number of legal and accounting issues that need to be addressed in doing so and these are beyond the scope of this book, but it is something that is attainable if the corporate occupier takes the right advice.

The process of preparing a company for disposal in an SPV may be something that will take some time. This will depend upon on what is required and also the focus the exercise gets from other parts of the business. Once prepared, the principal advantage will be that there will be no need to seek landlord's consent to the assignment of leases because those leases will already be vested in the SPV. Again completion of the transaction is a simple company sale and therefore it ought to be capable of being completed relatively quickly.

If the SPV only contains its own lease liabilities and there have been no previous inter group assignments or group company guarantees, there is a different dynamic in place. As a stand alone company with no links to the previous parent it becomes much more appealing to a purchaser because they will not have to consider privity issues in future negotiations with landlords and will be able to bring pressure to bear on landlords to accept surrenders in a way that they would be unable to do if the landlord had the security of a previous tenant and/or guarantor to call upon in the event of default by the current tenant.

That additional muscle for the purchaser is likely to flow through into the pricing of the legacy portfolio and the purchaser will be prepared to accept a lower payment from the vendor because they will be able to dispose of the lease liabilities at a lower price.

This approach can lead to an entirely different model altogether for the disposal of the surplus leases by the purchaser. If the SPV is sold to someone whose sole objective is to force landlords into accepting surrenders on the basis that there is limited funding available for the SPV being sold, it will be able to give the impression that landlords will have to compete for what little money there is for distribution. Such transactions have been carried out, although the sizes of the portfolios have tended to be small.

The number of operators in this marketplace is small but if the vendor is able to use it as a route, the price paid to the third party will be materially different from the IAS37 liability, something in the order of 50% to 80% less. Accordingly it is well worthwhile exploring what can be done with existing companies rather than automatically assigning to an SPV. Saving money is one thing, but the other very important benefit with this route is that it brings certainty to the corporate occupier. When it has been sold it is gone for good.

In reality, a clean dormant company disposal is unlikely. Usually a disposal is made up of a combination of dormant companies with property liabilities and surplus leases from operational companies where the leases are assigned to one of the companies being sold or an SPV.

2.5.4 Other options

For many corporate occupiers disposal of the portfolio is difficult and not viable. There can be a variety of reasons. For example:

- it could be that the surplus properties sit in the main trading vehicle or
- the corporate occupier does not have the cash available to pay a purchaser or
- there is some form of impediment to selling the company or
- the board does not want to spend the time and money structuring such a deal.

There are, however, other conventional methods of disposal available to the corporate occupier. Negotiations with landlords are all about perception, in exactly the same way as it is for landlords negotiating with tenants. When a tenant is looking to take a lease, negotiations usually involve other potential tenants in the bidding for the property and what is happening on rental values nearby. On surrenders it is usually the landlord who has the upper hand because the do nothing deal works in his favour. So generally the landlord has to be coaxed into agreeing a surrender deal. This is all about the balance of power, or more importantly the perceived balance of power, for perception can be more important than reality.

If a corporate occupier believes that a third party is about to take a much sought after building from right under its nose it is likely to react by improving the terms of its offer, often quite irrationally, as a means of securing the building. Landlords will therefore play on that to a greater or lesser extent, either just by raising the issue or at times creating a mythical prospective tenant.

There is the opportunity for a tenant to create a similar environment in respect of a surrender in certain circumstances. Where leases sit in either a dormant company or a company that is not trading well, there is scope to exploit the weak covenant of that company. This is the case whether the properties are operating or surplus properties, although the exact tactics, the way in which the landlord is approached and what is being sought as an outcome will vary according to the nature of the lease.

The passing rent, the open market value (OMV) and the yield drive the investment value of a property. The rent passing is a matter of fact as it derives from when the property was let or from a rent review. The OMV reflects the market place generally for that type of property in that location and is largely a matter of fact. The only other influence being will be the lease terms and the existence of restrictive covenants in the lease which will have a negative impact on the OMV. The yield is driven by the type and location of the property, the relationship of passing rent to OMV and when that might change. It is also influenced by the quality of the covenant of the tenant.

The assessment of covenant strength is arises from the credit rating of the company, which will come from the likes of Dun & Bradstreet who are continually monitoring the health of companies and issuing credit ratings. The better the credit rating for a corporate tenant, the lower the yield and therefore the higher the capital value. Very few landlords or their managing agents monitor what is happening with their corporate tenants and their credit rating. Indeed at times of bull markets when investors are chasing stock many investor landlords even at the point of acquiring a reversion will fail to properly research the strength of the tenant's covenant. If the agent's particulars say XYZ plc and it sees XYZ plc in occupation the prospective purchaser will carry out a search on that company. That tends to happen even where the tenant is a non-trading subsidiary and will have a materially different covenant strength to the parent company. For those that recognise the true covenant there is frequently an assumption that the parent will support the subsidiary because not to do so would breach the banking covenants of the parent. Such a universal assumption is dangerous at any time, but it can be particularly rash where a business is looking to dispose of surplus property or to maximise income from property.

This mismatch provides the tenant with leverage for dealing with the landlord either directly or indirectly. The direct approach is to highlight to the landlord the parlous state of the tenant's covenant. If the premises are wanted, the solution that benefits both parties is the assignment of the lease to a more substantial covenant in the group, in exchange for either variations to the lease eg a rent-free period or some other form of inducement. For a surplus property, a corporate occupier's objective is likely to be the surrender of the lease at a premium which is much lower than its provision for that property. There are a considerable number of nuances in how this should be dealt with; a well thought out strategy is needed to achieve results from this approach.

The indirect way of dealing with it is to involve a third party in some form of partnership, which can be an effective solution for surplus properties. It follows the same broad approach as selling a company subject to lease liabilities, but rather than spending time and effort resolving the credit enhancement scheme and all the other contractual terms, the sale is in effect a partnership approach to create the illusion that the company has been sold and that it will be put into insolvency unless the lease is surrendered. This has the benefit of removing the corporate occupier from the direct discussion with the landlord and therefore reducing the impact of any negative PR from such an activity. It is important however to conduct these negotiations very carefully. While bluff and counter bluff are very much part of the cut and thrust of commercial negotiations, no representations must be made that are not 100% accurate. If, in the course of the illusion, the corporate occupier or the third party makes statements which are not factually accurate, then any agreement may be set aside on the grounds of misrepresentation.

3 Occupation — Negotiations and Legal Agreement

3.1 Introduction

Once a corporate occupier has taken a decision to take on new space, it should give careful consideration to what tenure is required. The most secure form of tenure is freehold and in some circumstances (particularly in the industrial sector) a corporate occupier may elect to buy its premises. However, most corporate occupiers will take on space on a less secure form of tenure; the most common form being a lease.

As with most chapters of this book, the corporate occupier has two basic perspectives. In an operational context, the corporate occupier will be occupying the premises for the purpose of its business and the premises will be a factor of production. However, in a legacy portfolio situation, where the premises are no longer required for the purposes of the corporate occupier's business, it may be looking to sub-let the premises to offset its liability to pay rent to the head landlord.

A corporate occupier that is looking to take on space or alternatively to off load it to a sub-occupier will need to ensure that the most appropriate structure is put in place to suit its business needs. For example:

- if premises are needed by the corporate occupier as a temporary stop gap, security of tenure is unlikely to be important and an 'easy come easy go' arrangement may best suit its needs (ie a short excluded lease with a rolling break clause)
- if the corporate occupier will need to invest heavily in the fit out and has longer term aspirations for the property, security of tenure will be important and a longer term lease may be essential (ie a longer lease with security of tenure).
- if the corporate occupier wishes to sublet a surplus property, ideally it will want the lease to run as long as possible but it will (in most but not all cases) need to ensure that the lease is excluded from the security of tenure provisions contained in the Landlord and Tenant Act 1954 to enable it to hand the premises back to its landlord with vacant possession when the headlease expires.

3.2 Forms of occupation agreement

There are a number of different types of legal occupation of premises that corporate occupiers can be granted (operational context) or grant (legacy context):

- licence
- tenancy at will
- lease
- protected (fixed term/periodic)
- excluded (fixed term).

A corporate occupier needs to be familiar with each form of arrangement to enable it to determine which will best suit its business needs, and perhaps more importantly, which will not.

3.2.1 Licence

(a) Background

A licence does not confer any legal interest in land on the occupier. It is quite simply a permission from the owner of the premises without which the occupier's occupation would be unlawful and constitute a trespass. A licence is not protected by the Landlord and Tenant Act 1954 and with the possible exception of a tenancy at will (see 3.2.2), it is the least secure form of lawful occupation of land. As a licence does not constitute an interest in land it is a personal right which cannot be assigned by the licensor to a third party, and which does not bind a successor in title of the licensor.

In almost all circumstances this form of occupation will be unsuitable for the needs of the corporate occupier. This unsuitability arises not only because of the reasons given above, but also because of the difficulty and uncertainty of identifying as a matter of law whether a particular arrangement granted a licence, or whether it unwittingly granted a lease. However, despite its unsuitability for the commercial needs of landlords and occupiers of land, it is a form of occupation that a corporate occupier and its advisors may come across from time to time.

The licence/lease conundrum tends to arise where an occupier has gone into occupation of premises on an informal basis, or alternatively, where an agreement that purports to be a licence has been used by advisors who may be oblivious to the potential problems and uncertainty which such agreements can lead to.

(b) Distinguishing between a lease and a licence

The distinction between a lease and a licence can be of critical importance. In the commercial property context a licence confers no protection on the occupier under the Landlord and Tenant Act 1954. Once the licence is terminated the occupier must leave. In contrast, a tenancy (unless excluded or exempt from the security of tenure provisions) confers substantial statutory rights on a business occupier and can only be terminated in accordance with the 1954 Act (see Chapters 11 and 13).

This dispute is not one that has been restricted to the commercial context. In the last 20 years or so the debate has raged in the residential context. The motivation for the debate has been the same. If an agreement is a licence no security of tenure exists, whereas if it is a tenancy the occupier will be protected under residential statute law. The leading authority on how to distinguish between a lease and a licence was decided in the residential context by the House of Lords in *Street* v *Mountford* [1985] 1 EGLR 128. Lord Templeman in the course of his speech stated that a presumption that a tenancy has been granted will arise where three elements are present:

- the occupier has been granted exclusive possession of the premises
- for a definite term or period
- in return for the payment of rent.

A tenancy can arise either by an express agreement, or importantly, by implication of law by the payment and acceptance of periodical payments.

The presumption can be displaced by other evidence and Lord Templeman referred to a number of common exceptions where the grant of exclusive possession for a term at a rent will usually constitute licences and not tenancies, eg:

- where a potential purchaser of land is let into possession early
- where the grantor does not have the legal capacity to grant a tenancy.

At first glance it might be thought that a licence might suit the needs of a landlord which does not want to grant security of tenure. However, that is not necessarily the case. The problem is one of certainty. The question as to whether an occupational arrangement has granted exclusive possession can be difficult to determine and there are often respectable competing arguments. As such, purporting to grant a licence may merely result in uncertainty and costly litigation if and when the landlord seeks to terminate the agreement.

Where a written agreement describes itself as a licence, often the lawyers for the respective sides will embark upon a clause by clause analysis of the agreement looking for terms which are consistent or not consistent with a licence or tenancy. For example, it is often argued that the inclusion of the following terms are inconsistent with the grant of a licence:

- a forfeiture clause
- the reservation of a right to re-enter for inspection purposes and the like
- an alienation clause.

The rationale underpinning such arguments is that these terms are only necessary if the agreement in substance confers an interest in land (ie a lease) on the occupier as opposed to a personal right to occupy.

Conversely, it is often argued that agreements which contain the following clauses are only consistent with the occupation being on a licence:

- a mobility clause (ie a clause which gives the grantor the right to move the occupier into different premises)
- a clause that restricts when the premises can be used eg a stall in an indoor market.

The label used to describe the agreement is far from conclusive and often viewed with some suspicion by the courts. It was certainly given little weight in *Street* v *Mountford*. Furthermore, there have been a number of cases since where the agreement was described as a licence but where the court held in reality a lease had been granted.

However, the fact that the parties have chosen to describe the agreement as a licence can not be discounted altogether. In some cases, and in particular in the commercial context where the parties have received legal advice and bargaining positions tend to be more equal than the residential context, some weight may be given to this evidence. In *Manchester City Council* v *National Car Parks Ltd* [1982] 1 EGLR 94. the owner of land in the centre of Manchester and which was ripe for development granted

NCP a right to use the land for certain periods during the day for the purposes of a car park. The agreement was described as a licence. The Court of Appeal held that while the label which the parties have attached to an agreement is not conclusive, it should not be disregarded altogether. In this case there was a good commercial reason why the owner of the land did not want to grant a tenancy, namely, the future development potential of the land.

Following *Street* v *Mountford* there was some doubt about whether a label could have any weight in the lease verses licence debate. However the matter has recently been considered again by the Court of Appeal in *Clear Channel UK Ltd* v *Manchester City Council* [2006] 1 EGLR 27. In this case the agreement being scrutinised stated that it 'shall constitute a licence ... and confers no tenancy'. It the course of upholding the trial judge's view that licences had been granted, the Court of Appeal considered the label issue. It was stated that where a contract that was negotiated between two commercial parties of equal bargaining power and who had obtained legal advice contained a clause which set out the unequivocal intention of the parties as to its legal effect, the court would need some persuading that its true legal effect was different to the stated intention. Indeed, the Court of Appeal considered that it was 'surprising and unedifying' that a substantial and reputable commercial organisation that entered into a legal document after obtaining full legal advice should seek to argue that the document had a different effect to the intention expressed by that document.

Consequently, a carefully worded agreement with recitals setting out the commercial reasons why an agreement is a licence and not a lease and where commercial parties have been represented by solicitors stands much better prospects of withstanding a subsequent challenge. However, there will always be an element of doubt and for that reason a licence should be avoided unless there is no other viable option.

(c) Types of licence and their termination

The various types of licence which a corporate occupier may come across and the necessary requirements for their termination can be found in Chapter 11 at 11.2. Precedent letters terminating a licence can be found at 11.7.1 (licensee's letter) and 11.7.2 (licensor's letter). The important point from a corporate occupier's perspective is that a licence offers little in the way of security beyond any contractual termination provision or in the absence of such a provision, reasonable notice. In contrast, where a protected lease is terminated and a corporate occupier continues in business occupation of the premises, it will have statutory rights of continued occupation and renewal rights under the 1954 Act (see Chapter 13).

In a legacy situation, the risks involved from granting a licence are all too obvious. If the headlease requires any sub-occupation to be on an excluded sublease, the grant of a licence will in the case of most properly drafted alienation clauses put the corporate occupier in breach of covenant. Furthermore, if the arrangement is ultimately held by the court to be a periodic tenancy, and not a licence, the sub-occupier will have full protection under the 1954 Act notwithstanding the fact that the granting of the tenancy was unlawful. The corporate occupier may find itself in a position where it is unable to hand back the premises to the head landlord with vacant possession when the headlease expires. This could in turn lead to a substantial claim in damages (particularly if the head landlord had plans for the premises).

The most common situation where a corporate occupier may encounter a contractual licence to occupy is where it is granted occupational rights under a conditional agreement for lease. This device allows a corporate occupier to go into occupation (usually for fit out purposes) before a lease is ready for completion and whilst outstanding conditions are being fulfilled (eg obtaining superior landlord's consent). This is one of the rare occasions where a contractual licence may suit the business needs of the landlord and tenant and pose little risk to the landlord.

(d) Serviced offices

It is not uncommon to come across serviced office accommodation where licence agreements are used. However, it is extremely doubtful whether such agreements would be upheld as licences if they were seriously challenged. They usually contain a mobility clause (ie which gives the licensor the right to require the licensee to move to another office) and an unrestricted right of access (via a master key). It has been held that such clauses are capable of preventing exclusive possession being given to the occupant (see *Dresden Estates Ltd* v *Collinson* [1987] 1 EGLR 45). However, the courts will be vigilant to ensure that such clauses are not a sham. If, in reality, these clauses are never utilised by the licensor there is a significant chance that a court will ignore them and apply the presumption in *Street* v *Mountford*. As Lloyd LJ stated in *Dresden*:

> ... our decision... should not be regarded as providing a way around ... *Street* v *Mountford*. It will be in only a limited class of case that a provision such as is found in [here] will be appropriate. If it is included in an agreement which is not appropriate, then it will not carry the day.

3.2.2 Tenancy at will

This rather old fashioned means of occupying land arises where an occupier is allowed into premises, or continues to occupy premises following termination of an earlier unprotected occupation, and where no definite period of occupation or termination provisions have been agreed. It is by its very nature a precarious right. As its name suggests, it is an arrangement that can be terminated at the will of either party. No notice period is required.

There are essentially two types of situation where a tenancy at will can arise.

1. Expressly — A tenancy at will can be expressly granted by a written contract or a deed (see *Manfield & Sons Ltd* v *Botchin* [1970] 3 All ER 143 and *Hagee (London) Ltd* v *AB Erikson and Larson (a firm)* [1976] QB 209).
2. By implication and operation of law where:
   - a prospective tenant is allowed into occupation during negotiations for a lease: *Javad* v *Aqil* [1990] 2 EGLR 82 (see 3.3.2 below)
   - a tenant is allowed to remain in occupation after its lease and legal rights have expired (eg following the expiry of an excluded lease or where a tenant has failed to make a court application for a renewal lease in time): *London Baggage Co (Charring Cross) Ltd* v *Railtrack plc* [2003] 1 EGLR 141
   - a series of excluded leases have been granted but the tenant remains in occupation paying rent during negotiations for a further excluded lease: *Cardiothoracic Institute* v *Shrewdcrest Ltd* [1986] 2 EGLR 57.

Importantly, once a tenancy at will is terminated the occupier has no statutory rights under the 1954 Act which does not apply (see *Hagee* above). The termination of tenancies at will together with draft notices to quit can be found at 11.7.3 (tenant's notice) and 11.7.4 (landlord's notice).

A tenancy at will rarely satisfies the business needs of either the occupier (who will not relish the prospect of being evicted at the whim of the landlord and with little notice) or the landlord which will want to secure a regular income flow and not want the tenant to be able to terminate the agreement forthwith at any time it chooses. Consequently, most tenancies at will arise unwittingly and by operation of law.

Occasionally it has been suggested by commentators that a landlord can grant an express tenancy at will to avoid the consequences of the 1954 Act. For the reasons set out below (see 3.3.2), this practice is not without its risks.

3.2.3 Lease or tenancy

A lease is a contract between a landlord (the owner of the land or a longer leasehold interest in land) and a tenant (the occupier) whereby the landlord grants the tenant an exclusive right to possession of land for a definite period of time. It is one of the two legal estates in land which exist in English property law (the other being freehold).

There are various terms which are used in the property world which are interchangeable and mean the same thing: lease, tenancy, demise and letting. By the same token, a tenant is also known as a lessee and a landlord can be known as the lessor. If a tenant of premises wishes itself to lease premises (eg a corporate occupier in a legacy situation) it will sub-let the premises or grant a sub-lease or underlease. The ultimate occupier will be a sub-tenant or an underlessee. The intermediate landlord is also known as a *mesne* landlord.

Most leases that a corporate occupier will encounter are likely to be fixed term leases granted by a formal deed. However, it is possible for a periodic tenancy (ie monthly, quarterly or yearly) to be granted. A periodic tenancy can be granted expressly although such leases are usually granted informally by implication of law. For example, if a person is allowed into possession of premises and rent is demanded on a regular periodic basis, then depending on the circumstances there is a very good chance that the occupier will become a periodic tenant. If rent is invoiced and paid on a quarterly basis, it is likely a quarterly periodic tenancy has been granted. If rent is paid on a monthly basis then it is likely to be a monthly periodic tenancy. A periodic tenancy continues from period to period indefinitely until determined by proper notice. The termination of a periodic tenancy is discussed in detail at 11.3.1 and precedent notices to quit can be found at 11.7.5 and 11.7.6.

It is very important for a corporate occupier in a legacy situation to appreciate that by allowing a person into surplus premises on an informal basis it may be creating a tenancy by operation of law. Worse still, if that person occupies the premises for the purposes of its business, it could unwittingly grant a periodic tenancy with full security of tenure under the Landlord and Tenant Act 1954. It is therefore crucial that any legal paperwork is completed before a prospective sub-tenant is allowed into possession of the premises.

(a) Protected lease

When a corporate occupier takes space, or sub-lets surplus premises, it must decide whether the lease (or sub-lease) is to confer on the occupier security of tenure under the Landlord and Tenant Act 1954. This is considered in detail in Chapter 13, but in broad terms if the 1954 Act applies, and the tenant remains in business occupation of the premises on the expiry date of the lease, a statutory continuation tenancy will come into existence. A lease with 1954 Act protection can only be terminated in accordance with the Act and the tenant, if it so wishes, can seek a new lease. The landlord can oppose the grant of a new tenancy, but only on the basis of statutory grounds contained in section 30(1) of the 1954 Act.

If no action is taken, a lease for a term of more than six months will be within the 1954 Act. Equally, as indicated above, if a periodic tenancy is created by a business occupier going into possession and paying rent, it will automatically obtain security of tenure. If the parties intend to exclude a lease from protection they must grant a fixed term lease and follow the exclusion procedure set out below.

(b) Excluded lease

The parties can agree to exclude the security of tenure provisions from applying to the proposed lease. Such a lease is often referred to as an excluded lease. However, a tenancy (which is not exempt under section 43 of the 1954 Act) will only be excluded if the appropriate pre-tenancy procedure prescribed by the 1954 Act has been followed.

Prior to 1 June 2004, the exclusion procedure involved the proposed landlord and tenant making a joint application to the county court for an exclusion order. The process was for the most part a rubber stamping exercise and a good source of income for the courts. Importantly, the tenancy could not be granted unless and until the court had made the order authorising the parties to enter into the exclusion agreement.

The procedure changed on 1 June 2004. To exclude the security of tenure provisions after this date, the prospective landlord and tenant must comply or have complied with the new procedure before the grant of the lease. The new procedure is contained in section 38A of the 1954 Act and schedules 1 and 2 to the Regulatory Reform (Business Tenancies) (England and Wales) Order 2003. It is summarised below.

(i) *Standard exclusion procedure*

A warning notice in prescribed form must be served on the proposed tenant not less than 14 days before the tenant is granted a lease or enters into an agreement for lease. The notice warns the prospective tenant that it is being offered a lease without security of tenure and that it should not complete the lease unless it has taken professional advice. It also points out that "business tenants normally have security of tenure".

Provided that this notice has been served in sufficient time (ie not less than 14 days before execution of the lease), the next stage in the process is for the prospective tenant or a person duly authorised by him to do so, to make a declaration in a prescribed form. This standard declaration confirms that:

- it is agreed no security of tenure is to apply to the proposed lease
- the proposed landlord served the prescribed notice not less than 14 days before the date of the tenancy agreement (a copy of the notice must be attached to the standard declaration)
- the proposed tenant accepts the consequences of entering into the exclusion agreement.

Finally, the lease must contain a reference to the exclusion agreement or it must be endorsed on the lease.

Section 38(2) provides that any agreement to exclude security of tenure is void unless the procedure is followed to the letter. It is also essential that the warning notice and standard declaration are in the form prescribed by the 2003 Order or "substantially in the same form".

(ii) *Expedited procedure*

If the parties are anxious to proceed with the grant of the lease and do not wish to wait two weeks for the notice period to expire, an expedited procedure is available. This is contained in paragraph 3 of schedule 1 to the 2003 Order. The expedited procedure requires that:

- a warning notice must be served on the prospective tenant before the lease is granted or the tenant becomes contractually bound and
- before the lease is granted, the tenant, or a person duly authorised by him, must make a statutory declaration (again in a prescribed form)
- the lease must contain a reference to the exclusion agreement or it must be endorsed on the lease.

The statutory declaration will have to be sworn before an independent solicitor or someone else empowered to administer oaths. In the modern commercial world the "expedited procedure" tends to be used more than the standard exclusion procedure.

(iii) *Practical issues*

If a corporate occupier is taking space it is unlikely to be concerned about an exclusion agreement being void for failure to comply with the statutory procedure. After all, if the agreement is void, the corporate occupier will obtain security of tenure.

However, in a legacy portfolio situation, a corporate occupier will need to ensure that the procedure is followed carefully. Under the old exclusion procedure one issue which arose was whether a lease which contained a break clause was "for a term certain". In *Receiver for the Metropolitan Police District* v *Palacegate Properties Ltd* [2000] 13 EG 187 the Court of Appeal held that a lease which contained a break clause could be an excluded lease. There is no reason to suppose that the decision would be any different under the new exclusion procedure post June 2004.

Another issue which arose under the old procedure was whether the court order was invalidated, and the agreement for exclusion void, if after the court order and before completion, the parties changed the terms of the draft lease. This point was also considered in *Palacegate* and it was held the amendments would not affect the order provided that the amended agreement was substantially similar to the draft agreement that was before the court. Again it is thought that similar principles will apply under the new procedure, although this is likely to be less of an issue because there is no requirement to attach a copy of the draft lease to the warning notice. Further, if there is any doubt, the expedited procedure means that serving a fresh notice and making a further statutory declaration will be quicker than making a further application to the court for a second order under the old procedure.

A new issue that has arisen since June 2004 relates to guarantors of excluded leases. Often a lease will provide that if the lease is disclaimed by a tenant's liquidator the landlord can require the guarantor to take a new lease for the remainder of the contractual term. The obligation will usually provide that the new lease will be on exactly the same terms as the disclaimed lease and therefore will also be an excluded lease. There has been much academic debate about whether it is necessary to serve warning notices on potential guarantors at the same time as the proposed tenant. The concern is that if the warning notice is served much later, after the landlord has in fact required the guarantor to take a lease, it will be too late because at that stage the guarantor will already be contractually bound to take a lease. However, the approach taken by some firms of solicitors of serving a blizzard of notices before the original lease is granted seems equally unsound. One option may be to serve a warning notice immediately before formally requiring the guarantor to enter into a new lease, and then seek an order for specific performance if the guarantor fails to sign the standard declaration. This seems to be more logical than serving a nonsensical notice on a guarantor before the grant of the original lease. However it is not without its practical problems and there is no easy solution to this issue. Whether the issue will actually arise in practice remains to be seen. In most cases a guarantor that is on the hook to pay rent will be only too keen to obtain control of the premises so that it can occupy or relet to gain benefit from its payments.

It goes without saying that a copy of the warning notice and standard or statutory declaration should be kept with the lease for evidential purposes in case that a tenant subsequently tries to argue that a lease was not excluded.

3.3 The problems of early occupation and holding over

There are almost always pressures to allow a prospective tenant or sub-tenant to go into occupation pending the grant of a formal lease or sub-lease. The risks which flow from a tenant or sub-tenant going into early occupation are entirely those of the landlord or mesne landlord and therefore risks that a corporate occupier of surplus properties must be aware of. The problem arises because once heads of terms are agreed the parties believe the deal has been done. This is not the case. It may take many weeks (sometimes months) for the lawyers to negotiate the terms of the sub-lease. In addition, it may be necessary to obtain superior landlord's consent to the proposed sub-letting and there may be problems arising out of the requirements of the headlease alienation clause (see 5.6). It may also be necessary to obtain consent to proposed alterations. Throughout this process there may well be pressures to allow the sub-tenant into occupation. The corporate occupier will be keen not to lose the potential sub-tenant and will want the income stream to start as soon as possible to offset against head rent. Equally, the sub-tenant is likely to be keen to go into the space to commence fit out works or tailor the premises to its requirements. The same pressures exist on the expiry of an excluded lease in circumstances where the sub-tenant wants to remain in the property and take a new excluded lease.

There are essentially two problems to early occupation. The first issue is more technical than real. In most cases (but not all) parting with possession or allowing a third party into occupation will put the corporate occupier in breach of the alienation clause in the headlease (see 5.2.1). A landlord is unlikely to forfeit so the only real risk is that an aggressive landlord may issue injunction proceedings to reverse the unlawful alienation. This possibility is more theoretical than real. In any event, a landlord is likely to send a letter before action and at that stage the corporate occupier can require the prospective sub-tenant to move out.

The real problem of allowing a prospective sub-tenant into early occupation or allowing it to remain in occupation following the expiry of an excluded lease is that if the sub-tenant has exclusive possession and pays a periodic rent, there is a risk that it may unintentionally become a 1954 Act protected periodic tenant. This could cause significant difficulties when the headlease determines and the corporate occupier is unable to hand the premises back to the superior landlord with vacant possession because the sub-tenant has 1954 Act protection and refuses to leave. While these two circumstances are often described as the classic situations where a court will find that the occupier is a tenant a will with no protection (see for example *Javad* v *Aquil* [1990] 2 EGLR 82) these principles have limits and often negotiations can stall (particularly as the sub-tenant is in occupation and trading) and the temporary arrangement can continue for many years. Ultimately, it will always be a question of fact for the court, and as such, doubt can creep in.

3.3.1 Allow occupation at no rent

The traditional safe advice in these circumstances has been not to demand rent unless and until an excluded sub-lease (or a new excluded sub-lease in the case of a renewal) is put in place. The point about allowing occupation at no rent is that if no rent is paid, it will usually be impossible to calculate a period of the tenancy ie a definite date when the tenancy will end (which is critical for the existence of a tenancy — see *Lace* v *Chantler* [1944] KB 368). In the absence of express agreement as to the period for which the occupier is allowed into occupation, the way in which the period of the tenancy is ascertained is by reference to the payment of rent. If rent is expressed to be per annum, then the tenancy may well be an annual periodic tenancy. If rent is expressed to be per week, then the chances are that a court will hold that it is a weekly periodic tenancy. If no rent is paid, and there is no agreement as to the period of occupation, then it is very unlikely that a court will hold that there is a periodic tenancy. In those circumstances the arrangement is likely to be a tenancy at will or a licence terminable on reasonable notice.

If the occupation is to be for a short period this may be acceptable, but it is unsatisfactory and not without its drawbacks. If there is a significant delay in the lease or new sub-lease being negotiated, substantial arrears of rent can build up and in some cases the potential sub-tenant may change its mind and disappear without trace.

In the case of allowing a prospective sub-tenant into early occupation, this option should be the option of last resort, for example where a prospective sub-tenant is about to walk away and not prepared to entertain one of the alternative solutions suggested below. If the prospective sub-tenant is to be allowed into early occupation it is wise to document the arrangement before handing over the keys. A precedent early occupation letter can be found at 3.6.1 below.

Where an existing sub-tenant is holding over following the expiry of an excluded lease, it is prudent for a letter to be sent to the sub-tenant as soon as possible stating that it is a trespasser and that the landlord is entitled to possession. To soften the blow, a without prejudice letter (which can't be shown to the court if proceedings follow) can be sent to say that no immediate steps will be taken to secure possession so as to allow negotiations for a new excluded lease to be explored. If any rent is accepted this should be expressed to be as mesne profits or damages for unlawful use and occupation. Precedent open and without prejudice letters can be found at 3.6.2 and 3.6.3.

3.3.2 Tenancy at will

If the corporate occupier does not want to be at risk of losing rent should the prospective sub-tenant ultimately withdraw from negotiations, and the agreement for lease option referred to at 3.3.4 below is not viable (because the sub-lease is not in a final form or the prospective sub-tenant refuses to enter into such an agreement), the only other option is to grant a tenancy at will.

The leading case on the distinction between a tenancy at will (without security) and a periodic tenancy (with security of tenure) is:

Javad v *Aqil* [1990] 41 EG 61
Facts

- T had lost its place of business and needed to move into new premises quickly.
- L owned premises in Brick Lane E1 and entered into negotiations with T for a 10 year lease.
- On 25 June 1985 L allowed T into possession on payment of £2,500 (representing three months rent in advance) in anticipation of the parties being able to agree formal lease terms.

- On 27 June a draft lease was sent to T's solicitors. On 2 July it was returned to L's solicitors with amendments.
- Engrossments were sent out but the parties were unable to agree about certain outstanding matters.
- T paid further rent in November 1985 and January 1986.
- In February 1986 L required T to vacate the premises.
- T argued it was a periodic tenant with 1954 Act protection; L argued T had a tenancy at will which had been determined.

Held — Court of Appeal:

- A court can reasonably infer from one person allowing another into possession of land for a rent that the parties reasonably intended there to be a periodic tenancy.
- However, whether it is reasonable to make this inference will depend upon other material surrounding circumstances, in particular the existence of security of tenure conferred by statute.
- Where parties are negotiating the terms of a proposed lease, and the tenant is let into occupation or permitted to remain, in anticipation of terms being agreed, the fact that the parties have not yet agreed terms will be a weighty factor. The fact that rent has been paid does not mean that the parties intended there to be a periodic tenancy. That cannot be a sensible inference when considering the consequences that flow (ie 1954 Act protection) and that at the time of occupation the parties were still not agreed about all the terms.
- Entry into possession while negotiations proceed is one of the classic circumstances in which a tenancy at will may exist.
- In this case, T was allowed into occupation as an interim measure and at a time when all the terms of the proposed lease had not been agreed. It was a classic tenancy at will situation.

On the face of this decision it may be thought that the dangers of allowing a tenant into early occupation are relatively low. However, it must be remembered that ultimately the question will be one of fact. The Court of Appeal in *Javad* was at pains to make this point.

A further difficulty arises out of the following passage from the judgment of Nicholls LJ:

> But when *and so long as* such parties are in the throes of negotiating larger terms, caution must be exercised before inferring or imputing to the parties an intention to give the occupant more than a very limited interest...

The key point which arises out of the italicised words is that once the parties have ceased negotiations, it is arguable that what started life as a tenancy at will can transform into a periodic tenancy with protection. The very difficult question to answer is at what stage after negotiations end does it cease to be a tenancy at will and become a periodic tenancy: six months? one year? five years? This will always be a question of fact and extremely difficult to judge.

It is often thought that an express tenancy at will may be bullet proof. This is questionable. A court will be vigilant to ensure that the agreement is not a sham. For example, In *Hagee (London) Ltd* v *A B Erikson and Larson (a firm)* [1976] QB 209 Lord Denning MR stated:

> The court will look into it very closely to see whether or not it really is a tenancy at will, or whether it is a cloak for a periodic tenancy.

However, if it is one of the classic tenancy at will circumstances, it is more likely that such an agreement will be upheld.

If a corporate occupier decides to grant a tenancy at will, to minimise the risks of granting security of tenure it should:

- Use a standard tenancy at will document or a simple letter which contains recitals stating that a potential sub-tenant has been allowed into occupation or to remain in occupation as a tenant at will pending the successful negotiation of a (new) excluded sub-lease and/or the obtaining of superior landlord's consent and
- Ensure that negotiations are pursued diligently and monitored carefully. Responses from the prospective sub-tenant and/or superior landlord should be chased and all dealings documented carefully in writing. If at any stage negotiations falter or come to an end, the tenancy at will should be terminated
- Ensure that a tenancy at will does not normally continue beyond three to six months from the date that the tenant at will is allowed to enter into occupation.

3.3.3 Short fixed term lease

A corporate occupier which does not wish to take any risk whatsoever of granting a periodic tenancy but wishes to ensure immediate rental income could seek to grant a short lease that is exempt from security of tenure. Section 43 (3) of the 1954 Act provides that a fixed term tenancy for a term not exceeding six months does not confer security of tenure.

There are two points to watch out for:

- the lease must not contain an option to renew or extend the term and
- the tenant must not have been in occupation for a period which exceeds 12 months. This period includes any period a predecessor of the same business was in occupation.

To take advantage of this solution the parties have to agree to the terms of the short temporary lease. As this is for a short period, it may be that the prospective sub-tenant will be able to live with the terms proposed by the landlord and contained in the short lease. This will then give the parties an opportunity to agree the terms of the longer excluded lease and obtain the necessary consents. If the longer lease is ready for completion before expiry of the short lease, the excluded longer lease can be completed which will have the legal effect of surrendering the short lease by operation of law. Alternatively, if negotiations are not complete, a further short fixed term lease can be granted provided that the aggregated period of occupation does not exceed 12 months. Typically, the first lease is for six months and the second short lease is granted for a period of five months.

If a sub-tenant fails to enter into the longer excluded lease on the expiry of the second short fixed term lease, the corporate occupier should seek possession immediately and not accept any further rent. However, provided it takes these steps it will not be at risk by the tenant remaining in occupation after the 12 month long stop period. It was held in *Cricket Ltd* v *Shaftesbury plc* [1999] 2 EGLR 57 that any period which the tenant occupies as tenant at will after the expiry of the fixed term will not feature in calculating the 12 month period.

In many cases the thought of entering into two leases will not be attractive for the prospective tenant. Further, if the tenant is inclined to negotiate the terms of the short lease clearly this will not be a solution which achieves the purpose of immediate occupation.

3.3.4 Agreement for lease

By far the best approach is for the parties to enter into a binding agreement for lease which grants a temporary licence to occupy until conditions precedent are satisfied and the lease is granted, or until determined by one of the parties following the expiry of a longstop date.

This avoids the difficulties which are posed by the use of free-standing licenses and/or tenancies at will and has the added advantage that the potential sub-tenant is contractually bound to take a sub-lease in the event that the conditions precedent are fulfilled (eg on the obtaining of superior landlord's consent). If a potential sub-tenant has a free-standing licence or tenancy at will, it will be free to vacate and not take a sub-lease if it changes its mind.

The practical drawback is that it needs the parties to have negotiated fully the terms of the sub-lease as this will have to be appended to the agreement for lease. In the case of a new let this is not always practical. However, if the delays are external (eg outstanding landlord's consent for the transaction and/or alterations) this can be by far the best solution.

3.3.5 Conclusion

It will be appreciated that there is no simple one size fits all solution to this dilemma. The approach which a corporate occupier ultimately decides to adopt will be determined by its willingness to take a commercial risk, and its main objective. In most cases the benefit from using temporary express tenancies at will probably outweigh the risks particularly if the precautions identified above are followed. However, there must be awareness that these agreements are not bullet-proof and these agreements are from time to time challenged in court. It should also be remembered that any litigation which may follow, even if successful, is likely to result in the corporate occupier incurring some irrecoverable legal costs.

3.4 Lease negotiations and terms — operational perspective

A corporate occupier which is engaged in negotiating terms for a lease should refer to the Code for Leasing Business Premises in England and Wales 2007 prepared by the Joint Working Group on Commercial Leases *www.leasingbusinesspremises.co.uk*. The code has three constituent parts:

- Leasing Business Premises: Landlord Code
- Leasing Business Premises: Occupiers Guide
- Model Heads of Terms.

The code is likely to assist a corporate occupier in negotiating a lease not least by providing a useful checklist of the key topics which need to be considered and dealt with at the heads of terms stage and the negotiation of the lease terms stage. It may also prove of value when trying to persuade a landlord to relax its position on a point covered by the Landlord Code. It is important to note however that the code is not compulsory although a number of institutional landlords have endorsed it.

It is not the purpose of this chapter to provide a comprehensive checklist of negotiation issues (which is thoroughly covered by the Occupiers Guide referred to above) but rather to focus on a number of recurring legal and practical issues which are often overlooked by corporate occupiers or given inadequate consideration.

3.4.1 Cost of lease

When a corporate occupier considers taking on new space one of the most important factors will be how much that space will cost. It will influence the corporate occupier's decision whether to take on a lease of new space in the first place, and if so, which premises to take.

Often a corporate occupier will focus on the occupation rent alone. This is unlikely to give an accurate picture. There are likely to be hidden costs and it is important to tease these out. It is also sensible to try and estimate as accurately as possible the total cost of the lease over the entire lease term (or to a break date if one exists). This will enable a corporate occupier to compare the costs of alternative premises, and also compare leasing new space with alternative options such as the purchase of freehold premises.

To evaluate the total cost of a lease the corporate occupier should consider each of the following possible heads:

- tenant alterations/fit out costs
- rent — ie occupation costs
- service charges
- insurance premiums
- utility bills
- business rates
- ongoing maintenance
- repairing, reinstatement and redecoration costs at the end of the lease.

It is only when the corporate occupier has investigated each of these costs that it is likely to be in a position to compare different options and decide whether it can afford the space and whether it makes commercial sense. Some of these items will be within the corporate occupier's knowledge (eg utility costs). Others it will need to speak to the landlord or its agents about such as service charges. It will be particularly important to ascertain whether any substantial capital items are on the horizon (eg new lift or roof). The corporate occupier will also need professional assistance from a building surveyor to advise on the likely costs of the fit out, on going maintenance and/or the potential dilapidations liability at the end of the lease.

3.4.2 Lease length and security

When negotiating the contractual term of the lease (ie duration) the corporate occupier needs to consider carefully its business needs. If there is to be an expensive fit out then a longer lease term with security of tenure will be preferable (although it must always be appreciated that a landlord can oppose the grant of a new lease on certain statutory grounds — eg redevelopment or own use). If the premises are being taken for a new venture then security may not be as important as flexibility and the corporate occupier may wish a short lease and/or break rights.

Break clauses are a trap for the unwary corporate occupier and Chapter 12 deals with this thorny topic at some length. When negotiating a break clause the corporate occupier should try and ensure that the tenant's right to break the lease is unconditional, or if it is conditional, it is only conditional on serving a valid break notice and delivering up vacant possession on the break date. If the break right negotiated is conditional on absolute or conditional compliance with tenant covenants then there is a real risk that the break right will fail, and the corporate occupier will be saddled with the lease for the rest of the term. Paragraph 3 of the landlord code suggests that a tenant's break right should not be made conditional on compliance with covenants with the exception of the payment of rent and the handing back of vacant possession.

3.4.3 Alienation clauses

The negotiation of the alienation clause is of particular importance to a corporate occupier. A corporate occupier's need for premises can change very quickly as its business needs evolve. It is important therefore that it has the flexibility to assign or sub-let the premises without onerous conditions or lengthy legal disputes with its landlord.

Chapter 5 and in particular 5.6 deals with the difficulties that can be caused to a corporate occupier of surplus premises which are overrented and subject to onerous alienation conditions. It should resist any condition which prohibits sub-letting at a rent which is less than the higher of market rent or passing rent. Paragraph 5 of the landlord code suggests that only a condition that restricts a sub-letting to market rent at the time of the sub-letting is code compliant.

In relation to assignments, section 19 (1) of the Landlord and Tenant Act 1927 now permits a landlord to prescribe the circumstances in which it can refuse consent or the conditions which it can impose on granting consent. It is not uncommon to come across landlord alienation clauses that are bristling with restrictions and conditions. A corporate occupier should try and avoid these restrictions, or at any rate, keep them to a minimum. The landlord code suggests:

- that the alienation clause should not prescribe any specific circumstances for refusal with the exception of a requirement that in the case of an assignment to a group company of the tenant, the group company must be of at least equivalent financial standing (when assessed together with any guarantor) and
- an authorised guarantee agreement (see Chapter 10) should not be required as a condition of any assignment unless the assignee is of lower financial standing than the assignor (and its guarantor) or is an overseas company.

Ultimately it will be a question of bargaining position as to whether such restrictions can be negotiated out. However, if the corporate occupier does proceed with a restrictive alienation clause, it should do so in the full knowledge that it may have difficulties at a later stage if the premises become surplus to requirements and particularly if market rent falls below passing rent.

3.4.4 Repairing, redecoration and reinstatement covenants

Corporate occupiers rarely give much thought to dilapidations issues at the lease negotiation stage. If they do, and they have a good bargaining position, it may be possible to restrict any potential liability by reference to a photographic schedule of condition (see 14.2.5).

The first time most corporate occupiers consider dilapidations issues is when served with a terminal schedule by the landlord following lease expiry. However, to calculate the total cost of a lease a corporate occupier needs to have an indication of the likely costs of maintenance during the lease term having regard to the proposed repairing and redecoration covenants. A good building surveyor will be able to advise on this issue. He/she will also be able to advise on the likely cost of fitting out the premises and reinstating them at the end of the lease should the landlord require it. If the premises are multi-let or have common parts a building surveyor will be able to give some indication as to whether any large items of capital expenditure are likely in the foreseeable future which would be recoverable from the tenants by way of service charge.

The importance of taking advice on future dilapidations liability cannot be overstated. An obligation to keep in repair imposes an obligation on the tenant to put the premises in repair. This means that if a corporate occupier takes premises in poor condition (eg an old warehouse) it will be under an obligation to put the premises into a proper state of repair (see 14.2.1 (a)).

When it comes to the wording of a particular clause:

- It is important that the physical extent of the repairing covenant is clear: ideally the draftsman should visit the premises and be familiar with the layout of the demised premises, common parts and parts occupied by other tenants. In cases which do not warrant a lawyer inspecting the premises it is important that the building surveyor is given an opportunity to comment on the definition of the demises premises for the purposes of the repairing covenant.
- Words which can be interpreted as going beyond repair should be negotiated out if possible eg "condition"; "renewal"; "amend" and "rebuild" (see 14.2.1 (c)).
- If possible, a corporate occupier's advisors should seek to expressly exclude from the definition of "repair" works to defects which constitute or arise out of an inherent defect with the demised premises (see 14.2.1 (b)(iii)).

3.4.5 Service charges

There are two elements to service charge clauses which a corporate occupier of operation premises should be concerned with. First, it will need to be satisfied that the landlord is willing and financially able to provide the services contained in the lease. Significant problems can arise where a landlord gets into financial difficulties and fails to provide the essential services eg security, cleaning of the common parts or maintenance of lifts. Depending upon the identity of the landlord, it may be prudent to investigate the landlord's financial position before deciding to take the lease.

The second aspect of the service charge provision which is important is to gauge how much the corporate occupier is likely to have to pay both as an annual charge and also over the term of the lease. The landlord should be asked about this. One of the difficulties is that the reversion could be sold by the landlord at any time. Even if the current landlord has a relaxed approach to service charges there is no guarantee that its successor in title will take a similar view. It is therefore prudent to instruct a building surveyor to inspect the common parts to make an assessment of the prospects of significant capital expenditure arising during the course of the lease.

3.5 Lease negotiations — legacy perspective

If a corporate occupier is in the position of needing to sub-let surplus properties then its perspective is likely be very different. Ideally it will look to grant a sub-lease which ties the sub-tenant into paying rent up to a few days before the expiry of the headlease. It will be resistant to the sub-tenant having break rights and will look to pass on the full repairing and redecoration obligations which are likely to exist in the headlease. In all likelihood it will also be looking to ensure that any sub-lease granted is excluded from security of tenure provisions of the 1954 Act.

Again, the extent to which such aspirations can be achieved in negotiations will be influenced by the state of the market, the desirability of the premises and the keenness which the prospective sub-tenant has for taking the premises or the corporate occupier has for disposing of them.

If the premises are overrented and/or an inducement is necessary (eg a rent free period) care will need to be taken to ensure that the sub-letting does not breach the headlease alienation clause (see 5.6).

The purpose of sub-letting is to generate an income stream to offset against the head rent payable under the headlease. A corporate occupier of legacy premises will therefore need to think carefully about the financial strength of the proposed sub-tenant and whether it is likely to default. Where the proposed sub-tenant is a company, a company search should be carried out to ensure that it is registered and that the name it has given is correct. It is also prudent to consider its last three years accounts to ensure that it will be in position to pay the rent as it falls due and comply with the other tenant covenants.

If the sub-tenant is not a particularly strong covenant then the corporate occupier should consider additional forms of security in the event that the sub-tenant defaults:

- seeking a guarantor of stronger covenant strength
- a bank guarantee or bond linked to the sub-tenant's failure to pay rent, insurance and/or service charge
- a rent deposit.

If the sub-tenant is taking on a significant repairing obligation it is sometimes a good idea to take a rent deposit. If the sub-tenant fails to comply with its repairing obligations at the end of the lease this will allow the landlord to draw down on the rent deposit immediately and put the onus on the tenant to issue proceedings to recover these sums in the event of a dispute. This can be a very useful tactic where the amounts at stake are not substantial and the irrecoverable costs of going to court might not support going on the offensive.

3.6 Precedents

3.6.1 Early occupation letter — no rent

Dear Sirs

Re: Early occupation of Tumbledown House pending grant of excluded lease

We act for Getonwith IT Ltd who you are currently in negotiations with to take a sub-lease of Tumbledown House, London N1.

It is intended that the sub-lease will be excluded from the security of tenure provisions contained in the Landlord and Tenant Act 1954 (as amended).

We understand that you would like to go into immediate occupation of the premises pending the completion of the sub-lease for the purpose of carrying out works.

Our client is prepared to allow you to access the premises for the purposes of carrying out works but only on the basis of a bare licence which can be terminated [immediately] [on 7 days written notice] in the event that either our client decides that negotiations are not going to result in a completed lease or alternatively the superior landlord takes action against our client in respect of your occupation.

If you agree to this please sign the copy letter enclosed and return it to us whereupon our client will release one set of keys to you.

For the avoidance of doubt, both parties have had legal advice and agree that this letter constitutes the grant of a bare licence and does not grant a tenancy.

Yours faithfully

3.6.2 Holding over following expiry of court excluded lease — open letter

Dear Sirs

Re: Expiry of lease — Tumbledown House

We act for Getonwith IT Ltd your former landlord of the above premises.

On 24 March 2006 our client granted you a lease for a term commencing on 25 March 2006 and expiring on 24 September 2007 at a rent of £100,000 per annum. Your lease was excluded from the security of tenure provisions contained in the Landlord and Tenant Act 1954 (as amended).

Your lease has now expired and your continued occupation of the premises is as an unlawful occupier and is without our client's consent. Our client is entitled to claim mesne profits at the rate of [ ] per day from 24 September 2007 until vacant possession is delivered up.

Yours faithfully

3.6.3 Holding over following expiry of court excluded lease — without prejudice letter

Without Prejudice and Subject to Contract

Dear Sirs

Re: Expiry of lease — Tumbledown House

We refer to our open letter of today's date which you will appreciate our client has had to send to protect its legal position in the light of your continued occupation of the premises following the expiry of your excluded lease.

Notwithstanding the terms of the open letter, we understand that negotiations are continuing for a new excluded sub-lease. In the circumstances we are instructed to take no action to recover possession whilst negotiations are on going. However, please note that our client will require you to pay mesne profits at the rate to be charged under the new sub-lease ie [£] and these sums will be collected in arrears on completion, or if no completion takes place, when you are required to vacate the premises.

Yours faithfully

Service Charge 4

4.1 Introduction

In general terms a service charge is a payment by a tenant for services provided by its landlord under the terms of the lease. The exact definition of the services for which a tenant is required to pay will always be defined by the service charge clauses contained in the lease. In the UK the role of the service charge is predominately to maintain a multi-let property in the same way that a stand-alone building would be maintained by a tenant which occupies the building on the basis of a full repairing and insuring lease (often referred to as a FRI lease). Service charge regimes are not restricted to office blocks or shopping centres, but will apply to all sorts of premises ranging from industrial estates to mixed use properties — anywhere where there are common areas or facilities. A service charge regime is likely to be used in all circumstances where a tenant is not under an obligation to maintain plant, facilities or parts of the premises. The landlord will take on the role to provide the service or maintain the items, but will then seek to recover the costs from the tenant by way of service charge.

UK service charge provisions generally differ in comparison to other countries because of the way the service charge is structured. In the UK, with an onerous service charge clause, a landlord can be provided with the opportunity to recover all of its expenditure, even where an element of the costs arise from the landlord improving the building thereby providing a landlord with a clean investment. Elsewhere there tends to be more of an on-going investment by the landlord in its property. As a consequence the relationship between landlord and tenant in the UK is much more adversarial, and corporate occupiers have become wary of landlords' intentions when it comes to service charge clauses and expenditure. Most corporate occupiers will have come across a landlord seeking to improve premises at its expense. They are also likely to have come across managing agents who charge excessive fees and seek to recover these from the tenants.

In other countries many of the items included in the service charge will be seen as capital expenditure by the landlord and form part of its ongoing investment to maintain and improve the property. Indeed there is a compelling argument that the whole concept of the service charge is flawed. The landlord is the property professional and with a portfolio of investment properties is in a position to command best value when procuring services and maintaining the property. The landlord should have to invest to keep the property in good condition. Its return is the rental stream that is generated. Consequently, if the cumbersome process of the service charge regime was replaced by a simple obligation by the landlord to maintain the property and provide the services, and the rent was increased to cover the cost of that investment (ie the cost of maintenance and services provided) then the tenant would have certainty of costs (by avoiding uncertain and unpredictable service charges) and the landlord would be free to determine how best to invest his money in the property.

The development of the institutional FRI lease (with a 25-year term) at the start of the 1970s has been the biggest drawback for the Corporate Occupier in the UK, while at the same time it has been such a boon for the institutional landlord. However, a number of corporate occupiers who had freehold property portfolios have made considerable returns from sale-and-leaseback deals as a consequence of the resultant demand for good institutional grade investments.

Service charges have been a constant source of dissatisfaction and conflict between landlord and tenant over many years. Surveys have consistently shown that tenant satisfaction is low. When disputes have found their way to court, judges have occasionally been critical about the way in which the landlord's managing agents have approached their role. In *Princes House Ltd* v *Distinctive Clubs Ltd* [2006] All ER (D) 117 Jonathan Gaunt QC criticised the landlord's managing agents for simply being intent on recovering for the landlord as much as they could. He indicated that commercial tenants are entitled to expect professionals dealing with service charge recovery to perform their roles with diligence, integrity and independence. Unfortunately, he opined, such expectations can be misplaced. The judge also reflected that dissatisfaction was so high amongst tenants that there were signs that the commercial letting market was beginning to reject the inclusion of a full recovery service charge provision in new leases. A subsequent appeal by the landlord to the Court of Appeal was dismissed (see [2007] 27 EG 304).

In an attempt to address the problems and introduce best practice, the RICS has published a code of practice — *Service Charges in Commercial Property* which is effective from 1st April 2007. The Code of Practice is a remarkably thorough and clear document and if adopted by landlords (and tenants at renewal) would doubtless reduce the number of disputes and levels of dissatisfaction amongst tenants in relation to service charge recovery. In broad terms the service charge code:

- suggests the most desirable structure for service charge recovery
- sets out in detail how service charge regimes are to be operated to achieve best practice
- encourages better communication between the landlord (or its managing agents) and the tenant
- encourages transparency on costs
- promotes mediation and expert determination as alternative dispute resolution procedures
- encourages both parties to agree to old service charge clauses being updated on renewal.

The principal drawback to the service charge code is that there is no statutory requirement for the code to be taken into account in service charge disputes (see "All bark and no bite" by Stephen Jourdan EG Practice & Law 07/04/07). Consequently, landlords cannot be compelled to follow the service charge code. Furthermore, it appears doubtful whether a landlord or tenant will succeed in changing an existing service charge clause to bring it into line with the Service Charge Code on renewal against the wishes of the other party (see *O'May* v *City of London Real Property Co Ltd* [1982] 1 EGLR 76 discussed at 13.5.1 (c)). As against this, the service charge code is an extremely impressive piece of work which if adopted could eradicate many of the troublesome disputes which beleaguer a landlord as well as a tenant. Landlords may recognise that it is in their commercial interests to buy into the approach put forward by the service charge code.

The service charge code recognises that there will be a period of transition as leases expire at different times and for this period there may be dual service charge. One of the key elements which underpins the service charge code is 'transparency of costs'. Most corporate occupiers will be happy to accept changes to the service charge regime if they benefit from a more transparent arrangement that gives them value for money. The incentive for landlords to adopt the service charge code and on renewal to agree to updating the service charge clause is that tenants will be prepared to pay more rent for a lease which requires a transparent approach on service charges. The service charge code is referred to throughout this chapter and is a document that every CREM should refer to both as a means of trying to persuade landlords to adopt it, but also as a means of how they should operate a service charge regime where the corporate occupier is a landlord in a legacy portfolio situation. It is important that a corporate occupier tries to lead from the front on this not least because it will be in its commercial interests to do so. Contested service charges disputes can be a very significant drain on a CREM's time and a corporate occupier's resources.

4.2 The function of service charge

The fundamental purpose of a service charge is to maintain the premises and ensure that the building is kept in good condition throughout its life, and in doing so to be able to recover the cost of that upkeep from all of the tenants who occupy the building. This will apply equally across the spectrum from the landscaped areas of an industrial estate to the exterior of a shopping centre.

The costs involved in providing services are very substantial. For example, in 2005 UK businesses paid £4.42bn in office service charges (see David Barrass, "It's the devil's own job" *Estates Gazette* 16 September 2006). Achieving a sensible balance between quality and cost can give rise to very different views. The opportunity of gilding the lily on the landscaping of an industrial estate will be limited, as most tenants will start to wonder why they have fountains in place of the grassy knoll. However, the scope within large office buildings and shopping centres to improve the premises or to incur unnecessary costs is significant. Unfortunately the quality of the management of many properties is not good and the fact that it is the occupiers who are paying for the works being carried out seems to be lost on most managing agents. They also tend to overlook that it is the occupiers who are actually paying the managing agents fees, not the landlord. Many corporate occupiers feel that the properties occupied by them are seen as cash cows to be milked by managing agents and little regard is given to whether the work is really needed or whether it has been obtained at the best price. What has not helped is the practice of management fees forming part of the service charge costs. In general the managing agent's fees are geared to a percentage of the expenditure for the provision of services, or in some cases a percentage of the expenditure and a percentage of the rent. This provides an immediate conflict of interest as managing agents are being rewarded not on the basis of their performance in providing services, but instead on their ability to spend money.

Generally speaking, most corporate occupiers do not investigate in detail the service charge expenditure because they do not have the time — their focus instead being on the operational needs of the business. Once the money is spent, challenging the expenditure is difficult and costly. At that point the landlord will be looking to recover its expenditure, and the parties will tend to adopt adversarial positions.

As a consequence corporate occupiers have not always applied the rigour when looking at service charge costs which have been applied to contested rent reviews. In addition the ease of spending tenant's money has been helped by the growth of sinking funds, which are used to plan for large cost items such as heating system replacement and lifts. The existence of a sinking fund, provided it is used properly, can be advantageous because it allows the tenant an opportunity to vet expenditure before it is incurred and therefore provides the tenant with an opportunity to deal with such issues in advance of the spend. However, some managing agents may regard the money, which is already sat in a bank account, as belonging to the landlord, and therefore not consider the need to consult with the tenants as being particularly important. All too often with items of major expenditure, landlords fail to consult with, and engage the tenants in, the decision making process.

In theory, high service charge costs should impact upon the rental value of a building. However, that will only happen if all of the comparables of similar quality buildings have lower service charge costs, and most importantly, the point has been picked up by the rent review surveyor. At times of static or falling rents the impact of a high service charge on rental value will be minimal, indeed that is the time when landlords are probably planning or undertaking the maximum expenditure. However, when rents are rising tenants tend to ignore the service charge cost because rising rents generally reflects a period of economic growth. During those periods a business will be expanding and the effect on the business of higher costs tends to be ignored not least because the CREM will be focusing on other issues.

4.3 The operation of service charges

4.3.1 Management

The best practice approach to running a service charge regime is total transparency coupled with a recognition that the money, or at least a substantial part, has been paid by the tenants and the running of the building is a partnership between the parties. It also involves a duty of care owed both to the occupiers and to the landlord by the managing agent.

There must be policies and systems in place to encourage a sound management approach to running the property. These policies and systems should ensure that the correct procurement methods are used and that there is effective communication between the managing agent and the occupiers. Most corporate occupiers will be reasonable if they are properly treated and are kept informed and consulted on major issues. This means providing the tenant with the appropriate information promptly and without the corporate occupier having to ask for it, or indeed chase it.

Landlords should consider carefully the basis on which its managing agent is remunerated. Ideally the managing agent should not receive a fee based on a percentage of expenditure or on the amount of rent collected. It should not be rewarded for the amount of money it has spent. Consideration should be given to a fee based on fulfilling certain performance criteria. This could involve a fixed fee element together with a performance element over a fixed contractual period (say three years). It should also be driven by a tendered contract not an automatic appointment based on the fact that the agents introduced the investment to the landlord.

4.3.2 The lease clause

In the context of commercial leases there are no statutory rules which apply to service charges. The scope of each party's obligations will be determined by the words used in the service charge clause or clauses contained in the lease. Whether the service charge regime is onerous or not will be purely a matter of negotiation at the time that the lease terms are debated, and a product of the negotiation strengths of the landlord and the tenant.

A corporate occupier should focus carefully on the service charge provision being proposed by the landlord. It will need to have a clear understanding of what services are to be provided, the mechanism for payment of service charges, and what safeguards are built in from the tenant's perspective. It will want to ensure that the structure follows, as far as possible, that suggested by the service charge code.

If a landlord will not agree to a service charge code compliant clause, the corporate occupier needs, at the very least, to be aware of the potential liability under the service charge clause. This is likely to require advice from a building surveyor in addition to a commercial property lawyer.

Often the parties' lawyers do not carry out a site visit before a lease is granted. One area where it can be important is in relation to service charges. It can be important that the lawyers understand the common services and facilities. Drafting a service charge clause can be very difficult if the lawyer does not understand the interrelationship between the various parts of the property and the areas demised to the tenants. Relying upon plans or what they may be told by managing agents may not be sufficient. In most cases where the potential liability under the service charge is likely to be significant, a site visit is likely to be essential. Each building is unique and therefore each service charge clause should be tailored to fit that building. All too often a one size fits all approach is taken which can lead to expensive and protracted disputes at a later date.

The commercial property lawyer acting for the tenant when considering the service charge clause will be less concerned about what services are needed and more concerned about the manner in which the provisions operate. That too requires the solicitor to see the property to understand the building's configuration and the interrelationship of the services and the units.

When a potential dispute or issue arises, the starting point will always be the wording of the service charge clause. While there may be case law that is relevant, ultimately a court will look at the specific service charge provisions in the lease and decide the case on the basis of the evidence and the service charge clause.

4.3.3 Services provided

The principal obligation imposed on a landlord will be to carry out, or procure the carrying out of the services. The services which a landlord is required to provide will be defined by the lease. Some leases may distinguish between services which a landlord 'must provide', and those which it may provide. The purpose of the distinction is to give the landlord the opportunity to recover by way of service charge some items of expenditure which it may choose to carry out, without imposing a positive obligation to do so.

The services that can be covered by a service charge clause may be very extensive and cover not only maintenance but other items that may be necessary or appropriate for the property, such as child care. The service charge code contains a very thorough list of possible services.

Examples of items which may be defined as services include:

- cleaning
- fire protection
- security
- repairs
- landscaping
- maintenance of lifts
- heating
- maintenance of air conditioning
- energy
- water.

The amount of detail contained in the service charge clause will vary from lease to lease. There is no particular virtue in having very long lists identified. This is because most service charge clauses will have a catch all or sweeping up clause which is likely to provide that the landlord may provide any additional services that are capable or reasonably calculated to be for the benefit of the tenant and which are in keeping with the principles of good estate management. Trying to negotiate out specific items will delay completion of the lease and will be a waste of time if there is a broadly drafted catch all clause. The tenant has to look for protection elsewhere.

The service charge code stipulates that landlords should not seek to recover any initial costs in setting up the property via service charge. Set up costs should be part of the landlord's own costs. By way of example, the initial cost of a CCTV system should be borne by the landlord but thereafter the maintenance cost is a legitimate service charge cost, as is its replacement at the point it wears out. The code also provides that any costs which arise as a result of a failure by, or through the negligence of, the landlord or its managing agent should also be excluded. Furthermore, all costs associated with letting units, collecting rents and litigation should also be excluded and not be included within the service charge provisions. Again, as emphasised above, the service charge code is not underpinned by statute and if a landlord insists that a service charge provision include these items, beyond persuasion (and threatening to walk away) there is little a tenant can do.

When negotiating service charge provisions, a corporate occupier should be aware that:

- in relation to whether particular works constitute works of repair, the principles set out in Chapter 14 apply equally to service charge recovery. Importantly, care must be taken when including words such as "condition", "renewal", "amend" and "rebuild" which may make the obligation more onerous (see 14.2.1 (c))
- while a court may imply a term that the amount of service charge payable by a tenant should be fair and reasonable (see *Finchbourne* v *Rodrigues* [1976] 3 All ER 581), this will not always be the case, particularly where the service charge clause is detailed and provides for the landlord to be able to recover expenditure actually incurred. From a tenant's perspective, it is always better to ensure that a fair and reasonable clause is expressly contained in the lease.

4.3.4 Apportionment

The method of apportionment (ie the proportion that a particular tenant contributes to the total cost of services provided by the landlord) will be governed by the lease and usually reflects the nature of the building.

There are two principal approaches that are prevalent today:

1. rateable value
2. floor area, which can be the net area, gross area or a weighted floor area.

The most common apportionment is by reference to floor area. The problem with using rateable values is that every time there is a revaluation the apportionment has to be changed and this is unlikely to coincide with a year-end, which means that the apportionment has to change during the course of a service charge year. If the reassessments flow through at different times following revaluations then there is a continual process of changing the apportionment and the allocation of expenditure becomes difficult. This can be overcome by having the adjustment to the apportionment fixed to once a year at the start of the service charge year.

The use of fixed percentages was also very popular at one stage and had the great advantage of being simple to operate but fell out of use because it lacked the flexibility to respond to changes in circumstances. Problems frequently arose because of an incorrect calculation to begin with or because later tenants objected to paying more than an earlier tenant who had negotiated down the percentage it agreed to pay.

Where an apportionment is determined by reference to floor area, there are different methods of calculating the floor area depending upon the property use:

- offices — net internal area
- industrial and warehousing — gross internal area or occasionally gross external area
- retail — net internal area or weighted floor area, by reference to the zoning of the shop into 6m or 20ft zones, in the same way that retail rents are determined. The first zone — Zone A is worth twice as much as the next and so on. An apportionment is produced by analysing the shop in terms of Zone A (ITZA). In certain circumstances it may be appropriate to allocate a weighting to certain units possibly because they do not receive all of the services or because of the size of the unit they would pay a disproportionate amount of the cost. Shopping centres are a prime example of this where the formula of weighting will be similar to that which determines the rental levels. This will mean that the larger units such as department stores and supermarkets have a lower relative charge.

One of the thorny issues which must be considered when determining the appropriate apportionment for a particular tenant is who should and should not pay for specific services. For example, shopping centres units with only an external frontage will not benefit from escalators and the same level of security as those inside. Should they simply pay for the services that they directly receive or do all the services contribute to the attraction of the centre and therefore benefit all of the tenants? Another example is the office building. Does the ground floor tenant pay for the lifts? If not, does it pay more for the receptionist which it may make more use of? This argument if taken to its logical conclusion, would see only the top floor tenant in a block of offices paying for the repairs to the roof which is clearly not workable.

The argument for the tenants on the outside of the shopping centre is far stronger than for the office tenant. Clearly removing the issue by having inclusive rents would make life simpler and then the rent and the service charge would run together.

What is needed, therefore, is a logical sustainable and manifestly fair method of calculation. It must also be flexible to adapt to future changes of circumstances.

4.3.5 Creating a service charge

It is essential that whoever creates and runs the service charge does so from up-to-date information. The more services that have to be provided and the more complex the layout of a property, the more schedules of expenditure that will be required. The schedules will contain details of the individual categories of cost. Using the comparison between shopping centre and offices referred to above, the cleaning costs of offices will need to be charged back on a different basis to the cleaning costs of shopping malls. While overall taking one apportionment of the whole might produce nominal differences, tenants will want to be clear as to what costs have been incurred for different areas and ensure that they only pay for those on their part of the property. There is not necessarily a right or wrong way of creating a service charge; it is important to endeavour to find a fair and reasonable approach for a particular building.

All the information required needs to be presented in a format that is readily understandable by all the tenants. Before a new lease is drafted the landlord's management surveyor needs to have established what services are appropriate, how they are to be procured and how the costs should be allocated between the various parts of the property.

Shopping centres do give rise to their own issues, especially in terms of promotional activities. Promotional events will tend to benefit both the occupiers and the landlord, and consequently there may be a powerful argument for saying that promotional costs should be apportioned on a 50:50 basis to reflect that.

4.3.6 The budget

The simplest form of service charge regime seeks to recover expenditure over the previous quarter, or indeed as the expenditure is incurred. This simple format may be suitable for a parade of shops or a small industrial estate where all units are demised. However, once projected expenditure rises above a certain level and/or the services provided by the landlord are complex and extensive, landlords will look to recover the cost from the tenants in advance of expenditure to avoid the cost of funding and also to avoid the difficulty of spending sums up front and not then being able to recover those sums from the tenants. To achieve this, managing agents will need to prepare a budget of anticipated expenditure over a given period. Budgets of anticipated expenditure are generally prepared to cover a 12 month period. As with all aspects of service charge, the lease will set out the landlord's requirements for the preparation of an anticipated budget and its service on the tenants.

The budget should be prepared well before the start of the defined service charge year and will be based on the projected expenditure for the coming year. This should be provided to all tenants together with a schedule that specifies each tenant's on account payment. This will be a product of the total anticipated expenditure and the relevant apportionment for each tenant. Service charge on account payments are usually payable quarterly in advance. At the end of a defined service charge year the landlord's managing agent will carry out a reconciliation comparing actual expenditure against the projected costs. This will result in either the need for the tenants to make a further balancing charge to make up any shortfall, or for the landlord to give a balancing credit (or occasionally payment) if the landlord has spent less than anticipated in providing the services.

A corporate occupier with a legacy portfolio of premises that are surplus to its requirements may have to run the service charge regime itself. This usually arises because premises have been sublet to a number of different subtenants. To ensure full cost recovery and to match its own repairing obligations contained in the headlease, the corporate occupier will have to run the common areas and facilities as a service charge. A corporate occupier in this situation could charge inclusive rents and avoid the service charge approach. The benefit of this approach is that it avoids the hassle and attendant costs of running a service charge regime, and will avoid the disputes which tend to beleaguer service charge recovery. The potential risk in adopting an inclusive rent is that unexpected and significant expenditure may be incurred. Depending on the nature of the particular property, it may be a much lower risk option for a corporate occupier to run a service charge regime than it is to commit to inclusive rents and risk the direct and irrecoverable cost of unexpected repairs or maintenance. Nevertheless, in some circumstances an inclusive rent will be appropriate, although the alienation provisions in the headlease should be considered as they may require the sub-lease to be in substantially the same format as the headlease. A further factor to consider is that charging an inclusive rent to the sub-tenant may adversely affect the prospects of the superior landlord taking a surrender of the headlease at some point in the future, although this must be balanced against the easier and cheaper management of the property.

Ideally a budget will be based on the previous year's expenditure, with adjustments to account for inflation, new charges and one off costs. One-off costs incurred in the previous year clearly need to be stripped out to leave recurring costs only. Inflation should then be built in. This is unlikely to be uniform across all items, for example, utility costs have been rising ahead of inflation in recent years and so increases in these costs have been higher than most other costs. One off costs that are going to be incurred in the next service charge year need to be factored in to give the total budget. Corporate occupiers like to plan in advance for changes in expenditure and therefore having early notice of any likely increase in budget and potential shortfall is important. While this may not be required by the service charge provisions in the lease, it is good practice and should be adopted where practicable.

In preparing or reviewing a budget, care needs to be taken to ensure that there is no double counting. For example if a fire alarm is to be replaced in the next service charge year, the costs of maintenance from the previous year will need to be stripped out the following years budget if the cost of the installation includes the first years maintenance. Often the year leading up to such capital expenditure is characterised by high maintenance costs reflecting the poor condition of the equipment and the need to continually carry out repairs. Consequently, the timing of the replacement of a capital item should be considered carefully so as to avoid unnecessary repair costs being incurred in the years leading up to replacement. It is less palatable for a corporate occupier to pay the capital cost of replacement of an item where there have been expensive and frequent repairs in the recent past that with hindsight could have been avoided by earlier replacement.

The budget should be presented to the tenants in the form of a reasonably detailed breakdown between the various heads of expenditure. The right balance of detail will allow both sides the ability to understand and often resolve any issues. The service charge code suggests that there should be an ability to compare the proposed costs with comparable properties and identify specific comparators, for example, the energy cost per square foot for the premises.

Some service charge provisions will enable a landlord to introduce interim increases in the budget to take account of unbudgeted expenditure. This will involve a request for an increased on account payment. If such a provision is not included in the service charge clause then the landlord will have to bear the additional cost until the year-end reconciliation is undertaken. Unbudgeted expenditure should be at the landlord's risk so that any financial costs for the landlord funding the works should be no more than base rate. There is also a need to ensure that the monies held in the service charge account earn interest at a commercial rate.

4.3.7 Sinking funds

A service charge clause in a lease may provide for a sinking fund (otherwise known as a reserve fund or expenditure equalisation fund).The purpose of a sinking fund is to smooth out any major costs for a property so that tenants are not faced with widely fluctuating levels of service charge. It is also a means by which a landlord can ensure that monies are available, and within its control, when significant items of work become necessary eg replacement of a heating system in an office building. Most occupiers see the benefit of a sinking fund as being the ability to spread the cost of major capital works over a number of years. However, the landlord will not be able to demand advance payments towards a sinking fund unless the service charge provision in the lease provides for this.

Examples of heads of expenditure suitable for sinking funds include:

1. lifts, heating systems, electrical system and other mechanical services
2. internal and external decoration
3. planned preventative maintenance.

An area of concern for a tenant can be how sinking fund monies are held. Ideally the managing agents should fall under the jurisdiction of the Royal Institution of Chartered Surveyors and therefore be subject to a requirement to comply with their code of practice. Care needs to be taken here because some firms of agents may try to avoid code of conduct issues by placing monies into the accounts of different companies in circumstances where none of the directors are chartered surveyors. It is important to ensure that the service charge monies are clearly identified and held as client monies and referable to the leased property. From the tenant's perspective it is important that the lease specifies that the beneficial ownership of the money does not pass to the landlord but is held on trust for the occupiers of the building.

To achieve fairness, the lease should require a landlord to make its own contributions to the sinking fund in respect of any vacant units or in relation to any units where there is a dispute with a particular tenant and pending recovery from that tenant.

The terms of the bank account in which the sinking fund is to be held needs to be considered carefully. Where draw down from the fund is going to be infrequent; the landlord should secure the best possible rate of interest. There may be tax implications which arise on the interest paid and this should be considered with the tenants at the outset.

The current account for the service charge will need to be a cheque account in most cases, and should be configured to ensure that the benefit of any interest goes to the tenants and that there are no costs for drawing cheques.

An issue which can arise is what happens to a tenant's contribution to the sinking fund if the works are not actually carried out, and the expenditure incurred, before that tenant's lease expires. This will depend upon the wording of the service charge clause in the lease but, generally speaking, a landlord will not be entitled to demand a contribution to the sinking fund for works likely to take place after lease expiry (see by way of example *Brown's Operating System Services Ltd* v *Southwark Roman Catholic Diocesan Corporation* [2007] EWCA Civ 164). If the tenant has made a contribution and the works take place after lease expiry, the tenant will be entitled to a rebate.

4.3.8 Year end statements

Most, if not all, commercial service charge clauses will require a landlord to undertake a reconciliation of the service charge account at the end of a service charge year, often within a specified time. The clause may also stipulate that the reconciliation must be subject to an external audit of the expenditure and allocation of costs and the amount payable by the tenant to be certified by the auditor. Even if this is not the case, it is good practice for landlords to do this. While the costs will be passed on to tenants these should be relatively low unless it relates to a very substantial service charge account. Tenants may think this is a wasted cost, but in practice the audit is likely to assist them in verifying the expenditure.

The lease will usually provide for the amount payable by the tenant to be certified, and once certified to be final and binding on the tenant. As indicated above, the certification may be carried out by an external auditor. Alternatively, the lease may provide that it is to be carried out by the landlord's surveyor. There are two common legal issues that arise from certification clauses:

- whether the certification can be conclusive on questions of law and fact can be unclear and will largely depend upon the wording of the clause, the expertise of the certifier and the nature of the challenge
- where certification is required, it is likely that the obligation on the tenant to pay any shortfall will be conditional on a certificate being issued.

One of the vexed issues which is often debated is whether a service charge should be audited professionally or not. For a simple parade of shops this will be unnecessary but where, for example, the demised premises form part of a shopping centre with offices above, there may be benefits from a professional audit. A sceptical tenant may ask what auditors bring to the table. The answer should be clarity and independence. It is not the function of the auditors to catch out the managing agents; it should be to provide objectivity and a safety net because errors do occur.

However, all too often the auditor's role is simply one of ticking off invoices against a list of expenses and carrying out a similar exercise on the income side. This is often because their terms of reference are too restricted and there is a lack of clarity between the managing agent and the auditor as to what is going to be provided. What results is a fairly one dimensional exercise. More benefit would be gained if the service charge accounts were considered in the same way as a company's accounts, with a profit and loss account and a balance sheet together with comments on corporate governance and the like. More clarity would be forthcoming if a broader approach was taken with the service charge regime being regarded as a mini business. For the larger shopping centres or office complexes that is precisely what it is.

What approach should a tenant take in checking the reconciliation and in particular expenditure? At one end of the spectrum the tenant could simply check for arithmetical mistakes. At the other end of the scale it could try to obtain copies of all invoices paid by the landlord as part of the service charge account. Landlords will usually resist requests to provide copies of invoices because of the cost of management time taken with such work. A landlord may offer the tenant the opportunity to view the invoices at the landlord's office. However this is unlikely to offer the right environment for the exercise and should be resisted. An alternative approach may be to instruct one of a growing number of specialist surveyors or accountants who carry out a forensic audit of service charge accounts and charge a fee based on the savings made from overcharging by landlords.

4.3.9 Costs

Landlords should look to spend money in a prudent manner. As part of their best practice, landlords should approach service charge expenditure in the same way that it would approach its own expenditure with its own money. In particular they should look to obtain best value. Contracts for maintenance should be tendered regularly, and one-off costs such as redecoration should be properly specified and tendered competitively. The process of specification of contracts needs to be carefully considered to secure the best balance between on-going maintenance and having contracts with costs that are not appropriate.

4.4 Disputing service charge

4.4.1 The tenant's checklist

When a service charge reconciliation and/or service charge invoice is received there are a number of issues which a tenant, or its advisors, should consider and possibly investigate further.

(a) Procedural requirements

Most service charge clauses will prescribe a certain procedure which must be followed by the landlord. Some or all of these procedures may constitute conditions precedent (this will be a matter of construction turning on the precise wording of the service charge clause). If the procedural requirements are conditions precedent, and the landlord fails to comply with a certain requirement, this may mean that a tenant has no liability to pay the service charge demanded. This may be a matter which can be put right by the landlord very easily, eg where it has failed to provide the tenant with a copy of a certificate. However, if the lease requires certification by a certain date, and that date is time of the essence, then a landlord may lose its opportunity to recover the shortfall of service charge for that year if the date for compliance has already gone.

If the tenant has decided to investigate the costs in detail, there may be a contractual obligation on the landlord to provide all supporting documentation. The lease may also contain a dispute resolution clause which may dictate the tenant's approach to any challenge.

(b) The services provided

Each item of expenditure should be checked against the defined list of services which are contained in the service charge clause. There may be items of expenditure which do not fall within this list. If they do not, then they are not recoverable by the landlord. If the service charge provisions include a sweeping-up clause then it may be possible for an item to fall within that clause. However, such clauses are generally construed restrictively by the courts.

If the landlord is seeking to recover the cost of works, maintenance or replacement, it will be necessary to analyse carefully whether the works were capable of falling within those definitions. If the works were outside the scope of the defined services, they may not be recoverable.

Finally, the tenant will need to be satisfied that the service for which it is being asked to contribute towards was in fact provided by the landlord.

(c) A reasonableness test

A service charge clause may qualify recoverable expenditure by reference to a test of reasonableness. This tends to arise in one of two ways:

- the definition of services is restricted by the words "reasonably and properly incurred" and therefore focuses on the reasonableness of the landlord's actions in carrying out the work or
- the recovery provision stipulates that the landlord is entitle to recover the "reasonable costs" of providing the defined services.

Arguably, the first test is wider than the second in that it potentially extends to both the reasonableness of the landlord's decision to provide certain services (or carry out works), and also the reasonableness of the actual cost. In contrast, it may be argued that the second test merely addresses the reasonableness of the actual cost.

There can be no doubt that a service charge clause that contains an express reasonableness clause provides a tenant with more scope to challenge a landlord who appears to be improving his building at the tenants' expense. The most common source of frustration for a corporate occupier is where the landlord decides to carry out substantial works immediately before a lease expiry. The suspicion here is that the works are being carried out for the sole purpose of improving the landlord's prospects of re-letting the premises.

In *Scottish Mutual Insurance Co* v *Jardine Public Relations Ltd* [1990] EGCS 43 the landlord carried out long term repairs to the roof of a building. The tenant had barely a few months left on its lease. The landlord sought to recover a full contribution from the tenant and argued that notwithstanding the shortness of the tenant's lease, this constituted a fair proportion of the repair costs and the repairs had been reasonably and properly incurred. The court disagreed holding that having regard to the shortness of the tenant's lease, less extensive repairs would have been reasonable.

In *Holdings & Management Ltd* v *Property Holding & Management Trust plc* [1990] 1 EGLR 65 the Court of Appeal held that the correct test was not whether a prudent building owner bearing the cost himself might carry out the work, but rather whether the leasehold owners when considering their unexpired leasehold terms could be fairly expected to pay for such works (see also *Fluor Daniel Ltd* v *Shortlands Investments Ltd* [2001] 2 EGLR 103).

In the light of the above cases a court is only likely to consider that a reasonableness requirement has been fulfilled where the works are such as the tenants, given their length of leases, could fairly be expected to pay for those works.

(d) Expenditure

A rigorous service charge investigation will involve checking that the expenditure which appears in the reconciliation statement or service charge demand was in fact incurred by the landlord. The easiest way of establishing this is to check through the landlord's invoices. The corporate occupier will also want to verify the VAT position if the landlord is seeking to recover VAT from the tenants.

(e) Figures

The arithmetic should always be checked carefully. In addition, the corporate occupier should check that it has been charged the correct proportion. If the lease is vague and requires the tenant to pay a reasonable proportion, it should look to see if there is another way of calculating the proportion which could result in a significant saving. If the proportion is based on floor area then it should check that the landlord has used the correct figures.

If there are vacant units in the area covered by the service charge, or if there is a tenant withholding payments because of a dispute with the landlord, the tenant should ensure that it is not making up the shortfall which should be covered by the landlord.

4.4.2 Resolution of disputes

Traditionally a landlord faced with non payment of service charges had three options:

- send in the bailiffs to distrain against the tenant's property (provided that the lease reserved service charges as additional rent
- forfeit the lease or
- issue court proceedings for a money judgment and costs.

Sending in the bailiffs is probably the option of first choice for a landlord because it can be quick, cheap and downright effective (see 8.8). However, if it is contested it may become legally complex and it is not appropriate if the tenant has challenged the service charge demand. From a corporate occupier's operational perspective, therefore, it is important that any challenge to service charges is flagged up with the landlord as soon as possible. If, notwithstanding the existence of a dispute, the landlord still attempts to distrain, the tenant should write to the landlord and indicate that its action constitutes (or will constitute) wrongful distraint. The amounts at stake can be placed in an escrow account pending the outcome of the dispute.

A full blown county court trial will in most cases be unattractive to both parties. The process is slow, costly and to an extent unpredictable. If the issues at state are predominantly legal (eg construction of the lease) then a court determination may be appropriate as it may affect other tenants and be likely to recur year on year. However, a dispute about whether a certain item of expenditure has been reasonably incurred, or regarding the standard of workmanship, is unlikely to gain anything by being resolved by a judge who may have little or no previous experience in this area.

The modern approach to litigation involves a consideration of alternative dispute resolution procedures, with court determination being a last resort. The RICS has established its own Dispute Resolution Service which is particularly suitable for resolving most types of service charge disputes. There are two services which the RICS offer, both of which should be considered carefully by the parties about to engage in litigation as a result of a service charge dispute.

The first option to the parties is evaluative mediation. This involves an impartial but experienced mediator offering guidance and making settlement proposals to the parties. The mediation will involve all the parties to the dispute and its purpose will be primarily to find an acceptable solution to the dispute. The advantages of an evaluative mediation are:

- it is cheap, quick and can be very effective
- the mediator will be experienced in service charge matters and able to steer the parties towards a sensible solution
- it provides an opportunity to avoid misunderstandings and minor disagreements escalating.

If mediation is unsuccessful, the alternative to litigation is to have the dispute resolved by expert determination. This involves an impartial expert with experience of service charge disputes investigating the dispute, considering submissions from both parties and then reaching a decision which is final and binding on the parties. Expert determination is considered in more detail in the context of dilapidation claims at 15.2.5 (c).

In relation to court proceedings for a money judgment, the procedure is dealt with at 8.9.1. Suffice it to say that the parties should carry out a careful cost benefit analysis to ensure that the rewards of litigating a defended claim are sufficient to justify the risks of losing, and indeed the irrecoverable costs even if successful.

4.5 Benchmarking

An area of growing importance for corporate occupiers is benchmarking. Benchmarking is now being applied to service charge costs. This has become more sophisticated in recent years and there are various benchmarking services which look at the costs in a variety of ways, including per category and by property type as well as by location. Leading measures include OSCARTM, the Jones Lang LaSalle cost code, and ITOCC (the International Total Occupancy Cost Code from IPD). Benchmarking is likely to grow in importance as tenants increasingly look for cost savings.

These cost codes are also important to corporate occupiers as they provide basic information on costs in general and will be useful tools in assessing total occupancy costs for properties, and in determining the metrics for those buildings (see 2.2.3).

Dealings

5.1 Introduction

One of the key issues which a corporate occupier may face, is what it is permitted to do with a lease when the premises are surplus to its requirements. Can it dispose of the lease by assigning it to a third party? Alternatively, can it sub-let the whole or part of the property to a third party so as to extinguish or mitigate its liability under the lease? If so, what, if any, conditions must it comply with. Often the key issue is whether the landlord's consent is required, and if so, whether the landlord has reasonable grounds for refusing consent.

Generally a corporate occupier is keen to commence trading in a property as quickly at possible. Consequently it may not focus adequately in lease negotiations on the wording of the alienation clause (ie the clause which prohibits, permits or imposes conditions on what subsequent dealings can take place with the lease). This can, and has, led to significant difficulties for corporate occupiers. The most obvious example is the very restrictive conditions which were imposed on sub-letting in many leases granted in the 1980s and 1990s. Often leases were for lengthy terms (25 years) and prevented sub-letting except at a rent which is the higher of the passing or market rent. In a downward market, where rents have fallen, there is little opportunity for a corporate occupier to find a third party prepared to take an assignment of an overrented lease. It is more likely to take a new lease at market rent. However, sub-letting at below passing rent, in many leases, was not permitted. The difficulties and possible legal solutions are still being explored in court cases today. This is a topic we will deal with in more detail below.

5.2 When must a landlord be reasonable?

5.2.1 No alienation covenant

It is important to note that if there is no express alienation clause contained in the lease restricting dealing, the tenant is free to deal with a lease as it wishes and is entitled to charge, assign, sub-let or part with possession or occupation of the demised premises without the landlord's consent. The only restrictions on dealing will be those contained in the lease.

Most commercial leases will contain an alienation clause which restricts the tenant's dealings to a greater or lesser extent. The starting point is always, therefore, to look carefully at the lease to see what is or is not permitted, and whether the landlord's consent is required.

Careful examination of the alienation clause can sometimes pay dividends. The courts have consistently construed such clauses strictly against a landlord. So, for example, a covenant which provides that a tenant should not "assign underlet or part with possession" of premises does not prevent the sharing of those premises with a third party, nor does it prevent parting with "occupation" of the premises (eg by the grant of a licence as opposed to a sub-lease). In *Akici* v *Butlin* [2006] 1 EGLR 34 the Court of Appeal held that there had been no parting of possession (and therefore no breach of covenant) where an unrelated company used and occupied the premises but where invoices and utility bills were paid by the tenant and he retained a key and entered when he wanted.

In *Chaplin* v *Smith* [1926] 1 KB 198 the tenant assigned his business to a company controlled by him and the company carried on business from the premises. The rates were paid by the company which also appeared as the occupier in the rating valuation list. The tenant retained a key and was a managing director of the company. It was held that there had been no parting of possession.

The only circumstances where a corporate occupier may come across a tenancy with no restrictions on dealings is where a tenancy has come into existence accidentally eg where an implied periodic tenancy is created by the payment and acceptance of rent.

5.2.2 *Absolute prohibition*

Where the lease contains an express covenant against charging, assigning, sub-letting or parting with occupation or possession, the landlord is entitled to refuse consent to one of those modes of dealing even if his refusal is unreasonable.

The prohibition may prevent all or some of the modes of dealing. For example, it is common to find an absolute prohibition against sub-letting part of premises, even where there is a qualified covenant governing other modes of dealing.

A corporate occupier's only hope when faced with an absolute prohibition is that the landlord can be persuaded to waive the covenant (invariably for a premium or an increase in the rent). This can then be formally documented as a specific or general variation of the alienation clause depending on what is negotiated. From the corporate occupiers perspective a permanent change is much more valuable than one linked to a specific tenant. Having to re-negotiate with the landlord if a sub-tenant breaks the lease, defaults or does not renew will at best lose time and hence money, and at worst there is always a risk that the landlord will not co-operate the next time around.

A further way in which an absolute prohibition may arise is where the alienation clause imposes mandatory conditions on the circumstances in which a dealing can occur. This most commonly arises in the context of sub-letting where the right to sublet may be restricted by reference to rental levels, or the terms contained in the proposed subletting. If the proviso is not satisfied and the condition fulfilled, then the tenant will have no right to grant a sublease. This is dealt with in more detail below (see "Entitlement to seek consent").

5.2.3 *Qualified covenant*

A qualified covenant prohibits dealings or a specific type of dealing except with the landlord's prior consent (usually in writing).

A common example of a qualified covenant is a clause prohibiting a tenant from assigning, sub-letting or parting with possession without the prior written consent of the landlord. Usually such a covenant will also specify that the landlord's consent is "not to be unreasonably withheld or delayed".

A corporate occupier is likely to encounter alienation clauses that have elements of both absolute and qualified restrictions on dealings.

(a) Landlord's consent not to be unreasonably withheld

This qualification arises by one of two possible routes:

- *An express covenant in the lease*
  For example

  > not to assign, sub-let or part with possession or occupation of the whole of the premises without the prior written consent of the landlord, such consent not to be unreasonably withheld or delayed.

- *Implied covenant by section 19(1) of the Landlord and Tenant Act 1927*
  This section was enacted to turn qualified covenants (where landlords could unreasonably refuse consent) into fully qualified covenants (where landlords can only refuse consent where it is reasonable). This is achieved by implying the words "not to be unreasonably withheld" into the qualified covenant. Consequently, all alienation covenants which require landlord's consent are now subject to a requirement that its consent cannot be unreasonably withheld.

  It is important to note that while a similar proviso is implied into a covenant prohibiting improvements without prior consent, no such proviso is implied into a covenant prohibiting the change of user. Consequently, it is still possible for a landlord to frustrate a proposed transaction if the new tenant wishes to use the premises for a user which the lease prohibits. This again emphasises the importance of negotiating flexible terms at the outset.

(b) Landlord and Tenant (Covenants) Act 1995

Section 19(1) of the Landlord and Tenant Act 1927 applies in a modified form to "new tenancies" as defined by the Landlord and Tenant (Covenants) Act 1995 (ie tenancies granted after 1 January 1996). With new tenancies the parties are able in the alienation clause to expressly provide for:

- the circumstances in which a landlord may refuse consent to an assignment (eg where the proposed assignment is to a limited company which is not registered in the United Kingdom), and/or
- conditions on which consent will be granted (eg the requirement for a guarantor where the proposed assignee is a limited company).

If there is subsequently a dispute between the landlord and tenant, the landlord will not be regarded as unreasonably withholding its consent or giving consent subject to an unreasonable condition if the circumstances relied upon, or the conditions imposed, were specified in the alienation clause.

However, section 19(1) will not assist if the alienation clause is drafted in a way which allows the landlord to be the arbiter of whether the circumstances exist unless:

- the landlord's power to determine the matter is required to be exercised reasonably or
- the tenant has a right to have such determination reviewed by a person independent of both landlord and tenant.

5.3 What is landlord's consent? — the unintentional licence

Landlord's consent to a particular dealing will in most cases, in practice, be recorded formally in a deed executed by all parties (ie a licence to underlet or licence to assign). However, this is not always legally necessary, and will depend upon what the alienation clause actually requires.

Once a landlord has given its consent and that has been acted upon, it cannot be withdrawn, even if the landlord subsequently unearths facts which would have justified the withholding of consent. It is important, from the landlord's perspective therefore, that consent is not given inadvertently.

However, if the tenant misrepresents the details of the proposed dealing, or obtains the landlord's consent by fraud, the licence or its consent can be set aside and the assignment or subletting will then be treated as if no landlord's consent had been given.

5.3.1 What does the covenant say?

It is important to read the alienation covenant carefully to see what, if any, formality is required. Older leases do not always require the landlord's consent to be contained in a formal deed of licence. A typical clause may provide:

> Not without the previous consent in writing of the landlord ... to assign, underlet or part with possession.

A letter from the landlord or its duly authorised agent giving consent will be sufficient in such a case. This is often overlooked by the landlord's agents and is a trap for the unwary. By the same token, it may present an opportunity for a tenant who is in a hurry to assign or underlet.

5.3.2 Consent "Subject to Licence"

Importantly, the courts have held that a letter from the landlord or its agents giving consent "subject to licence" (ie conditional upon the parties entering into a formal licence) may constitute consent even though the writer of such a letter patently did not intend to grant consent in that document. In other words, the use of these words alone will not prevent a letter from constituting consent in certain circumstances.

See *Prudential Assurance Co Ltd* v *Mount Eden Land* [1996] 74 P&CR 377

Facts

- T applied for consent to carry out alterations.
- The lease provided that alterations could not be carried out without the "previous written consent" of L.
- L's agents wrote to T in the following terms:

> *Subject to licence*
>
> *I refer to your request to consent to alter the external appearance of the ... building.*
> *I can confirm that the Freeholder... gives consent for the works subject to the following conditions:*
>
> 1. *A formal licence is entered into by your client...*
> 2. *[T] will pay [L's] costs...*
> 3. *[T] are to obtain the necessary consents and statutory approvals.*

- T obtained the necessary consents and wrote to L confirming the conditions were agreed.

Held — by the Court of Appeal

- L had provided its "previous written consent". No formal licence was required.

This principle has now also been applied to alienation clauses (see *Next* v *National Farmers Union Mutual Insurance Co* [1997] EGCS 181 and *Aubergine Enterprises Ltd* v *Lakewood International Ltd* [2002] EWCA Civ 177).

In short, if the alienation clause does not require formality, the subject to licence condition will not prevent consent from being granted.

A landlord (which will include a corporate occupier who has sublet) and its advisors should now take greater care in its initial response to an application for consent stating clearly:

- whether the letter does or does not constitute consent for the purposes of the lease
- if the alienation clause requires a formal licence, that the landlord will be enforcing this obligation.

To prevent unintentional consent being granted, a landlord or its professional advisors would be wise to include the following standard paragraph in all correspondence leading up to the formal licence:

> This letter does not constitute consent to the proposed [assignment/underletting] and consent will only be given, if appropriate, by completion of a formal licence executed as a deed on terms to be agreed between us.

5.4 A landlord's statutory duty

The Landlord and Tenant Act 1988 imposes a statutory duty on a landlord to give consent within a reasonable time, or provide written reasons demonstrating that the refusal or conditions attached to consent are reasonable. Failure to do so constitutes breach of statutory duty and can give rise to a claim in damages.

5.4.1 When does the 1988 Act apply?

It only applies to a covenant not without the consent of the landlord to:

- assign
- underlet
- charge
- part with possession.

It does not apply to:

- absolute covenants against alienation (ie total prohibition)
- qualified covenants against change of user
- qualified covenants against alterations or improvements.

5.4.2 What is the landlord's duty?

Under section 1(3) of the 1988 Act, where a landlord receives a written application for consent, he owes a duty to the tenant *within a reasonable time*:

1. to give consent, except in a case where it is *reasonable not to give consent*
2. to give the tenant written notice of his decision whether or not to give consent specifying:
    - if the consent is given subject to conditions, those conditions (which must be reasonable)
    - if a consent is withheld, the reasons for withholding it.

The landlord must, therefore, deal with the tenant's application and provide a decision within a reasonable time.

The key points are set out below.

- If the landlord gives a notice refusing consent within a reasonable time, it must give reasons.
- Importantly, the landlord will not be permitted to put forward reasons in subsequent proceedings justifying a refusal unless those reasons were contained in his refusal notice/letter (ie put forward in writing within a reasonable time).
- A failure by the landlord to give a written decision to the tenant within a reasonable time will be treated by a court as a refusal to give consent with no reasons and will make a landlord liable to pay damages to the tenant for breach of statutory duty. It is no defence for the landlord in subsequent proceedings to show that it had reasonable grounds for refusal of consent (see *Footwear Corporation* v *Amplight* [1999] 2 EGLR 38). The moment that the landlord gives written reasons for its refusal, time will stop and the reasonable time period will expire. If those reasons are bad, the landlord will not be allowed to rely upon fresh reasons that are good which emerge from subsequent correspondence with the tenant or its advisors (see *Go West* v *Spigarolo* [2003] 1 EGLR 133).
- If the tenant issues court proceedings, the landlord has the burden of proving:
    1. if he gave consent, he did so within a reasonable time
    2. if he gave consent subject to conditions, those conditions were reasonable
    3. if he did not give consent, the refusal was reasonable.

When a corporate occupier receives an application from a sub-tenant it is therefore of great importance that it considers and deals with a completed application swiftly and if it decides to refuse consent, or impose conditions, it must carefully draft the decision letter to ensure all reasons relied upon are set out in full.

5.4.3 When does time start to run?

The clock does not start and time begin to run against a landlord until a completed formal application is received from a tenant. Informal exchanges between the landlord and tenant or their respective agents will not start the clock (but these may affect what is later determined to be a reasonable period in which to respond).

The assessment of whether a reasonable time has expired will be made at the time that the tenant claims that it has expired.

5.4.4 What constitutes reasonable time?

What amounts to reasonable time is a question of fact and will depend upon the circumstances of a particular case. The Law Commission Report that led to the 1988 Act suggested 28 days was a reasonable period for a landlord to consider a tenant's application for consent. However, this recommendation did not find its way into the 1988 Act.

In *Norwich Union* v *Shopmoor Ltd* [1991] 1 WLR 531 Sir Richard Scott V-C indicated that a landlord was required to deal with a tenant's application "expeditiously" and "at the earliest sensible moment". What do such statements mean in practical terms? There are now a considerable number of cases in which this issue has been considered. For example:

- In *City Hotels Group* v *Total Property Investment* [1985] 1 EGLR 253 a delay by a landlord in pursuing enquiries and making a further request for information over a period of two months was held to be unreasonable.
- A delay of two to two and a half months in replying to an application was held to be unreasonable in *Midland Bank* v *Chart Enterprises* [1990] 2 EGLR 59.
- In *Dong Bang Minerva (UK)* v *Davina* [1995] 1 EGLR 41, the court indicated that consent ought to have been given "at the latest" within one month of an application by the tenant for the grant of a five year (1954 Act excluded) subletting of part.
- In *Footwear Corporation* v *Amplight Properties* [1999] 2 EGLR 38, it was held that a delay of nine weeks (some seven of which had been spent in negotiations) was excessive.

While these authorities shed some light on the question, they should be treated as guidelines only. The courts have avoided laying down any benchmarks for the same reason that the Law Commission's recommendation of 28 days was not enacted. What is a reasonable period will vary from case to case. It will be influenced by the following points.

- *The terms of the transaction*: in the case of an underletting, the landlord is entitled to know the principal details and true nature of a proposed underletting; if the transaction is unusually complex, the landlord may be allowed longer to consider it. In *NCR Ltd* v *Riverland Portfolio* [2005] 2 EGLR 42, which involved a complex underletting (with potential damages of £3m if consent was unreasonably refused), the Court of Appeal adopted a more landlord friendly approach and emphasised the importance (to both parties) of the landlord making a good decision as opposed to a rushed decision. The landlord was entitled to consider the serious financial and legal implications of a refusal with its advisers, and if necessary, report to the relevant board. In that case three weeks was insufficient time.
- *The amount at stake and the relative financial positions of the parties*: a court is likely to be more lenient to a landlord with meagre financial resources faced with an application to sublet from a major covenant than it would be where the facts are the other way round.
- *The time of the year at which the application is made*: common sense dictates that the landlord may be less well equipped to deal with an application when it arises, for example, in the middle of August or at Christmas time (when many professional advisors may be away).
- *The personal circumstances of the parties*: if, for example, the landlord is an individual who is on holiday for a month at the time the application is made, a court is unlikely to hold that time should start running until his return.
- *Any previous knowledge by the landlord of the tenant's proposals*: from a tenant's perspective therefore, one approach is to start the flow of information to the landlord as soon as possible, rather than

waiting until everything is together. While the drip feed can be disruptive and some landlords will refuse to look at things until they have received everything, it is a way of starting to pressurise the landlord in situations where time is of the essence and some information is not yet ready to be sent to the landlord. This does need to be thought out and it is important that the major requirements are provided at an early stage. For example, sending through the last three years accounts whilst waiting for trade references is preferable to waiting until the latter are obtained before providing the landlord with anything.

While no hard and fast rules can be extracted from decided cases, it is clear that a reasonable period in respect of a straightforward application will be measured in days and a complex application will be measured in weeks rather than months. In most cases the 28 day period suggested by the Law Commission will be a good starting point.

5.5 How reasonable must a landlord be?

5.5.1 The reasonable landlord on the Clapham Omnibus

When a court considers whether a landlord's refusal to grant consent to a transaction is reasonable or unreasonable, it must consider the application through the eyes of the particular landlord involved and consider what a reasonable landlord would have done in the same circumstances.

The court must not substitute its own view for that of the landlord. To put it another way, a landlord will discharge the burden of proving that a refusal of consent is reasonable if it can show that some landlords, acting reasonably, might have refused consent for the reasons given, even though some other reasonable landlords might have given consent (see Carnwath LJ's judgment in *NCR Ltd* v *Riverland Portfolio*).

5.5.2 An approach to reasonableness

In *International Drilling Fluids* v *Louisville Investments (Uxbridge) Ltd* [1986] 1 All ER 321 the Court of Appeal laid down some important guidance on how to approach the issue:

- the purpose of a qualified covenant is to protect a landlord from having his premises used or occupied in an undesirable way or by an undesirable tenant or assignee
- a landlord is not entitled to refuse his consent on grounds which have nothing to do with the relationship of landlord and tenant, the lease and the leased premises
- it is not necessary for a landlord to prove that the conclusions that led him to consent were justified, if they were conclusions that might be reached by a reasonable man in the circumstances
- it may be reasonable for the landlord to refuse consent if the proposed assignee/subtenant is going to use the premises in an undesirable way, even if the lease does not prohibit that use
- while a landlord usually only needs to consider his own interests, if the prejudice likely to be suffered by the tenant is disproportionate to the prejudice likely to be suffered by the landlord, it may be unreasonable for the landlord to refuse consent.

In *Straudley Investments Ltd* v *Mount Eden Ltd* [1996] EGCS 153 the Court Appeal added two further guidelines:

- it will normally be reasonable for a landlord to refuse consent or impose a condition if this is necessary to prevent his contractual rights under the lease from being prejudiced by the proposed assignment or sublease
- it will not normally be reasonable for a landlord to seek to impose a condition that is designed to increase or enhance the rights that he enjoys under the lease.

Whether a landlord's consent is being unreasonably withheld or not is a question of fact which needs to be considered in each individual case and considering how the courts have dealt with a particular issue in the past may not be determinative. Nevertheless decided cases do provide some assistance in showing how the courts have approached certain recurring issues.

When considering a tenant's application, a landlord may consider refusing consent on one or more to the following grounds:

(a) Can the proposed assignee/subtenant pay?

A landlord can reasonably refuse consent when references or accounts provided by the proposed assignee/subtenant cast doubt on its ability to pay the rent or discharge its obligations under the lease as they fall due. In considering this issue a landlord will look at:

- the proposed tenant's references and accounts
- its financial standing and prospects
- the extent of the obligations under the lease
- the financial health of any guarantors offered.

Although it has been judicially frowned upon, a commonly used rule of thumb is to look at a proposed tenant's profits to see if they are at least three times the rent under the lease. If the proposed tenant satisfies this test, it is likely that a landlord's objection on this ground is unreasonable. However, it is doubtful whether a landlord's invariable practice of refusing consent if this test is not satisfied, will be upheld by the court (see *Footwear Corporation* v *Amplight Properties* [1999] 2 EGLR 38).

Arguably there should be a difference between what the landlord will accept in the circumstances of an assignment compared to a subletting. With the latter the superior landlord does not have any direct contact with the subtenant and the tenant remains in place. The superior landlord's only risk is that the tenant becomes insolvent or looks to restructure. However, there is rarely a difference in approach in practice and often superior landlords will take issue with accounts submitted by a prospective sub-tenant on a small property in circumstances where the tenant is a major international company and the risk to the superior landlord is non-existent.

The poor financial position of an assignee (and guarantors) cannot be remedied by a guarantee given by the outgoing tenant. The landlord is entitled to be satisfied that the tenant in possession can discharge rent and obligations under the lease. However, an assignee with a poor financial standing (eg a new company) can bolster its position with its own guarantors provided they are of sufficient financial standing.

(b) Covenant strength

This is a growing and fruitful area for landlords as a result of recent case law and the Landlord and Tenant (Covenants) Act 1995.

A landlord may refuse consent if it reasonably concludes a proposed transaction will reduce the value of its reversion. Traditionally, however, the courts have treated such claims by landlords with some caution and they have usually failed unless the court has been satisfied that there was some prospect of the reversion being sold (and a diminution in value occurring).

In *International Drilling Fluids Ltd* v *Louisville Investments (Uxbridge) Ltd* the landlord's valuer suggested that an assignment of the lease to a tenant which intended to use the premises for serviced office accommodation would result in a lower price if the landlord were to sell its reversion. However, as there was no evidence that the landlord intended to sell, the reduction in the paper value, was not a real detriment to the landlord and did not justify refusal of consent.

In *Pondersosa International Developments* v *Pengap Securities (Bristol) Ltd* [1986] 1 EGLR 86 the position was different. The landlord was a single project development company that intended to sell a retail development (which included the tenant's lease) once the development had been completed. The tenant was a large international company and a good covenant. The proposed assignee was a new company with little share capital. The proposed guarantors did not improve the weak covenant strength. The landlord refused consent to assignment on the basis that this would adversely affect its prospect of sale and result in a lower price than if the transaction was structured as a sublease. The landlord did indicate that it would consent to a sublease. The tenant argued this was unreasonable because, as a matter of law, the tenant would continue to be liable for rent if the assignee defaulted under the principle of original tenant liability. The judge rejected this argument and accepted the landlord's expert evidence that it is a widely held view in the property market, and in particular among investors, that, after an assignment of a lease, the identity of the original tenant is of no interest. The judge said that the landlord had to:

> ... live in the real world and to take the market as it finds it, not as lawyers might wish it to be.

The issue has recently been reconsidered again in a case that has been regarded by many property professionals to have improved the landlord's lot.

NCR v *Riverland Portfolio No 1 (No 2)* [2005] 2 EGLR 42

Facts

- T occupied the premises on a 25 year lease with upwards only rent reviews, and with six years of the term remaining.
- The current rent was £710,000 pa which significantly exceeded market rent (£19.30 psf compared with a market rent of £16 psf).
- The lease prohibited subletting except at the higher of market rent or current rent.
- T proposed a sublease to ST which reserved the market rent but on the basis that T would pay to ST a reverse premium of £3m as an inducement.
- L refused consent on the basis (i) the unusual nature of the terms and (ii) the inadequate covenant strength of ST.
- T issued proceedings for unreasonably withholding of consent.

Trial judge held

- L's refusal was unreasonable.
- ST's covenant strength was irrelevant because the proposed transaction was an underletting and T would remain liable under the headlease.

Court of Appeal held (allowing the appeal)

- The same general principles applied to applications to sublet and assign. The trial judge had set the standard of reasonableness too high.
- L was entitled to refuse consent on the ground ST and its proposed guarantor lacked covenant strength. This decision was based on L's expert evidence which showed:

  (1) at expiry of headlease, ST would have a right under the 1954 Act to renew its lease
  (2) ST would become the direct tenant of L
  (3) the weakness of ST's covenant strength would be likely to diminish the value of L's reversion by in excess of £500,000.

Riverland is an important departure from early cases and perhaps suggests a more landlord-friendly approach. It has clarified that there is no distinction to be made between the principles to be applied to applications for subletting or assignment. It is also important to note that the landlord succeeded even though there was no evidence before the court that the landlord intended to sell its reversionary interest, in contrast to the earlier cases.

Arguments concerning covenant strength are now much more likely, not only as a result of *Riverland*, but also in the case of a new lease under the Landlord and Tenant (Covenants) Act 1995. In the case of a new tenancy (ie granted after 1 January 1996) an assignment of the lease will release the former tenant from its continuing liability under the lease, unless it is reasonable to require the tenant to enter into an authorised guarantor agreement. An assignment may therefore be of more concern to a landlord, particularly where this would result in a dilution of covenant strength.

In the 30th Annual Blundell Lecture, Timothy Fancourt QC suggested that a landlord is entitled to take into account the extent to which it is likely to be disadvantaged by an assignment and the loss of the tenant's covenant when considering whether to refuse consent, or in requiring an AGA as a condition of its consent. He also suggests that a court when considering reasonableness ought to be more astute to protect a landlord's legitimate interests given the impact of the 1995 Act.

(c) Collateral advantage

A landlord is not acting reasonably in refusing consent if it does so with the intention of improving or enhancing its position, or obtaining a commercial benefit outside the contemplation of the lease.

It has been held to be unreasonable to refuse consent:

- to prevent a comparable transaction which might adversely affect rent review of other property the landlord owns in the neighbourhood (*Norwich Union Life Insurance Society* v *Shopmoor* [1991] 1 WLR 531)
- to try and force the tenant to surrender the lease (*Staudley Investments Ltd* v *Mount Eden* [1997] EGCS 175)
- except on terms that required the tenant to agree a variation to the user covenant.

(d) Breaches of covenant

Landlords regularly withhold consent on the basis that the tenant is currently in breach of covenant. While this is a factor that a reasonable landlord can consider, it is rarely reasonable for a landlord to refuse consent on this ground.

A landlord cannot reasonably refuse consent if the benefit to the landlord is greatly and disproportionately outweighed by the detriment to the tenant. Whether refusal is justified will depend upon:

- the degree of seriousness of the breach and
- whether the landlord's position will be prejudiced by the transaction going ahead.

A landlord would not usually be justified in refusing consent:

- On the basis that the tenant is in breach of its repairing covenants provided that the breach is not serious or requiring urgent remedy, or the proposed assignee has the wherewithal to carry out repairs (see *Farr* v *Ginnings* (1928) 44 TLR and *Staudley Investments* v *Mount Eden Land Ltd* [1997] EGCS 175). If the breach of covenant is a continuing breach, the landlord's consent to assignment will not prejudice its rights to forfeit and it is not open to a landlord to refuse consent on the grounds that if granted it could constitute a waiver of the landlord's right to forfeit the lease.
- Where there is an outstanding service charge dispute (*Beale* v *Worth* [1993] EGCS 135).
- Where the tenant is in breach of an alterations covenant and the breach would be easily remediable at the end of the term.

However, the position may be different if:

- The landlord reasonably anticipates that the transaction will result in a breach of covenant. In *Ashworth Frazer* v *Gloucester City Council (No 2)* [2001] 1 EGLR 15 the House of Lords held that a landlord would generally be justified in taking into account an anticipated breach of covenant by the assignee. In that case the user proposed by the assignee was in clear breach of the user clause in the lease.
- There are extensive and longstanding breaches of the repairing covenant, it will generally not be unreasonable for a landlord to refuse consent unless it can be reasonably satisfied that they will be remedied by the assignee/subtenant (see *Orlando Investments Ltd* v *Grosvenor Estates Belgravia* (1990) 59 P&CR 21 and *Crestfort Ltd* v *Tesco Stores Ltd* [2005] 3 EGLR 25).

5.6 Entitlement to seek consent

5.6.1 The problem

There will be occasions where a landlord does not object in principle to the transaction, but is unhappy with the detailed terms. This is a recurring problem for corporate occupiers of surplus properties.

Often, in older leases, the alienation clause may contain a restriction on subletting in the following or similar terms:

> ... the Tenant shall not be permitted to underlet unless granted at the best rent obtainable in the open market without taking a premium or other capital consideration or, if greater, the rent payable under this lease...

Other variations may include:

> ... any permitted underlease shall be granted subject to like covenants contained in this lease except as to the rent reserved and the length of term ...

> ... and shall be excluded from the security of tenure provision contained in the Landlord and Tenant Act 1954

The effect of these covenants is:

- if the proposed sublease does not satisfy these mandatory restrictions or conditions the landlord is not required to consider the application for consent. This is because the lease prohibits subletting *unless* these conditions are (or would be) fulfilled (see *Crestfort Ltd* v *Tesco Stores Ltd* [2005] 3 EGLR 25). The landlord can, therefore, refuse consent, however unreasonable that refusal may be
- the mandatory restriction on rent, (which was a very common restriction in leases granted in the 1980s and 1990s), prevents a tenant from granting a sublease at a rent which is below the passing (ie the rent reserved in the headlease) or market rent, which ever is higher.

A mandatory rental restriction can be very problematic for a corporate occupier of surplus premises where the lease has many years to run to expiry. In a downward market, where the rent reserved by the lease is significantly higher than market rent (a common phenomenon with long leases granted in the 1980s and 1990s with upwards only rent reviews), a corporate occupier is unlikely to find a third party which is prepared to take an assignment of the overrented lease. The only option to a corporate occupier in those circumstances is to try and sublet the premises. Any rental restriction on subletting however could prevent this.

If the mandatory rent restriction is merely a requirement that any subletting must not be at less than market rent, a tenant may take some comfort from the decision in *Blockbuster Entertainment Ltd* v *Leakcliff Properties Ltd* [1997] 1 EGLR 28. If the tenant and subtenant strike a genuine deal after the premises have been properly marketed, then this is the best evidence that the rent is a market rent. That is not to say, however, that the rent proposed by the tenant will not be challenged by the landlord, and that the matter will have to be determined at a court hearing. The problem for the corporate occupier is that this will cause uncertainty, significant irrecoverable legal costs and delay, and invariably the loss of the tenant.

If there is a restriction on underletting at less than market or passing rent (which ever is the higher), and the passing rent is significantly higher than market rent, a corporate occupier is unlikely to find a subtenant at that higher rent. In the absence of an incentive, it is very unlikely that a subtenant will want to pay more than market rent for its premises. This is a much more serious problem and could sterilise the property until market rent improves, or the lease comes to an end. Traditionally, the solution adopted by tenants was for the sublease to stipulate a rent at the passing rent, but for the tenant to enter into a side deed or collateral arrangement with the subtenant which had the effect of reducing the amount of rent actually paid by the subtenant to the tenant. Often this deed was not disclosed to the landlord, which in turn, seldom enquired whether such arrangements existed.

Unfortunately, this pragmatic solution suffered a substantial set back as a result of the well know case of *Allied Dunbar Assurance plc* v *Homebase Ltd* [2002] 2 EGLR 23.

Facts

- T applied for consent to underlet to ST.

- The alienation covenant in the lease prevented underletting unless:

    1. the rent was not less than full market rent (without taking a fine or premium)
    2. the underlease contained covenants in the same terms as the lease
    3. the rent review provisions in the lease were mirrored in the underlease.

- T entered into an agreement for the grant of a sublease with ST which complied with the strict requirements of the lease; but T and ST entered into a collateral deed which provided:

    1. a rent that was significantly below passing rent (and the landlord argued below market rent)
    2. an indemnity from T to ST for the rental difference between the agreed rent contained in the side deed and that expressed to be payable under the underlease
    3. an indemnity from T against the cost of complying with the repairing covenants contained in the underlease insofar as they required ST to put the premises in any better condition than evidenced by a schedule of condition.

- The obligations contained in the collateral deed were expressed to be personal to T and ST and were expressed not to affect L

Held by the Court of Appeal

- The sublease and the collateral deed had to be read together when considering whether the mandatory conditions had been satisfied. The court had to approach the issues as if all the terms had been contained in the underlease and not in two separate documents.
- When considering the arrangement as a whole (including the personal concession), the proposed underlease terms breached the underletting conditions.
- Despite the suggestion that the side deed was personal to T and ST; it could have an adverse impact on L's position, namely:

    1. the level of rent which a court may fix on lease renewal under the 1954 Act (the sublease was not to be excluded from the provisions of the 1954 Act)
    2. the terms of new lease which a court may determine on renewal
    3. by affecting T's ability to pay rent and comply with its covenants
    4. if a notice were served under section 6 of the Law of Distress Amendment Act 1908.

5.6.2 The impact of Homebase

It is perhaps no exaggeration to say that the approach of the Court in *Homebase* was a cause of great concern for corporate occupiers with surplus properties and their advisors. Its immediate impact, it was thought in some quarters, threatened to sterilise a significant category of property. With hindsight, this was perhaps an overstatement of the position. It has, however, left many corporate occupiers of surplus properties with a potentially very difficult obstacle to underletting and the prospect of increased costs in implementing any transaction.

It is now unlikely that a personal concession between a tenant and a subtenant contained in a side letter or deed, in the circumstances discussed above, will satisfy mandatory underletting conditions. Furthermore, if a tenant decides to proceed by way of a side deed, generally speaking it will be extremely unwise to proceed without disclosing the side deed to the landlord when seeking consent.

A tenant and subtenant who do decide to take their chances and proceed without disclosing a side deed must consider the following points carefully:

- A landlord is entitled to be told the precise nature of the proposed subletting (see *Fullers Theatre and Vaudeville Co Ltd* v *Rofe* [1923] AC 435). If the landlord grants consent after being told only part of the story, the consent will be treated as not having been given. Equally, if the consent is obtained by fraud, it can be set aside by the landlord.
- If the conduct could be classified as fraudulent misrepresentation, there could be serious implications for any company director and professional who may have authorised it. Equally, any failure to disclose a side deed in these circumstances could potentially leave a solicitor acting for the tenant susceptible to disciplinary action. If a tenant refuses to agree to the side deed being disclosed, a solicitor should consider very carefully whether s/he can continue to act on the transaction.

The real risks of proceeding without prior disclosure lie in the potential remedies available to a landlord if it discovers the existence of a side arrangement and that the transaction proceeded without its proper consent. Tenants and subtenants have, on occasions, taken a bold approach and proceeded on the basis that either it is unlikely that a landlord will take action (because it is against its commercial interests eg forfeiture), or if the landlord does take action, nothing significant will happen because it has suffered no loss or nominal loss. This is an unwise assumption, as Tesco's solicitors discovered in *Crestfort Ltd* v *Tesco Stores Ltd* [2005] 3 EGLR 25, after advising Tesco that once the sublease was in place there was little the landlord could do about it short of forfeiting the lease which it was unlikely to do. A tenant should very carefully evaluate the potential remedies available before adopting a bullish approach.

1. *Forfeiture*: if the landlord does not want the premises back, the tenant is a good covenant and/or the lease is overrented, the landlord is unlikely to forfeit the lease. However, from the perspective of the ingoing subtenant (which may spend a good deal of money fitting the premises out and have no where else to go) this is a significant risk. If the landlord did forfeit the lease (and consequently the underlease), a court is unlikely to grant relief from forfeiture in view of the deliberate deception and flagrant breach of covenant, which the subtenant will be aware of and party to (see *Crown Estates Commissioners* v *Signet Group plc* [1996] 2 EGLR 200).

2. *Injunction reversing the subletting*: until recently this was a remedy often overlooked (although successfully used by the landlord in *Hemingway Securities Ltd* v *Dunraven Ltd* [1995] 1 EGLR 61). This has now changed following the case of *Crestfort Ltd* v *Tesco Stores Ltd* [2005] 3 EGLR 25. In *Crestfort* Tesco requested landlord's consent to sublet to Magspeed a warehouse that was surplus to its requirements. The lease contained mandatory conditions on subletting. Those conditions were not fulfilled. Furthermore, Tesco was in serious breach of its repairing obligations, and also in breach of its insuring obligations. The landlord withheld consent, but in flagrant breach of covenant Tesco proceeded with granting the sublease. The court held that Tesco had knowingly in breach of contract granted, and the subtenant unlawfully accepted, a sublease. It ordered that the sublease be surrendered and the subtenant was therefore required to vacate the premises. It was clear that the court was unimpressed with the calculated and deliberate breach of covenant by Tesco, and the advice given by its solicitors. The publicity which the *Crestfort* case generated means that this is now a remedy which will feature prominently in the minds of the landlord and its advisors when considering how to respond to an unlawful subletting.

3. *Award of damages for breach of covenant*: the traditional view of tenants was that a landlord, in these circumstances, would suffer no loss as a result of the breach, and therefore, any damages awarded would be nominal. The reason for this view was that the direct relationship between the landlord and the tenant would remain following the grant of the sublease, and the tenant would continue to be liable to the landlord to pay rent and observe the covenants in the headlease for the remainder of the contractual term. It was also thought that it would be very hard for the landlord to prove diminution in value of its reversion so as to give rise to damages. This view is no longer sound. The modern approach of the court is to award compensatory damages either in addition to the grant of an injunction, or as an alternative (see *Jaggard* v *Sawyer* [1995] 1 EGLR 146 and *Experience Hendrix LLC* v *PPX Enterprises Inc* [2003] FSR 46). This approach was confirmed in the *Crestfort* case, where in addition to granting a mandatory injunction reversing the unlawful subletting, the judge also awarded compensatory damages. Compensatory damages can be very considerable and it is important to note:

    - when awarding damages, the judge is entitled to award such amount as the landlord might reasonably have demanded at the date of the breach of covenant for relaxing the mandatory conditions on underletting (eg to permit an underletting at a rent below market or passing rent, or allowing the underlease to impose a diluted dilapidations obligation on the subtenant)
    - in *Crestfort*, although the court did not ultimately determine damages, one of the agreed alternative basis of damages was that Tesco pay the landlord 25% of the value of the benefit of a relaxation of the mandatory condition to Tesco which equated to a quarter of the rental for the whole period of the subletting (which in that case would have resulted in damages of £120,000). This is a far cry from the days of nominal damages.

Clearly, a corporate occupier who proceeds in a cavalier way now does so at significant risk of the arrangement being unravelled by the court, and/or faces a potentially substantial claim in damages. However, the position is not necessarily as bleak as it may at first appear. There are potential solutions and strategies that can be used in certain circumstances to overcome the *Homebase* problem.

5.6.3 Solutions and strategies

Whether a solution or strategy exists that is likely to result in a successful subletting despite mandatory restrictions will depend on a number of variables. The starting point is to study the alienation clause very carefully to determine precisely what can and cannot be done. This is a point which was emphasised in *Homebase* and was re-emphasised in *NCR* v *Riverland No 1* [2004] EWHC 921 (Peter Smith J). It will also depend on the particular landlord and whether it can be persuaded that it would not be in its interests to be difficult, and ultimately whether the landlord would be prepared to litigate the issue and risk significant irrecoverable costs. In many cases, a cost benefit analysis may not support the landlord risking litigation even if it has been advised it has good legal grounds to oppose the proposed subletting. It will also depend on the subtenant, and whether it is prepared to enter into more complex legal structures than would be the case if it were taking a new lease.

When faced with a *Homebase* problem, the following potential solutions and strategies may be available to a corporate occupier.

(a) Subletting with a reverse premium

Although in the immediate aftermath of *Homebase* there was some doubt, it has now been held by the Court in *NCR* v *Riverland No 1* [2004] EWHC 921 (Peter Smith J) that the payment of a reverse premium by the tenant to the sub-tenant as an incentive to accept a sublease at passing rent (ie above market rent) did not fall foul of the principle in *Homebase*.

In *Riverland* the passing rent under the lease was £710,000 pa compared to a market rent of £585,000 pa. The lease imposed a mandatory restriction not to sublet at less than market or passing rent (which ever was the higher). It also prohibited the tenant from "taking a premium or other capital consideration" for the sublease. The tenant proposed to grant an underlease to the subtenant at passing rent (£710,000 pa) but with the tenant paying a reverse premium of £3m to the subtenant. The bulk of the premium was paid into an escrow account and released to the subtenant in five equal payments. The court held that although the two agreements had to be read together, the lease did not prohibit a genuine reverse premium, and it did not result in a letting below passing rent.

Riverland is much needed good news for corporate occupiers. However, it is not necessarily a complete answer and it is not for the fainthearted tenant or subtenant. There are potential drawbacks.

- Whether or not this is a solution will turn on the specific wording of the alienation clause, and the precise circumstances of the case (in some respects it is very difficult to reconcile *Homebase* and *Riverland*). For instance, in *Crestfort* the lease contained a mandatory restriction that prohibited the tenant from "accepting or *paying*" a premium. The granting of a sublease at a reverse premium would not be an option, therefore, with a restriction of that nature.
- There will be a cashflow issue for the tenant if it is required to pay out the full amount of the premium at the date that the sublease is granted.
- There will be a credit risk issue caused by the payment of a lump sum. If the tenant pays the full premium to the subtenant, and the subtenant defaults (or worse becomes insolvent), the tenant could suffer significant losses. Equally, if the tenant makes periodic payments and becomes insolvent, the subtenant stands to lose out. The most acceptable solution is likely to be the payment of the bulk of the reverse premium into an escrow account, with periodic payments being made to the subtenant on rent days. This will leave the tenant exposed to the risk of losing one premium instalment, in the event of the subtenant defaulting, although that can be covered by only making payment on production of receipts.
- This is not a practical solution where one of the key issues relates to dilapidations. If the premises are not in a good state of repair, and the subtenant wants protection, this approach will require a quantification of the amount which the subtenant may be required to pay either discharging its full repairing covenant, or to pay by way of damages in lieu of carrying out the works. This is equally unattractive from a tenant's perspective as the subtenant may not actually carry out any repairs, or pay any damages (eg if there is ultimately a defence under section 18 of the Landlord and Tenant Act 1927).
- There is a legal debate about whether the arrangement must be disclosed to the landlord at the application stage. Where a reverse premium was paid to an assignee in *Kened* v *Connie Investments Ltd* [1997] 1 EGLR 21, the court held that the tenant was not required to provide the landlord with the details. It is unclear whether this principle would extend to these circumstances. The safest course is to disclose the existence of any side arrangement and argue about it before completion.
- The corporate occupier will need to retain its provision for the monies that have to be paid out as part of the rent subsidy. With an assignment there is only a contingent liability but with the

subletting of an over-rented property there is potentially a greater risk and the corporate occupier should consider carefully whether to retain a provision.

(b) Indemnity from related company

A second device for side stepping *Homebase* that now has judicial support is by a third party indemnifying a subtenant in respect of the discharge of a full repairing covenant (as opposed to one qualified by a schedule of condition). In *Crestfort* it was proposed that the sublease would contain a full repairing lease, but that a sister company of the tenant would indemnify the subtenant against costs in excess of those needed to maintain the premises in their current condition. The court held that while an indemnity from the tenant would fall foul of the decision in *Homebase*, an indemnity provided by a third party (albeit related) company was permissible. If the tenant has a substantial sister company, this may be an effective way of dealing with a dilapidations problem. In most instances the headlease dilapidations will be dealt with by the tenant and therefore it is unlikely that the indemnity from the sister company is likely to be called upon. However, what is less clear is whether this devise could be used to side step a rental restriction. If the sister company makes periodical rebates to the subtenant, which are funded by periodical payments from the tenant to the sister company, there must be a real risk that a court considers this is merely an artificial device to lower the rent actually paid by the subtenant.

In terms of provision, as there is a liability a provision will be needed for the commitment, although depending on the group structure, it is possible that the related company indemnity could be classified as a current obligation and therefore be viewed as an operating cost. This may allow the original tenant company to avoid making a provision, but how that then flows through to the impact on the overall company is open to debate.

(c) Assignment with a reverse premium

If the proposed sub-letting is for the remainder of the lease term, and the landlord is, or is likely to be, resistant to the grant of a sublease, one option that may be available to the parties is to restructure the deal as an assignment with a reverse premium.

This will only be an option if there are no restrictions on assignments contained in the lease. While the lease may not permit an assignment for a premium (ie capital sum), it is unlikely that a lease will prevent an assignment for a reverse premium. If it does, this device will not provide a means around the *Homebase* problem.

The sticking point is likely to be dilapidations. The tenant could bring the premises up to scratch before assignment. Alternatively the assignee could accept an additional amount to account for a likely claim by the landlord for terminal dilapidations. The problem with this approach is that it may be very difficult at that stage to assess what, if any, damages a landlord may seek at lease expiry for dilapidations. A corporate occupier is unlikely to want to pay out any money for dilapidations on surplus premises, particularly in circumstances where there may be no claim by the landlord on lease expiry (eg where there is a possibility that the landlord may redevelop or refurbish the premises).

An alternative is for the tenant to indemnify the assignee (fully or up to an agreed limit) against any claim from the landlord for dilapidations over and above the deterioration in the premises since the assignment (evidenced by a schedule of condition). While this may seem a fair and equitable approach at the date of the assignment, if no agreement can be reached at lease expiry, it could give rise to a fairly complex tripartite dispute between the landlord, assignee and tenant. A further alternative, if the premises are in poor condition, is for the tenant to bite the bullet and offer a full indemnity to the assignee and recognising that this is a necessary cost to avoid the premises being sterilised. Of course, the indemnity will be worthless unless the tenant is good for the money, and the assignee is only likely to accept an indemnity from a tenant (or related company) that is likely to have the wherewithal to discharge it at lease expiry.

Such an arrangement should be formally documented by:

- a licence to assign from the landlord
- a formal deed of assignment
- a side deed.

The precise nature of a side deed will depend upon the circumstances and the nature of a particular deal. However, in a typical side deed a tenant will agree to:

- pay the assignor a reverse payment by instalments, linked to proof of payment of rent by the assignee
- to indemnify the assignee (either wholly or partially) for any claim by the landlord for breaches of the repairing, alteration and redecoration covenants.

The key issue, which a tenant will face, is whether it should disclose the side deed to the landlord when applying for consent. This will be particularly relevant if the landlord has already refused to consider an application for subletting because the proposed rent was below passing rent. It is the authors' view that a side deed, which provides for the payment of a reverse premium, does not need to be disclosed to the landlord. It is not relevant to the landlord's decision whether or not to grant consent. The landlord is concerned with the identity and character of the proposed assignee, not the detailed terms of the assignment (see *Kened Ltd* v *Connie Investments Ltd*). The absence of the need to disclose the side deed will, in some cases, make this option more attractive to a tenant. It allows a corporate occupier to offload an overrented onerous and surplus lease swiftly without the risks and potential litigation of going down the sublease route.

The potential drawbacks are listed below.

- *Covenant strength*: the landlord is likely to take a closer look at the covenant strength of an assignee than a subtenant (although following *Riverland* some landlords may now be just as vigilant with a subletting).
- *Loss of control:* if the assignee defaults or becomes insolvent, and the former tenant remains liable under the lease or an AGA, the former tenant will have to either wait until the landlord serves a notice under section 17 of the Landlord and Tenant (Covenants) Act 1995 to obtain an overriding lease, or wait until a notice of disclaimer is served and make an application for a vesting order. During the period of the delay, the former tenant is likely to be liable for arrears of rent, service charge and insurance premium that accrue under the lease.
- *Shorter duration*: problems will arise, and the structure may not be suitable, if the proposed tenant does not want to take a lease for the remainder of the lease term. It may be possible to incorporate a 'put and call option' which allows the assignee to call for the former tenant to take an assignment back at a given date. However, this level of complexity may be a step too far for most occupiers and it is, of course, subject to landlord's consent in the future. There will, always

therefore be a risk that the then landlord will withhold consent and the assignee will remain on the hook for the remainder of the term.

In the event that the assignment takes place and the premium is paid there is only a contingent liability and as such no provision is needed.

(d) Persuasion

If none of the three options suggested above are suitable, there is one other strategy that might work. In some cases landlords can be persuaded to take a more flexible approach to rental restrictions on subletting.

A tenant adopting this strategy may wish to raise some or all of the following points when disclosing a proposed side deed and making an application for consent to sublet at passing rent.

- The determination as to whether a personal side deed breaches a pre-condition to subletting is a question of construction of the particular alienation clause to establish the intention of the original landlord and tenant. *Homebase* is no more and no less than a decision on the particular clause in that lease. A different court is entitled to reach a different view on a differently worded clause.
- *Homebase* concerned a sublease that could potentially be renewed under the 1954 Act. While this is not a complete answer, if the restrictions in question required the underlease to be excluded, the rental to be not less than passing rent (this was not the case in *Homebase*), and there was no restriction on the terms contained in the sublease, it is possible that a court could take a different view. It may be possible to distinguish *Homebase*.
- Even if there is no condition requiring that the sublease be 1954 Act excluded, if this were to be offered, it would remove one of the significant concerns raised by the Court of Appeal. There would be no risk of the landlord on renewal being saddled with a direct tenant (ie the former subtenant) with a right to a concession rent. There would be no renewal.
- One of the concerns often expressed by landlords (and indeed referred to by the Court of Appeal in *Homebase*) is that it has a commercial interest in the sublease rent because it is an obvious comparable on the headlease rent review. This is an old fashioned view that is overplayed by landlords. If, as was the case in *Homebase*, the premises were marketed over a period of 18 months and the proposed sub-rent was negotiated at arm's length with the only potential subtenant who came forward, and that rent was substantially below passing rent, then the obvious conclusion to draw is that unless the market improves dramatically, the headlease is overrented and any imminent rent review of the headlease is likely to result in a nil increase.
- The fact that the landlord stands in the way of the granting of the sublease is not going to improve its position on rent review. Indeed, the modern view is that in a falling market with overrented premises, a passing rent restriction is likely to be regarded as an onerous provision on rent review with a depressing effect on rent. This will particularly be the case if the tenant is not permitted to mitigate the position by use of a personal side deed. In effect, the passing rent restriction, in these circumstances, constitutes a total prohibition against subletting in a falling market.
- If the landlord still remains unconvinced on the rent comparable issue, the tenant can offer to enter into a confidentiality agreement with the subtenant to try and avoid the rent becoming a comparable (although the subtenant would have to disclose such a document if ordered to do so by the court), and also agree with the landlord not to refer to the side deed in subsequent rent reviews of the headlease.

- The landlord should be reminded of the fact that the British Property Federation has named a number of its members who are prepared to offer flexibility on restrictive subletting conditions and have entered into a voluntary code to provide greater flexibility. It may be that the landlord is itself a signatory and it is worth checking on the British Property Federation's web site *http://www.bpf.propertymall.com*. Furthermore, paragraph 5 of the Commercial Lease Code 2007 provides that if subletting is allowed, the sublease rent should be the market rent at the time of subletting.
- The tenant can remind the landlord of its duties under the 1988 Act and that if it refuses consent, it will face costly court proceedings and the possibility of a substantial claim for damages.
- Finally, the tenant can point out that the effect of the landlord's stance will be to cause the premises to be empty for the remainder of the term. In some cases, where the rent is high and the corporate occupier is not a large company, if there is no means of mitigating its loss by subletting, this could affect the corporate occupier's ability to fund the head rent, and ultimately affect its solvency.

While some landlords faced with the above points will still withhold consent, others may reach a different conclusion after considering what objective it is seeking to achieve, and carrying out a cost benefit analysis to see whether the prospects of success justify the risks of losing. Further, in some cases there is great merit in a corporate occupier in tandem (and without prejudice) offering to pay the landlord for a variation to the alienation clause rather than take the risks involved with paying a reverse premium to a tenant.

While there is no one single solution, there are steps that a corporate occupier can take to address this vexed and reoccurring issue. It can be avoided altogether if a corporate occupier refuses to accept such restrictive conditions when negotiating the lease. In view of the difficulties that have arisen over the last 20 years, it would be unwise to enter into such clauses lightly in the future.

5.7 The unreasonable landlord — the tenant's remedies

The 1988 Act imposes a clear and unequivocal duty on a landlord to act swiftly and reasonably. If a landlord withholds or delays consent a corporate occupier needs to know:

- what action it can take
- how long will the process take and
- how much will it cost.

The remedies available against a landlord can be both potent and swift (although there may be delay if the matter leads to contested court proceedings). In the majority of cases a firmly worded letter before action reminding a landlord of its duty under the 1988 Act and the damages that may flow from its delay will be sufficient to elicit a swift consent.

If this does not achieve the desired result the corporate occupier has two possible remedies:

- issue court proceedings
- defiance — proceed without consent.

5.7.1 Issue court proceedings

The most common solution a corporate occupier will consider is the threat and possible issuing of court proceedings against the unreasonable landlord. When considering proceedings there are a number of practical and legal issues which it should have regards to.

(a) Declaration

A tenant can issue proceedings in the county court or, in some cases, the High Court, seeking a declaration that the landlord has unreasonably withheld consent, or has granted consent subject to an unreasonable condition. If the tenant's application is successful, the transaction can proceed without the need for the landlord's consent, or compliance with the unreasonable condition.

(b) Damages

A tenant can also claim damages for any loss that it has suffered as a result of the landlord's breach of statutory duty. If the proposed assignee or subtenant is still around and willing to proceed at the date of the court hearing, a tenant is likely to be principally concerned with obtaining a declaration to allow the assignment/subletting to proceed. If the proposed assignee/subtenant has disappeared and found alternative premises by the date of the court hearing, the tenant is likely to seek damages. However, in certain circumstances the tenant may wish to claim both a declaration and damages, for example where a tenant is unable to collect sub rent for period of delay caused by the landlords unreasonable withholding of consent.

If a court decides that the landlord is in breach of statutory duty by unreasonably withholding consent, the tenant must establish that it has suffered a loss, and that loss was caused by the unreasonable delay or refusal. If the proposed transaction failed to proceed for other reasons (eg because the proposed assignee/subtenant found alternative accommodation at a lower rent), then the tenant's claim for damages will fail.

A tenant can potentially recover the following losses.

- *In the case of a proposed sub-tenancy*, the loss of rent, service charges and insurance premiums that may have been payable under the sub-tenancy (for a period a court considers appropriate having regard to the tenant's prospects of subletting the lease to a new subtenant).
- *In the case of a proposed assignment*, any premium the tenant may have received if the assignment had gone ahead and/or the continuing rental and other liabilities (again having regard to the tenant's prospects of assigning the lease to a new assignee).
- Business rates.
- Abortive legal/professional costs of the proposed transaction.

The tenant must take reasonable steps to mitigate its losses. It is important, therefore, that the premises are marketed as soon as possible after the failed transaction and that this is evidenced carefully. The longer a tenant seeks damages for, the greater scrutiny a court is likely to give to the tenant's attempts to remarket the premises.

The following further points may be raised by a landlord seeking to minimise damages.

- *Break clause*: if the proposed transaction was a sub-letting and the proposed sub-tenant was to have a break clause, the court will be able to take into account the possibility that the sub-tenant would have exercised the break clause thereby limiting the losses suffered by the tenant.
- *Early occupation*: often the proposed assignee or subtenant is let into early occupation without the consent of the landlord and therefore in breach of covenant. If the landlord becomes aware of the breach it can take steps to forfeit the lease, or alternatively seek an injunction enforcing the covenant. If the landlord takes no action, early occupation is unlikely to cause the tenant's claim for damages to be extinguished or reduced. Even if the landlord counterclaimed for damages for breach of covenant, those damages are likely to be nominal, particularly if the premises were substantially over-rented, because it would have been against the landlord's financial interest to forfeit the lease. In the circumstances, the landlord is unlikely to have suffered any loss, and the breach is largely irrelevant.
- *Failure to notify the landlord of the need for urgency*: (eg the existence of a long stop date). Occasionally landlords argue that if the tenant had warned the landlord about an approaching long stop date, the licence may have been expedited and the loss avoided. There is nothing in this point. The landlord must deal with a tenant's application in a reasonable time. Whether a tenant perceives an application to be urgent or not is of no relevance except where he extends the period by, for example, telling the landlord that "there is no particular urgency". There is no principle that a landlord should devote more resources to an application mainly to suit a tenant's perceived urgency. Nevertheless, it is always prudent for a tenant to notify a landlord of any particular urgency.

The corporate occupier should be aware of the possibility that a court may award exemplary or punitive damages (in addition to ordinary damages) if it takes a particularly dim view of the landlord's conduct. In *Mount Eden Land Ltd* v *Folia Ltd* [2003] EWHC 1818 the court raised the possibility that a tenant could be awarded exemplary damages if the landlord had deliberately withheld consent for its own purposes and for reasons which it could not properly advance on the question of reasonableness.

In *Design Progression Ltd* v *Thurloe Properties Ltd* [2004] 1 EGLR 121, the court took the view that the landlord had pursued a deliberate and obstructive policy designed to prevent an assignment of the lease for its own purposes. The court therefore awarded £25,000 to the tenant for exemplary damages in addition to the award of damages to compensate the tenant's losses.

The threat of an award of exemplary damages may act as a powerful deterrent to a well advised landlord who is withholding consent for an ulterior motive.

(c) Delay

A serious drawback to a tenant pursuing an application for a declaration is that if it is defended, a court will need to hear oral evidence (including in most cases expert evidence) and the process from start to finish is unlikely to take less than four–six months, and could take longer, particularly if the decision is appealed.

It may be possible, in some circumstances, for a tenant to obtain an order for an expedited hearing that may speed up the process.

In many cases the issue of court proceedings is likely to result in a speedy resolution of the problem. The fear of incurring substantial costs (some of which are likely to be irrecoverable even if successful), and possibly exemplary damages, is likely to dissuade many landlords from defending. In the remaining cases, the longer the case runs on, the more likely it is that the proposed assignee/subtenant will look for alternative premises. To this extent court proceedings to allow the transaction to proceed, as opposed to recover damages, are something of a blunt instrument.

If it appears that proceedings are likely to become bogged down, it may be sensible for the tenant to suggest an early mediation to try and break the deadlock quickly. The problem for the corporate occupier is that a potential assignee or subtenant will be unlikely to await the outcome of court proceedings because the time it is likely to take to resolve the court proceedings will be far longer than the desired occupation date. The only time this may not be the case is when the premises are specialist in nature, or the assignee/subtenant is a special purchaser.

(d) Binding contract

A further possible solution, which may offer a tenant some protection, is to require the proposed assignee/subtenant to enter into a binding contract subject to obtaining landlord's consent with a longstop date after which either party can rescind the contract. This has a number of advantages from the tenant's perspective:

- If the longstop date is sufficiently far away, it will give the tenant the reassurance of knowing that the proposed assignee/subtenant cannot walk away before the proceedings are disposed of. However, as mentioned above the assignee/subtenant may be resistant to being tied in to a date that is too far away because of its business needs to go into occupation.
- If the longstop date is approaching it may provide the tenant with good grounds for arguing that there should be an expedited hearing.
- If the longstop date passes, and the proposed assignee/subtenant terminates the contract, it will be much more difficult for a landlord to argue that it was not its delay/withholding consent that caused the tenant's losses (which is a common defence used by a landlord).

Obtaining landlord's consent from a landlord who is intent on being difficult is often about putting as much pressure on the landlord as possible. If a contract exists this should certainly be disclosed to the landlord. The landlord will then be aware of the extent of any potential claim for damages, which the tenant may bring if the contract is rescinded as a result of consent being unreasonably withheld. In a surplus property situation with many years to run on the lease and where the premises are hard to let, it may be unlikely that another potential occupier will come forward and consequently the potential damages may be very substantial. The potency of this point should not be overlooked and should be stressed to the landlord at every opportunity.

(e) Costs

It is a well known fact that the costs of litigating can be very substantial. It is therefore very important that a corporate occupier whether as tenant or landlord (in a legacy portfolio situation) carries out a very careful cost benefit analysis. Most experienced property litigators should be in a position to give a reasonably accurate estimate of the cost of going to, and the various stages leading up to, trial.

It is very important that the parties should bear in mind that even if successful, a party is unlikely to obtain a costs order that will result in 100% recovery of its legal costs in bringing, or defending, a claim. Typically, a successful party that is awarded costs on a standard basis recovers between 50% and 60% of its actual legal costs.

From a landlord's perspective therefore, the key question is this: is the objective that it is seeking to achieve (eg preventing the grant of a sublease to a weak covenant which it perceives may have an adverse affect on the value of its reversionary interest) such as to justify the risks of continuing to withhold consent? If the landlord's legal costs of defending are likely to be in the region of £40,000–£50,000, even if successful it is likely to lose £20,000–£25,000 in successfully defending court proceedings. If it loses, in addition to paying its own costs (£40,000–£50,000), it is likely to be ordered to pay approximately a further £20,000–£25,000 in costs to the successful tenant. This may be higher if the court awards costs on an indemnity basis (see *Redevco Properties* v *Mount Cook Land Ltd* [2002] All ER (D) 26 where the court considered the landlord's tactics to be so unreasonable to merit disapproval by the award of costs on a higher basis). In addition, the landlord may face a substantial claim for damages.

In the example given, the landlord may consider costs of £20,000–£25,000 if it wins, or £60,000–£75,000 (plus any damages that may be awarded) if it loses, does not justify seeking to avoid what may be no more than a paper loss if it has no immediate plans to sell its investment (reversionary) interest.

A tenant should carry out the same exercise. If the rent is high, and the lease has a number of years to run, the risks may well justify the issue of proceedings. The risks could be less for a tenant, not least because there will be no counterclaim for damages.

If the parties go to mediation, a skilled mediator is likely to carry out a cost benefit analysis similar to the one above. This exercise can be a compelling reason to reach an agreement.

5.7.2 Defiance — complete the transaction without the landlord's consent

An alternative to potentially expensive and protracted litigation is to proceed without landlord's formal consent. At common law, if a landlord unreasonably withholds consent the tenant is released in respect of that particular transaction, from its obligation to obtain consent from the landlord. The tenant is, as a matter of law, entitled to complete the transaction.

There are two practical problems to this approach.

1. If the tenant's judgment (or that of its advisors) is ultimately proved wrong and the landlord's refusal held to be reasonable, the lease may be forfeited. There is no guarantee that an assignee will obtain relief from forfeiture where there has been a "deliberate breach" (see *Crown Estate Commissioners* v *Signet Group plc* [1996] 2 EGLR 200). In most cases a corporate occupier of a lease of surplus premises will be unconcerned about this possibility, indeed would probably welcome it, but the assignee/subtenant will be very concerned about it. However, a landlord who does not wish to bring the lease to an end may apply for an injunction to reverse the alienation to which both the tenant and proposed assignee/subtenant would be parties, or bring a claim for damages.
2. In most cases an in-going assignee/subtenant will insist on a formal landlord's consent before completing the transaction (to avoid buying into an immediate dispute with its new landlord) and/or risking being evicted and incurring the losses and wasted expenditure that that would entail.

In most circumstances these problems will deter a prospective assignee/subtenant but occasionally this may be an appropriate course of action (eg in a company group re-organisation where the lease is to be occupied by another group company and where the lease does not allow group sharing).

5.8 A practical guide to landlord's consent

The purpose of this section is to provide a step by step practical guide to a corporate occupier when making an application for consent in a typical transaction, or alternatively, in responding to one as a landlord in a legacy portfolio situation.

5.8.1 Tenant — read the alienation clause

The starting point is always to study the detailed clauses of the lease carefully:

- is the proposed dealing permitted?
- is landlord's consent necessary ?
- if so, are there any conditions that must be fulfilled before an application can be made (eg a proviso that only permits a subletting at passing or market rent and/or on an excluded sublease)?
- does the proposed transaction require any other consents from the landlord (eg to a change of user or to carry out alterations)?

5.8.2 Tenant — formal application

Time will not start to run against a landlord unless and until it is served with a complete application from the tenant that includes all the necessary information and supporting documents.

As a matter of good practice this should:

- provide sufficient details of the proposed transaction to enable the landlord to make an informed decision. If a subletting is proposed, in most cases it would be sensible to include a copy of the draft sub-lease or the heads of terms
- identify the proposed sub-tenant/assignee and provide names and addresses
- provide evidence of the proposed sub-tenant/assignee's financial standing:

  (a) three years audited/ management accounts
  (b) trading references
  (c) bank, solicitor or accountant references
  (d) previous landlord's references
  (e) in relation to a start up company, a business plan.

- if the matter is urgent, a statement to this effect
- if the tenant and subtenant have entered into a conditional contract with a longstop date, the details of that should be provided
- an undertaking to pay the reasonable and proper costs of the landlord in dealing with the application.

The corporate occupier should put itself in the shoes of the landlord before making any application and weigh the likely views of the landlord upon receiving the paperwork. What are the likely issues going to be? Is the covenant strength good enough or should some form of guarantee or a rent deposit be offered? If these things have been covered objectively then it should speed up the process of the landlord's decision-making. Overlying this will be issues around the landlord's attitude generally and whether the landlord will look to use the application as a negotiating tool on something else, such as a rent review. While the landlord cannot use outstanding reviews as a lever, it may well do so as a delaying technique, which will put the corporate occupier under pressure. The solution is to ensure that everything has been thought through and to leave the landlord with no room for manoeuvre.

There has been some debate in the cases as to whether time will start to run against a landlord before a costs undertaking is given (see *Dong Bang Minerva* v *Davina Ltd* [1995] 1 EGLR 41). Whilst the strict position may be that time will only be suspended in circumstances where the landlord has made a reasonable request for an undertaking, the safest course to avoid the argument is to give an undertaking in the application letter. However, most property lawyers will want to obtain an estimate from the landlord's surveyor and take instructions before giving an undertaking capped at that level. This can delay the consideration of an application considerably, and therefore, if achieving a rapid consent is the principal objective, it is important to give a general cost undertaking in the initial application. Conversely, if cost is of paramount importance, seeking an estimate will be more appropriate.

5.8.3 Precedent — formal application letter

Dear Sirs,

Re application for consent to assign [sublet]

We act for and on behalf of Twaddle Solutions (No 1) Ltd your current tenant of 65 Mancky Road, London, EC1, under a lease dated 22 December 1985 between Diffy Cult Ltd and our client ("the Lease").

Please treat this letter as a formal application on behalf of our client for consent to assign [sublet] the Lease to Fly By Night Ltd pursuant to Clause 2(18) of the Lease.

[We enclose a copy of the draft sublease/heads of terms. The principal terms are:]

Fly By Night Ltd is a well-established marketing company and we enclose:

1. up to date company search showing its current address and company number
2. copies of its last three years audited accounts
3. a recent bank reference.

Please note that Fly By Night Ltd has indicated that it must complete this assignment within the next five weeks or it will pull out of negotiations and look elsewhere. As you are aware Twaddle has been looking to assign or sublet this lease for a number of years and therefore we would be grateful if you could deal with this application as a matter of urgency.

We undertake to pay the landlord's reasonable and proper legal and surveying costs in considering this application [up to the sum of £] whether or not the matter proceeds to completion.

Your faithfully,

5.8.4 Landlord's response

On receiving an application the landlord, and its advisors, need to consider:

- whether the tenant has supplied sufficient information for it to properly consider the application
- making an insolvency search against the proposed assignee/sub-tenant
- whether the proposed assignee/sub-tenant is in good financial health or whether it would be appropriate to consider requiring a guarantor of adequate means, and/or insisting on a rent deposit deed
- obtaining valuation advice if there is a concern that this transaction might lead to a dilution in covenant strength to ascertain whether there would be a diminution of value of the landlord's reversion
- whether there are any serious breaches of covenant
- whether the proposed use of the premises would be detrimental to the landlord's interests and/or in breach of covenant
- whether it would be appropriate to grant consent but impose conditions.
- whether the landlord needs to make an application to a superior landlord or to notify a mortgagee.

If the landlord has insufficient information, or wants time to consider the application with its professional team, it may be appropriate to write a holding letter.

5.8.5 Landlord's holding letter

Dear Sirs

Re application for consent to assign [sublet]

We act for and on behalf of Diffy Cult Ltd and have been passed a copy of your letter requesting consent to assign [sublet] the Lease.

We will be reporting to our client and making a recommendation once we have had an opportunity to fully consider your client's application. To that end:

1. Please confirm the intended user of the premises after the proposed assignment [subletting] takes place.

2. Fly By Night Ltd's accounts show a trading loss of £4m for the last financial year, is it your client's intention to offer a guarantor?

3. [Please confirm that the entire terms of transaction will be contained in the sub-lease you have provided for approval and that there are no side deed or agreements.]

Please note that neither this letter nor any correspondence between our respective clients or us constitutes consent to the proposed assignment [subletting]. If consent is given, this will only happen by completion of a formal licence to assign [sublet] executed as a deed on terms to be agreed between the parties.

Yours etc ...

5.8.6 Landlord's decision letter

If the landlord has sufficient information then it must make a decision, and communicate that decision to the tenant or its advisors (depending where the application has come from) with reasons as soon as possible.

In view of the importance of the decision letter (not least in defending any subsequent court application — the landlord will only be entitled to rely upon reasons contained in a decision letter sent to the tenant within a reasonable time), the scattergun approach to service may be appropriate (i.e. to be sent by ordinary post; recorded delivery and fax). It would also be prudent to telephone the tenant or its advisors to ensure that it has arrived safely. The onus will be on the landlord to satisfy the court that a decision letter was served on the tenant within a reasonable time of receiving the tenant's application.

Dear Sirs

Re application for consent to assign [sublet]

We act for and on behalf of Diffy Cult Ltd and have been passed a copy of your letter requesting consent to assign [sublet] the Lease.

Our client has considered your application very carefully. However, it has decided to refuse consent to assign [sublet] on the following grounds:

1. Fly By Night Ltd is a fledgling company that made a substantial financial loss in its last full trading year. You have been unable to demonstrate, by past accounts, references or otherwise, that if the transaction proceeded it would be able to pay rent as it fell due or that it would be in a position to comply with other tenant covenants. Furthermore, you have failed to offer any or any adequate guarantor(s).

2. There are currently very serious outstanding breaches of your repairing obligations. We notified you of these breaches on [ ] and you failed to carry out the necessary repairs. As a result, the state of the premises has deteriorated to a condition which requires extensive works to be carried out. For the reason set out in 1 above, we do not believe that Fly By Night Ltd would be in a financial position to carry out the necessary repairs if the transaction went ahead.

3. Whereas Twaddle Solutions (No 1) Ltd is an established company which would be recognised by an investor purchaser as a strong covenant, Fly By Night Ltd has no track record and will be regarded by a potential purchaser as being of weak covenant strength. We have been advised by our valuer that should the transaction proceed, Diffy Cult Ltd's reversionary interest will suffer a substantial diminution in value. Prior to Twaddle Solutions (No 1) Ltd's application, Diffy Cult Ltd was considering selling its reversionary interest and it is likely to be put on the market very shortly.

4. [The proposed underletting is at a rent which is below passing rent. Clause 2(18) of the lease provides that the tenant can only sublet the premises if the proposed sublease is at the higher of market or passing rent. As this condition is not satisfied, Twaddle Solutions (No 1) Ltd is therefore not entitled to apply for landlord's consent.]

We appreciate that Diffy Cult Ltd's decision may be unwelcome, but it is prepared to consider any fresh application which Twaddle Solutions (No 1) Ltd may make if it is able to put forward any further evidence or solutions which may address the reasons given above.

Yours etc ...

5.8.7 Tenant's letter before action

If a corporate occupier considers that a landlord is unreasonably withholding consent, the next stage is to send a letter before action.

Dear Sirs

Re application for consent to assign [sublet]

We act for and on behalf of Twaddle Solutions (No 1) Ltd your current tenant of 65 Mancky Road, London, EC1, under a lease dated 22 December 1985 between Diffy Cult Ltd and our client ("the lease").

By clause 2(18) of the lease, our client covenanted not to assign the whole of the demised premises without the written consent of the landlord such consent not to be unreasonably withheld or delayed.

On [ ] we made a formal application for landlord's consent to assign the lease to [ ]. All the necessary information has been provided to you in respect of this straightforward assignment [sublease] and we have provided an undertaking to pay the landlord's reasonable and proper legal and surveying fees that are reasonably and properly incurred.

Under section 1(3) of the Landlord and Tenant Act 1988 the landlord is under a duty to give consent to this assignment within a reasonable time unless it is reasonable not to give consent. The landlord is also under a duty to serve a written notice of his decision whether or not to give consent and if the consent is withheld, the reasons for withholding it.

You have to failed to give consent within a reasonable time [and you have failed to provide a written notice giving your reasons for withholding consent] [we do not accept that the objections set out in your written notice are reasonable and form a proper basis for withholding consent because:]

It is clear from the circumstances outlined above that the landlord is in breach of its duties under Section 1(3) of the 1988 Act. In the circumstances, unless we receive your client's written consent within 7 days of today's date, our client will be left with no alternative but to pursue its legal remedies which are likely to include (but are not restricted to) the issuing of court proceedings seeking a declaration that the landlord's consent has been unreasonably refused or delayed [and/or that it is seeking to impose an unreasonable condition], together with a claim for damages, interest and costs.

For the avoidance of doubt, these premises have been surplus to our client's requirements and vacant for a considerable period of time. If the proposed assignee [subtenant] does not proceed because of the delay caused by your failure to provide consent, our client is likely to suffer considerable losses (including, but not limited to, loss of rent to the end of the lease) which it will seek to recover from you by way of damages.

We look forward to hearing from you within the next seven days.

Yours etc ...

5.8.8 Issue court proceedings

The final stage is to either proceed without landlord's consent, or issue court proceedings. The practical aspect of each option is considered at 5.7 above.

5.8.9 Practical tips

(a) Landlord

- Deal with the tenant's application promptly. In particular, if further information, references or financial accounts are required from the tenant, request these quickly. In a straightforward case one month is likely to be a reasonable period in which to respond.
- If consent is to be withheld, or granted subject to conditions, set out in detail in response to the application all factors which have influenced the landlord's decision. This must be in writing.
- To avoid the risk of granting a licence in correspondence unintentionally, make sure that all pre-licence correspondence explicitly states that a licence will not be given except in a formal deed and that any initial response is not consent for the purposes of the alienation covenant.

(b) Tenant

- When applying for consent include all relevant information concerning the proposed transaction (eg with a proposed sublease enclose a copy of the heads of terms, or if available, a copy of the draft sub-lease. Also include any financial information or accounts which will enable the landlord to consider the covenant strength of the assignee or subtenant.
- The tenant must undertake to pay the landlord's reasonable costs. Failure to do so may mean that the clock does not start to run against the landlord.
- Respond fully and promptly to requests from the landlord for further information where those requests are reasonable so as to avoid any suggestion that the delay has been caused by the tenant's failure to respond.
- If the landlord delays, remind the landlord of its duties under section 1(3) of the Landlord and Tenant Act 1988 and draw the landlord's attention to any likely losses that may occur should the delay lead to a loss of a potential assignee or sub-tenant.

6 Tenants' Alterations and User

6.1 Alterations — introduction

It is rare for a corporate occupier to move into a property and be totally happy with what is there. Generally something does not suit its operational needs. The need for a corporate occupier to make alterations to a property will range at one end of the scale from minor items, such as changes to partitions to improve the amount of light in an office, through to major adaptation projects, which may entail virtually rebuilding the entire property. The cost implications of such changes will vary considerably. However, it is not only the direct and immediate costs that are of importance, but the total cost of the works which usually involve hidden costs that tend not to be identified clearly when the decision to carry out works is taken. That 'failure' can be in part the consequence of ignorance but also it can be a symptom of how difficult it can be to ascertain the total cost, in particular predicting whether the current or future landlord may require re-instatement.

6.2 Alterations — legal framework

The legal framework within which a corporate occupier will need to operate is governed by the lease terms and by statute. A corporate occupier that occupies freehold premises is likely to face considerably less restrictions than where premises are occupied under a lease.

6.2.1 Alterations clauses in leases

It is rare to find a lease that contains no restrictions on a corporate occupier's ability to alter the premises. However, the alterations clauses in leases vary considerably. The flexibility that a clause will offer is generally dependent upon the length of the lease and the nature of the property. The position can range from complete flexibility to one of total prohibition. For example, a ground lease on an industrial unit for 99 years is likely to offer greater freedom for the tenant than a two-year lease on a retail unit in a shopping centre.

There are three types of restriction which a corporate occupier may encounter (each of which is considered below), although a lease may contain a hybrid clause composed of each type. For example, an office suite in a multi-let building may have an absolute prohibition on the tenant carrying out structural and external works; allow the tenant to make changes to demountable partitioning without landlord's consent; but in respect of all other proposed alterations, require landlord's consent, not to be unreasonably withheld.

(a) Absolute bar

In short-term leases alterations are often forbidden in all circumstances. This provides the landlord with total control, but fetters a corporate occupier's ability to adapt the building to suit its operational needs.

In normal circumstances in the case of leases of more than five years this would reduce the appeal of the property to occupiers, and if a landlord insisted on such a term it would have an adverse affect on its ability to let the property. It would also reduce the amount of rent a tenant would be prepared to pay and hence would materially affect the capital value of the building. However, there can be circumstances where such a restriction is seen as an advantage or a necessity. The most obvious example of this is where the building is listed and where the ability to alter the premises is severely constrained by planning legislation. By including an absolute prohibition in the lease the landlord is reinforcing compliance with planning legislation. Corporate occupiers who take space in such buildings must do so with the knowledge of this restriction and the benefit to them of being in that type of building will have to outweigh the disadvantages of not being able to adapt the space to their needs.

An absolute restriction is most often encountered on short-term lettings especially in the retail sector, for example, lettings of premises for the three months leading up to Christmas. It may also be encountered where a landlord is itself hindered by operational requirements, for example, kiosks on railway platforms.

(b) Qualified restriction

This is the most common type of restriction and allows works to be undertaken provided that the tenant obtains the prior consent (usually in writing) of the landlord. Sometimes that consent will be subject to an express requirement that the landlord's consent should not be unreasonably withheld. In many cases there will be a combination of qualified and unqualified restrictions. Some alteration clauses will allow certain works to be carried out without the need for landlord's consent. A common example arises in the context of office suites where the tenant is given the flexibility to make changes to non-structural internal partitions without consent, but other works require landlord's consent.

In the industrial sector alterations tend to be to the fabric of the building and consequently alteration clauses tend to reflect the broader needs of occupiers of that class of property.

Retailers who operate a chain of shops will in the main have their own fit-out design that is put into every unit (and which will tend to include the shop front, the walls, ceiling and the like). Consequently, leases of retail units tend to reflect the shell finish and the alterations clause is likely to focus more on the prohibition of structural alterations. While this tends to be the approach for stand-alone retail units, landlords of shopping centres are notorious for having strict policies on alterations. Landlords of shopping centres prefer to grant consent to the original fit-out and thereafter exert a considerable degree of control. Unfortunately, such restrictions are often drafted at a time when the shopping centre was originally let and consequently the restrictions are put in the lease without considering whether it continues to be in the right location and of the right type and quality to merit such a restriction being included.

(c) Unrestricted alterations

If a lease has no covenant restricting alterations, a tenant has free reign to carry out alterations to the premises provided that the alterations do not constitute acts of waste by the tenant. It is very rare to come across a lease with no restrictions on alterations, even with ground leases, because the landlord will want to protect the value of his investment. Unfettered alterations could see the building knocked down and not replaced with something comparable, and may have an impact on the rent achieved at review.

6.2.2 Other lease clauses

The corporate occupier must be aware that even if the alterations clause is sufficiently flexible to allow works to be carried out, there may still be other lease clauses to consider which may impact upon whether the works can proceed, or the total cost of the works.

(a) Services

Some leases include a separate clause for alterations to the services and in particular works to the electrical system. If a corporate occupier is, for example, putting in some new non-structural partitions, which do not require consent under the alterations clause, it may well fall foul of a prohibition not to amend the electrical system. This could arise because the design for the new partition includes a power outlet in the studwork or would involve a change to the light switch. Both of these works will entail the contractor amending the electrical system and hence breaching the clause.

(b) Consents

Consideration must be given to determine whether consents from external parties will be required. Most leases will provide that the tenant has to comply with statutory requirements. Examples of consents that may be required include:

- planning permission
- building regulations
- CDM regulations
- party wall consents and awards
- rights of light issues.

(c) Third parties

The need to consider third parties consents also arises because often proposed alterations will infringe the rights of others.

A good example of this is the installation of air conditioning. Under a qualified alterations clause the landlord's consent will be required for the works in the suspended ceiling and also to the electrical system which powers the air conditioning units. In addition, consent will have to be sought to run pipes and cables from the ceiling units to the condensers, using the risers in the building. As the condensers are likely to be roof mounted, landlord's consent will be required for these as well. It is unlikely that the lease will provide the tenant with the ability to put the condensers on the roof and the landlord can therefore unreasonably refuse consent if he/she so wishes. Furthermore, pipe work may need to go through the demise of another occupier and therefore approval will be needed from that party as well. For these reasons a corporate occupier taking a lease and proposing to install air conditioning units should make sure that a licence for alterations is completed, and third party consents obtained, at the same as the lease is executed so as to avoid any problems.

(d) Yield up

The yield up clause is generally overlooked by tenants, especially at the point when they price the cost of the work and the benefits that will derive from it. Often the yield up clause will either require, or provide the landlord with the ability to require, the tenant to re-instate the premises to the original pre-alterations condition. Alternatively, the re-instatement provision may be contained in the original licence for alterations.

6.2.3 The Landlord and Tenant Act 1927

The contractual position governing alterations is enhanced by the Landlord and Tenant Act 1927. The provisions contained in the 1927 Act are often overlooked or not used by corporate occupiers but can be of considerable assistance. The 1927 Act addresses three potential issues concerning alterations:

- it grafts onto a covenant that improvements cannot be carried out without consent, a proviso that the landlord cannot unreasonably withhold consent: section 19(2)
- it allows a business tenant in certain circumstances to seek the court's authority to carry out improvements which are prohibited by the lease or for which the landlord has refused consent: section 3(4)
- it provides a business tenant on quitting premises with a mechanism to recover compensation for improvements carried out during the lease term: section 1.

(a) Implied proviso to qualified covenant

Where a lease provides that a tenant cannot make an improvement without the landlord's consent, section 19(2) of the 1927 Act implies a proviso that the landlord's consent should not be unreasonably withheld. The word improvement is not defined by the 1927 Act and it will be a question of fact as to whether a proposed alteration constitutes an improvement. Strictly speaking, therefore, it is possible that a proposed alteration may not be an improvement and section 19(2) does not apply. However, this possibility may be more theoretical than real. It has been held that whether a proposed alteration is an improvement must be considered from the point of view of the tenant (see *Lambert* v *Woolworth* [1938] Ch 883). In most cases, therefore, a court is likely to consider that a proposed alteration is an improvement and imply the proviso contained in section 19(2).

There are a number of restrictions to the implied proviso:

- by definition, it will not assist where there is an absolute covenant against the tenant carrying out alterations
- it will not assist if some of the works need to be carried out on premises outside the demise; here the landlord has an absolute right to refuse consent

- it does not prevent the landlord from charging a reasonable amount for granting consent to cover damage to, or diminution in the value of, the premises or neighbouring property
- it does not prevent the landlord from recovering its legal or other expenses in dealing with the application for consent and
- where the improvement does not add value the landlord may require the tenant, if it is reasonable to do so, to give an undertaking to reinstate the premises at the end of the lease.

If the alterations clause is conditional, and either section 19(2) implies the proviso or alternatively there is an express requirement for the landlord not to unreasonably withhold consent, in the event of refusal or delay, the tenants can seek a declaration from the court that the landlord is unreasonably withholding consent. The approach of the court as to whether the landlord is unreasonably withholding consent is likely to be the same as in the case of an alienation clause (see 5.5 above). While recent case law on this topic is thin on the ground, in Lambert Slesser LJ indicated:

> ... many considerations, aesthetic, historic or even personal, may be relied upon as yielding reasonable grounds for refusing consent ... The wider the connotation given to the idea of improvement, the more necessary it may be that the landlord should have his protection.

It is clear from *Lambert* that unlike a dispute in the alienation clause arena, the onus of proving that the landlord has unreasonably withheld consent is firmly on the tenant.

A more recent case where this issue arose was *Iqbal* v *Thakrar* [2004] 3 EGLR 21. In *Iqbal* the tenant wanted to convert ground floor premises (which he held on a 999 year lease) into a restaurant. The lease contained a covenant that prevented alterations without the landlord's consent which was not to be unreasonably withheld. The landlord objected on two grounds. First, he argued that the use would diminish the value of the residential upper parts. This ground was held to be unreasonable. However, his second ground was that the tenant's proposals could give rise to structural problems. The tenant maintained that the landlord should not have refused consent on this basis, but rather granted consent conditional on ensuring the works were done without causing structural problems. The Court of Appeal disagreed. The tenant must make its proposed alterations sufficiently clear to the landlord to enable it to make an informed decision.

(b) Court's authority to carry out improvements

The 1927 Act provides a mechanism for a tenant to seek to recover compensation for improvements in certain circumstances and where a certain procedure has been followed. An incidental benefit of carrying out this procedure is that under section 3(4) of the 1927 Act a tenant can seek the court's authority to carry out improvements which:

- are prohibited by the alterations covenant or
- where the landlord has refused consent to proposed improvements.

Consequently, if a tenant has not been able to persuade a landlord to consent to proposed alterations, this procedure may be the only means by which the tenant can put pressure on the landlord. Alternatively, in the case of a conditional alterations covenant, it may be a more cost effective and a more potent solution than issuing proceedings seeking a declaration that the landlord has unreasonably withheld consent to alterations. This can be a very powerful stick with which a corporate occupier can beat a landlord.

To pursue a claim for compensation for improvements, a tenant must follow the procedure set down in sections 1–3 of the 1927 Act. This procedure is split into two essential parts. The initial steps must be taken before the works are carried out and involves obtaining authorisation to carry out the works. The final part of the process takes place much later and involves making a claim for compensation when the lease is coming to an end.

The initial procedure involves the points set out below.

(1) The tenant serving on the landlord a written notice of its intention to make the improvements together with a specification of works and plans which clearly show the proposed improvements and the part of the premises affected. A copy of a draft notice can be found at 6.6.1 below.
(2) The landlord is entitled to object provided that it serves a notice of objection within three months. A precedent notice of objection can be found at 6.6.2.
(3) If the landlord fails to object within three months of the tenant's written notice to make improvements, the tenant can go ahead with the works even if they are prohibited by the alterations clause and/or the landlord has already refused consent.
(4) If the landlord wishes to carry out the proposed works itself and charge a reasonable increased rent to the tenant to reflect those works, it can offer to do so. However, if the tenant does not wish to go down this route, it is not obliged to accept the landlord's offer and can withdraw its original notice (see *Norfolk Capital Group* v *Cadogan Estates* [2004] 2 EGLR 51).
(5) If the landlord does serve a notice of objection within three months, the tenant may make an application to court requesting that it certify that the proposed improvement is "a proper improvement" and authorise the tenant to carry out the works. Usually the appropriate court will be the county court closest to the premises.
(6) A court will only certify in the tenant's favour if:
    - the proposed improvement is calculated to add to the letting value of the property at the end of the tenancy and
    - it is reasonable and suitable to the character of the property and
    - it will not diminish the value of the property of the landlord or a superior landlord and
    - the landlord has not offered to do the work in return for a reasonable increase in rent (unless the landlord offered but failed to actually carry out the improvements).
(7) Once the works have been carried out, the tenant can require the landlord to certify that they have been duly completed. If the landlord refuses, the tenant can apply to the court for a certificate. Whilst the certificate is not essential for claiming compensation, it is useful for proving at a later date both the work was done and when it was completed.

The authorisation procedure has two possible benefits for a corporate occupier. If no notice of objection is received, or a court certifies that the works are a proper improvement it may carry out works which are prohibited by the alterations clause or for which the landlord has refused consent.

Alternatively, it may result in the landlord offering to carry out and then rentalise the works. This provides a corporate occupier with an opportunity to mitigate capital expenditure on large-scale improvements, such as building an extension to a factory. Rather than funding the capital expenditure itself, the landlord pays for the work and charges the corporate occupier a market rent for the improvement. This offers a number of advantages:

- it preserves capital
- it avoids the need for a licence for alterations and

- it removes the possibility of being required to remove the extension at the end of the lease by virtue of a reinstatement clause.

(c) Compensation claim procedure

The authorisation procedure referred to above paves the way for a tenant which has carried out improvements to claim compensation after the lease has ended. However, this does not guarantee the compensation will be payable, merely that a claim can be made. However, if a tenant fails to comply with the authorisation procedure, it will not be entitled to make a subsequent claim (section 3(5) of the 1927 Act).

To claim compensation at the end of the term the tenant must:

(1) serve a written notice on its former landlord within strict time-limits:
   - where the lease is terminated by notice to quit, within three months of the date that the notice to quit is served
   - where a section 26 request is served under the Landlord and Tenant Act 1954, within three months of the landlord's counter-notice, of if not given, three months from the latest date that a counter-notice could have been given
   - where the lease expires by effluxion of time, not earlier than six nor later than three months before the contractual expiry date
   - where the lease is brought to an end by forfeiture, within three months of the date of the re-entry or the possession order.

(2) The notice must be in writing signed by the tenant or its agent and must include the following details:
   - the name and address of the tenant and the landlord
   - the address of the premises
   - the nature of the business carried on at the premises
   - a concise statement of the claim
   - particulars of the improvements (ie nature of works; location in premises; cost and date completed)
   - the amount of compensation claimed.

(3) If the parties cannot agree the claim, the tenant must issue a claim form in the county court.

The time-limits are strict and the court has no power to extend time.

The amount of compensation which a tenant can claim is the residual value of the improvement at the end of the lease from which the landlord benefits. Section 1(1) of the 1927 Act provides that the compensation shall not exceed:

- the net addition to the value of the property as a whole resulting directly from the improvement or
- the reasonable cost of carrying out the improvement at the end of the lease, less any cost of putting the improvement works into a reasonable state of repair (except to the extent that this is covered by the tenant's repairing obligations under the lease).

An important factor in assessing the level of compensation is the future use of the property. If the property is to be demolished or redeveloped, or alternatively structural alterations are to be carried out that will render the improvements nugatory, then compensation for the improvements will be nil, because they will be of no benefit to the landlord and have no value.

Other drawbacks to a potential compensation claim are:

- a tenant cannot claim compensation for an improvement which it is contractually required to carry out and a carefully drafted lease may avoid the possibility of a compensation claim by including a clause which obliges the tenant to carry out any proposed improvement to which the landlord agrees
- if the lease or licence for alterations requires the tenant to re-instate the premises at the end of the lease this will defeat a claim for compensation because the landlord will gain no residual benefit or value from the improvements
- it requires the corporate occupier to serve its claim letter *before* the expiry of its lease within strict time-limits and consequently unless the date is carefully diarised there is a real risk that a potential claim will be lost by the failure to serve a claim letter within the time limits. The natural time to consider serving a claim letter would be on expiry of the lease by which stage it will be too late. It is a curious requirement not least because although a claim letter is required before lease expiry, no payment is due unless and until the tenant has quit the premises.

Tenants very rarely bring compensation claims for improvements against landlords at the end of the term. It is for this reason that the Law Commission Report in 1989 Compensation for Tenants Improvements recommended that the compensation provisions be abolished. These recommendations have not been implemented and the procedure remains on the statute book. A number of reasons were suggested why tenants fail to bring claims. The principal reason suggested by the Law Commission was that most leases contain standard provisions which ensure a claim is likely to fail (eg a requirement to reinstate). It was also suggested that most tenants are only likely to carry out improvements for which they are able to write down during the period of the lease. It was further considered that ignorance of the procedure perhaps also accounted for its lack of use.

There will be occasions where a corporate occupier may be able to recover compensation at the end of the term which at the very least justifies the service of a notice. Of much more importance is the ability to put pressure on a landlord to do the works itself, or alternatively, to overcome a prohibition contained in the lease or a specific objection by a landlord to proposed alterations and obtain court authorisation to do the works.

(d) Assessment of options

The 1927 Act rarely gets used because of the reasons referred to above and the cumbersome nature of the process. A far more efficient approach is to have a dialogue with the landlord to seek to persuade it to carry out the proposed works. If the landlord refuses, then the 1927 Act procedure can be followed if there is sufficient value to the tenant to warrant the time and expense involved in pursuing the process and carrying out the improvements. Alternatively, if the landlord agrees to carry out the works, then the terms of the lease can be amended by a deed of variation and the change in rent agreed.

In assessing whether to carry out the works proposed by the tenant, the landlord will consider factors such as:

- the costs of the works, including all fees
- the amount of floor area created

- the rental value of the space created
- the likely increase in the asset value for the landlord.

These are the elements that the landlord is going to need to consider, for example, in the case of an extension to an industrial unit. It will then be a question of comparing the costs incurred with the return that the landlord is going to achieve.

The calculation becomes more complicated with a refurbishment of an office building because a landlord will be concerned about the amount by which the rent and asset value will be increased if it carries out the works. It may be harder to persuade the landlord to carry out the works unless there is a significant change in the value of the building.

A key determinant in a landlord's decision will be what lease length remains. If a short term remains the landlord will need to consider what a potential future tenant might want with the property and whether the proposed works add value for all tenants or just the current one. With longer unexpired lease terms the decision is likely to be influenced by the return the landlord will get and the security of the income stream.

6.3 Landlord's perspective on proposed alterations

A corporate occupier will take an entirely different view of proposed alterations to its landlord, and the corporate occupier needs to understand how the landlord will perceive any application for alterations. A corporate occupier should seek to anticipate and address likely concerns the landlord might have before the application for alterations is made. This is likely to speed up the process of getting the works done and reduce the costs incurred in achieving consent.

Examples of the considerations a landlord will take in to account include:

- *Fees*: Is the tenant going to meet all of the landlord's costs?
- *Rental value*: Will the works add to or detract from the rental value? Will there be an impact on the rent at the next review?
- *Capital value*: Will the works add to or detract from the capital value? What is the benefit of the landlord doing the works and charging an additional rent in exchange for doing so?
- *Compensation*: It is possible that the corporate occupier will make an application for compensation under the 1927 Act? What will the level of compensation be if such a claim is made?
- *Leverage*: Can the application be used to force the tenant on something else? For example to settle an outstanding rent review.
- *Impact*: Will there be an effect on the running of the building and other tenants? For example, if a tenant is proposing to install air-conditioning that will increase the consumption of electricity. Who will meet that additional cost?
- *Physical appearance*: What are the aesthetics of the proposal internally and externally? Landlords will be concerned about the impact that internal alterations might have on the interior. For example, partitioning can be seen externally and if there is an unsympathetic design which results in partitions finishing in the middle of a glazing pane rather than against the framework, the appearance can be very poor.
- *Structural implications* — What impact is the proposal going to have on the structure? This is likely to be an issue if the tenant is intending to create an opening through a structural wall, or install an internal staircase between two floors of offices, or by installing a mezzanine floor in an industrial unit.

- *Services*: What services will be affected by the works? Will the new partitioned offices require changes to the fire alarm?
- *Compliance*: Will the proposal get consent from building control?

The primary focus for the landlord is likely to be around value, but the landlord will also be concerned to some extent about the impact the proposed alterations will have on the management of the property.

6.4 Practical considerations for the corporate occupier

6.4.1 Considerations

Before making a decision to carry out alterations to premises and approaching the landlord, the corporate occupier needs to consider a number of matters.

In the event that the corporate occupier does not have a property professional working as a CREM, the first step is to find someone who is able to act as the intelligent client. It must be someone who understands the issues that are involved in the process but who can also see the corporate occupier's needs in the context of the overall scheme. This is unlikely to be someone who will be part of the design team because they will be too focussed on the scheme itself to ask the difficult questions. All of the issues that surround the proposed works need to be considered before it gets beyond the concept stage. Key considerations include:

- how much will it cost?
- how long is left on the lease?
- what revenue will be generated from the scheme?
- is there an alternative solution? If so, how much will that cost in comparison?

Creation of a project team should, if properly constituted, lead to a balanced view of the proposal. The people on the team will depend on the scale of the works. For some simple partitioning it will probably be the CREM, the facilities manager and the head of the department for whom the works are being carried out. If, for example, the proposed works involved an extension to a factory then a much larger team will be needed:

- project manager
- architect
- structural engineer
- services engineer
- quantity surveyor
- operations manager
- factory manager
- representative from the finance department.

(a) Documenting the works

It is essential the proposed works be properly authorised and formally documented. Without a formal licence for alterations in place disputes are likely to arise between the parties, especially at the time of rent reviews, lease renewals or dilapidations claims. The consent granted by a landlord will vary from a simple letter of consent to a detailed licence for alterations, which includes specification, drawings and plans of the works.

There is an inherent problem for corporate occupiers because generally they are under pressure to start the works as soon as possible. Often, therefore, a corporate occupier is happy to proceed on the basis of an informal indication the consent is agreed in principle but will only be granted in a formal licence for alterations which will follow. This puts the corporate occupier at risk either from the landlord raising issues about the project at a later date when the works have already been carried out or because the landlord's failure to grant immediate consent results in the works having to be put on hold and penalty costs being incurred with the contractor. The corporate occupier needs to be conscious of the amount of time that obtaining consent will take.

Ideally the CREM needs to engage his landlord counterpart as early as possible and ensure that the landlord's surveyor is taken through the proposal and any possible concerns are dealt with before the formal process starts. This will avoid abortive costs in having to take into account issues that the landlord raises at the application stage. It is easier for these to be covered at the initial design stage than at the point when the job has been tendered and all the working drawings have been issued. It is important to ensure that any licence for alterations includes "as built drawings" because of the effect it will have at the time of rent reviews, the exercise of a break option and the expiry of the lease. All too often the drawings that are contained in the licence do not reflect the works undertaken, with changes to partition layouts and the like.

If works are carried out without landlord's consent and/or formal documentation being put in place, there is a risk that at review the landlord will seek to rentalise the improvements. If there is no documentation to prove matters one way or another, the landlord may seek to gain competitive advantage. An example which occurs frequently is where a corporate occupier installs a mezzanine floor in an industrial unit and the landlord at a subsequent rent review seeks to incorporate the additional floor area into the total floor area of the building. If the tenant is either unrepresented or badly advised it may find itself paying for rent on the improvement it has created. In the example of the mezzanine floor, it is rare to find the additional floor has been created by adding to the building. There tends to be a free standing lightweight structure installed, that may be tied into the structure to stop movement, but can be readily removed without damage to the fabric. Clear consent and formal documentation would remove this problem.

If the corporate occupier adopts the authorisation procedure under the 1927 Act there will not be a formal licence for alterations. However, the documents generated by that procedure should be kept safely with the lease and other important documents concerning the property.

(b) Building regulations

Generally speaking, the larger the project is the more probable it is that building control consent will be required. However that does not mean that minor alterations do not need consent. The subject matter is too extensive to be covered here, but the CREM needs to ensure that the question of consent is properly dealt with by the project team.

(c) CDM regulations

The Construction (Design and Management) Regulations 1994 were introduced to try and reduce the number of injuries and fatalities in the construction industry, and as such have placed specific duties on all parties involved in the construction process (eg clients, designers and contractors) in respect of health and safety. The 1994 Regulations created the position of planning supervisor and require a health and safety plan and health and safety file to be kept during the construction process.

The 1994 Regulations also impose a number of specific duties on the client (in this case the tenant carrying out the works) or its duly appointed competent person. The function of the planning supervisor is to co-ordinate all of the health and safety aspects of the project from design to completion, and in particular the health and safety plan prepared before construction commences.

(d) Party wall issues/rights to light

Larger projects may involve work that interferes with neighbouring property and give rise to party wall matters. There is a legal procedure for dealing with such issues, although in the main the impact tends to be restricted to Central London and the larger cities. In a similar vein, substantial alterations may raise issues concerning rights to light which need to be resolved. However, if a right to light issue does arise this will generally be the result of major changes being proposed to the fabric of the building and it does raise the question as to whether a corporate occupier should be doing this unless it owns the freehold of the property.

(e) Planning permission

Corporate occupiers occasionally think that the only time planning permission is required is for large-scale schemes. This is not the case and it may be that relatively small works will need planning consent. An obvious example is the installation of condenser units for air conditioning where in all probability planning permission will be needed for the units, especially when they are located at ground level.

(f) Tax

Consideration should be given to the tax implications of the works and the possible benefits that can be gained by structuring the works in a particular way. In particular a corporate occupier should take advice on capital allowances and the impact on the cost if the property is elected for VAT.

(g) General considerations

It is important that a corporate occupier carries out an assessment of the total cost of the alterations over the lifetime of the works, which will include running costs. For example, with the installation of air conditioning units regard should be had to the following items.

- Installation costs: This needs to include all the fees involved and the costs of appropriate consents.
- Running costs: Maintenance of the equipment and increased costs for power.
- Reinstatement costs: A lease or licence for alterations will usually provide that the tenant has to reinstate the premises to their pre-alterations state at the end of the lease. The cost of this can be substantial and is rarely assessed by a tenant at the decision stage. A good example of this is the cost of works to suspended ceilings.

- Any additional rental value that could be incurred on review: Usually not a problem if the appropriate consents are obtained, but if the proper procedure is not followed the corporate occupier could be required to pay an additional rent in respect of alterations which it has paid for.

A typical trap arises with partitions. At the very least, the compliance with reinstatement will involve removing the partitions and costs can escalate quickly if:

- partitions run through a suspended ceiling as this may mean that a new ceiling is required
- electric wiring runs through the partitions, and when the partitions are removed the cabling in the floor also has to be moved and, as a consequence of working on the electrical system, the circuitry of the entire premises needs to be brought up to current standards.

It is important for a corporate occupier to consider the alternative options that are available.

- Carrying out works to someone else's property where there is a finite involvement by the corporate occupier is not necessarily the best approach.
- For extensive works, a key question has to be whether there may be premises available that are better suited to its operation without the need to undertake extensive works.
- Does it really need such an expensive solution? Tenants devise very complicated fit-outs because of perceived business needs. These are expensive to install and equally as expensive to remove but can be mitigated if a different approach is taken.
- A key element is reinstatement. Will it have to rip out the alterations on expiry of the lease or will the works remain in place? The tenant needs to consider the life span of the works and what value they may add to the property. If value is added to the property the landlord is more likely to agree to no reinstatement provisions in the lease.

6.4.2 The reluctant landlord

Where premises let to a corporate occupier are surplus to its needs, one of the best ways of mitigating the position is for it to sublet the property to a third party. However, this can create problems if the subtenant wants to undertake alterations, such as those set out below.

- Unless it is a very short-term sub-lease, it is likely to contain a qualified alterations clause. Consequently, it will be difficult for the corporate occupier to prevent the subtenant from carrying out works.
- The corporate occupier is faced with the possibility that the subtenant may disappear before the expiry of the sub-lease and leave the corporate occupier with a reinstatement liability under the headlease, and the prospect of no recovery against the sub-tenant.
- If the sub-tenant is still around and worth suing at expiry of the sub-lease, the corporate occupier will need to require the subtenant to reinstate the premises to the pre-alterations position. However, this can become factually and legally complex if both the head and sublease are co-terminus because the reinstatement obligations may be different and it may be very difficult to apportion liability and damages. Equally, the headlease re-instatement obligations may provide the sub-tenant with a defence based on causation.

6.4.3 Summary

In summary, a corporate occupier will secure significant benefits by:

- determining its accommodation needs before it chooses premises
- selecting premises that do not require significant works or premises that can be adapted easily and cost-effectively
- securing lease terms that avoid prohibitive constraints or conditions
- exploiting opportunities to transfer some or all of the cost to the landlord
- ensuring that all works undertaken are formally and comprehensively documented.

6.5 User restrictions

6.5.1 User clauses

Most leases will contain a user clause which restricts the use to which the tenant can put the premises. The extent to which the tenant's use is restricted will vary according to the landlord and the tenant's requirements and their respective negotiating positions at the time the lease was granted. A corporate occupier will generally want as much flexibility as possible so that in the event of a change in its business environment there is the ability to adapt the nature of its operation without the need to revert to the landlord for consent. Landlords can be inflexible in granting changes of use and either such consent will not be granted or the landlord may insist on receiving a premium for the privilege of varying the lease terms. This issue can also arise if the corporate occupier needs to assign or sub-let if the premises become surplus to its requirements. While the landlord may not be able to reasonably refuse consent to assign or sub-let, it may nevertheless be able to frustrate the transaction if the landlord can lawfully refuse consent to the change of use intended by the incoming tenant.

Securing a wide user clause tends to be easier for stand-alone properties and for offices and industrial units, although for the latter group there tend to be restrictions in the case of certain types of industrial use. User clauses in leases of retail units will generally be more restrictive, especially in shopping centres where landlords seek to control the mix of uses within the centre. A tenant mix policy will usually benefit the tenant while it is trading because if properly implemented it will avoid the problem of too many retailers in one category being in a centre, or inappropriate uses being introduced at a later date. For example, in a centre that is predominantly occupied by fashion retailers, the introduction of a hardware store might have an adverse effect on the retailers either side of the hardware unit.

Landlords will want to control the use of their premises to ensure that there is no adverse impact on asset value. In the retail environment that means ensuring that the right type of retail use is in the centre, or in an industrial context, that there is not a switch to heavy industrial use, which may have an unacceptable impact on the property. For example, it may lead to the installation of heavy equipment, which could have an impact on the floor slab, or the production process may produce gases which are detrimental to the building fabric and corrode the cladding.

There are instances where user restrictions are not a result of the landlord trying to control the property but are caused by external factors. A good example of this can be a business park where the restrictions have been imposed as a condition of development planning conditions (eg some business parks outside Cambridge have seen such restrictions contained in leases).

The use clauses in a lease can occasionally define the permitted use by reference to the Town and Country Planning (Use Classes) Order 1987, rather than a specific use. This provides consistency of definition and a degree of flexibility for the corporate occupier to change uses within the class or classes specified in the user clause.

6.5.2 Landlord's consent

As with alienation and alterations, user clauses can take the form of an absolute restriction (ie the clause allows no possibility of change) or a qualified restriction (ie the clause allows change provided landlord's consent is obtained). However, whereas section 19 of the Landlord and Tenant Act 1927 implies a proviso that landlord's consent shall not be unreasonably withheld to alienation and alteration covenants which contemplated landlord's consent, there is no such implied proviso in relation to a user restriction. Instead, section 19(3) of the 1927 Act implies a different proviso that:

> ... no fine or sum of money in the nature of a fine, whether by increase of rent or otherwise, shall be payable for granting consent; but this proviso does not preclude the right of the landlord to require payment of a reasonable sum in respect of any damage or diminution in value of the premises or neighbouring premises belonging to him and any legal or other expenses incurred in connection with such licence or consent.

When considering a user restriction the following propositions apply.

- The proviso does not apply to absolute prohibitions; a landlord can lawfully demand a premium for a variation in the lease terms to permit the change of user sought by the tenant.
- If the user clause simply provides that a change cannot take place without the landlord's prior consent (ie there is no express proviso providing that the landlord should not unreasonably withhold its consent); the landlord can arbitrarily refuse consent without the risk of any challenge.
- However, if the landlord demands a sum for diminution in value of the premises, and this is challenged, the court has jurisdiction to assess what is a reasonable sum. Importantly, once this has been determined, the landlord must grant consent on payment of this sum.
- If the user restriction expressly provides that the landlord shall not unreasonably withhold consent, then if the landlord refuses consent, the tenant can apply for a declaration that the landlord has unreasonably refused consent. The burden of proof will be on the tenant to establish the landlord's decision was unreasonable. Alternatively, if the landlord has unreasonably refused consent, at common law the tenant is released from the need to obtain the landlord's consent and can proceed. However, defiance is not for the faint hearted and if a court subsequently disagrees, the tenant may lose the lease through forfeiture, or find itself at the wrong end of an injunction to restrain the unlawful user.

6.5.3 Impact on rent review

As with many aspects of landlord and tenant law, the legal implications of one issue can have ramifications on an entirely different issue which may arise in the future. Consequently, while a tenant may find a restrictive user covenant (e.g. an absolute restriction) problematic when attempting to assign or sublet, the restricted user clause may reduce the rent payable at rent review, because the market for a lease with such restriction is limited. However, this argument is unlikely to apply in a lease of a shopping centre unit where all the units are similarly restricted. Whether a particular user clause will impact at review will depend upon the particular wording of the clause and the nature of the property.

6.5.4 Keep open clauses

In the retail context it is not uncommon for a lease to contain a positive obligation on the tenant to keep the premises open and trading at specific times. Such clauses are often inserted into the leases of anchor tenants in shopping centres. Anchor tenants are important to a landlord as they draw in business to the smaller shops in a shopping centre. If the anchor tenant ceases trading this can have a dramatic effect on other retailers and in turn on the landlord's reversionary value. However, at a time when a retailer may be suffering from poor trade generally and there is little prospect of finding a replacement tenant, it can be the only solution. The courts have tended to award damages against a tenant that ceases trading in breach of a keep open clause rather than force it to continue trading. However, damages are unlikely to be an adequate remedy. In *Co-operative Insurance Society Ltd* v *Argyll Stores(Holdings) Ltd* [1987] 2 WLR 898 the Co-op were so incensed with its tenant's wanton breach of such a clause it argued all the way to the House of Lords that a mandatory injunction should be granted requiring the tenant to recommence trading. While enjoying brief success in the Court of Appeal, the House of Lords held that an injunction should not be granted in these circumstances, particularly where the effect would be to require the tenant to trade at a loss.

6.6 Precedents

6.6.1 Tenant's notice of intended improvements

NOTICE OF TENANT'S INTENTION TO MAKE IMPROVEMENTS

To: [NAME AND ADDRESS OF LANDLORD] "The Landlord"
From: [NAME AND ADDRESS OF TENANT] "The Tenant"

1. By a lease dated [INSERT DATE] the Landlord demised to the Tenant premises known as [INSERT ADDRESS] "the Premises" for a term of [INSERT LEASE DURATION].
2. The Tenant hereby gives the Landlord notice that it intends to make improvements to the Premises in accordance with the specification attached which provides details of the intended works and the plan which identifies the parts of the premises which will be affected.
3. For the avoidance of doubt this constitutes a notice of intention to make an improvement for the purposes of section 3(1) of the Landlord and Tenant Act 1927.

Signed .... Dated ....

6.6.2 Landlord's notice of objection

LANDLORD'S NOTICE OF OBJECTION TO PROPOSED IMPROVEMENTS

To: [NAME AND ADDRESS OF TENANT] "The Tenant"
From: [NAME AND ADDRESS OF LANDLORD] "The Landlord"

1. On [INSERT DATE] the Tenant served the Landlord with a notice of intention to make improvements to the premises known as [INSERT ADDRESS] demised by a lease dated [INSERT DATE] between [INSERT PARTIES].
2. The Tenant is hereby given notice that the Landlord objects to the execution by the Tenant of the improvements detailed in the specification and plan attached to the Tenant's notice.

Signed .... Dated ....

Rent Review 7

7.1 Introduction

A rent review clause is included in a lease to allow a landlord to have the benefit of future increases of rent (and also increases in the value of its investment), while at the same time allowing a tenant to have a lease duration which provides adequate security of tenure. A rent review is therefore a valuation exercise that brings the current rent payable by the tenant (often referred to as passing rent) into line with the market rent of the premises. Lease durations until the late 1960s had been relatively long, typically with terms of 21, 42 and 99 years and with rent review cycles which tended to be 21 years or longer. With inflation becoming a major factor in the 1960s, lease durations dropped to 25 years and the rent review cycle shortened initially to seven and then to five years. This resulted in the institutional full repairing and insuring lease of a 25-year term, with five yearly upwards only rent reviews (often know as the FRI lease).

More recently corporate occupiers in some sectors have favoured the flexibility of shorter leases (typically 10–15 years), and in the retail sector some retailers now seek leases of five years. However, rent review cycles have tended to remain five yearly (although occasionally they can be three yearly). With leases of five years or less, the parties usually do not feel the need to include a rent review clause.

The period from the 1970s through to the early 1990s saw the structure of rent review and the rent review clause evolve and a substantial body of case law was generated as a result. While there are still points of law occasionally emerging on the subject, the basic principles are relatively settled, with new cases tending to involve the courts construing the wording of specific rent review clauses, rather than deciding any new points of principle. There are a number of books dedicated to this subject which address it in considerable depth. The purpose of this chapter is to provide an overview of the pertinent issues for the corporate occupier, and not to provide a detailed analysis.

7.2 The key components of determining rental value

While a rent review involves a valuation exercise, and in particular the determination of the open market rent on the rent review date, it can be a curious exercise with the valuer taking into account some elements which are real, and some which are unreal and hypothetical. The review process generally looks at the actual building but refers to a hypothetical lease and assumes a hypothetical landlord and hypothetical tenant. The review clause will set out assumptions, which the valuer is to assume when valuing the rent, and disregards, which he is to ignore. The assumptions and disregards will not necessarily reflect the reality of the lease itself.

When interpreting a rent review clause, great care must be exercised in referring to previously decided cases. In practice, no two rent review clauses are identical, and the construction of a particular word or phrase is unlikely to be assisted greatly by considering how an early judge interpreted a different clause in a different lease. However, inevitably, decided cases have given rise to some general principles and presumptions as to how rent review clauses should be interpreted. However, those principles and presumptions are by no means consistently applied in practice, and care should be exercised when relying upon them.

There are essentially two schools of thought when a problem arises in construing a rent review clause. The strict literal approach is based on the freedom of contract principle which is underpinned by the belief that the parties to a contract are free to agree whatever terms they choose and the courts should not re-write or misconstrue the bargain which was agreed if it was clearly expressed in the contract however harsh the result may be. Only if the result of the words used would be absurd can the court conclude that such a result could not have been the intention of the parties.

The second school of thought, which is more interventionalist, is based on what has become known as the presumption of reality. Here, in the case of ambiguity, the courts lean towards a construction of the rent review clause which gives effect to 'the presumed commercial purpose of rent reviews' ie a fair and proper increase in rent for the premises. A good example of a case in which the presumption of reality was applied is *British Gas Corporation* v *Universities Superannuation Scheme Ltd* [1986] 1 EGLR 12. It fair to say that in recent years the presumption of reality has tended to prevail, although this has not always been the case and the presumption must not be followed slavishly (see *Canary Wharf Investments* v *Telegraph Group Ltd* [2003] EG 132). It should certainly not operate in a way that defeats the clear words of the rent review clause.

7.2.1 The basis of valuation

Leases contain many terms to define market rent such as best market rent, open market rent,'best rent reasonably obtainable. Landlords will look to exploit the wording to try and include special purchasers to boost the rent over and above what was originally agreed between the parties. For the vast majority of properties there will be little difference which ever term is used.

The valuation method which should be adopted in a given case is a matter of expert valuation opinion and will vary according to property type and user. The process of valuation is essentially undertaken by analogy. A very clear and readable description of this exercise was given by Lewison J in *Marklands Ltd* v *Virgin Retail Ltd* [2004] 2 EGLR 43.

Lewison J pointed out that a valuer will look for a comparable transaction as close as possible to the hypothetical lease. He will assume that if the hypothetical transaction were to go ahead in the real world it would command a similar value. However, because it will be impossible to find an identical comparable, the valuer will have to make adjustments to reflect the differences between the hypothetical lease and the comparable. These will include differences in size, location, lease terms and the like. To overcome the difference in size of the property between the comparable and the hypothetical transaction, the valuer will divide the rent for the comparable by the area of that property to produce a rent per square foot (a process which is known as devaluation). The valuer can then apply the rent per square foot devalued from the comparable to the hypothetical transaction by multiplying the rate per square foot by the area of the demised premises. However, as part of the valuation process, the valuer may consider it appropriate to make further adjustments for size, for example, where the demised premises are significantly bigger, or the comparable is significantly smaller, so as to avoid producing an extremely high rent. This method is often referred to as the 'overall' method of valuation.

With retail properties, valuations tend to be carried out on the basis of the more sophisticated zoning method.

Lewison J summarised this method as follows:

> ... This method notionally divides a shop into parallel zones (usually of 20ft), starting at the street frontage and working backwards to the rear of the shop. The zone nearest the street is called "zone A", the one immediately behind it is called "zone B", and so on. At some point, however, usually at around zone C, the valuer stops zoning and classifies the rest of the shop as "remainder". Zone B is taken to be half the value of zone A, zone C as half the value of zone B, and so on. This process is known as "halving back". An alternative way of looking at this method of valuation is for the valuer to divide the area of zone B (as opposed to its value) by two, the area of zone C by four and so on. The resulting area is expressed "in terms of zone A" or "ITZA". A zone A rate per square foot can then be applied directly to the area ITZA.

Lewision J then went on to explain:

> ... One of the theoretical underpinnings of this method of valuation is that a shop with a wider street frontage (and, hence, a bigger shop window) is more valuable than a shop of the same area but with a narrower shop front and a greater depth. Even this method of valuation may require adjustments to be made over and above the process of devaluation. A common adjustment would be an increase in the zone A rate to take account of the frontage; that is, a shop window at the side of the shop.

While the zoning method will almost always be used with high street shops and standard shopping centres, department shops and non retail premises tend to be valued using the overall method. One further possible method of valuation is profits valuation which is based on the profit a tenant would be likely to generate from occupying the premises and then considering how much a tenant would be prepared to bid for the property. This method of valuation tends to be used in the leisure industry.

7.2.2 The premises

The valuer will have to identify precisely what premises he is valuing. This will usually be the demised premises (although not necessarily so) and the valuer will have to look at the lease to determine what property was let. The property may be defined by a plan, and it is obviously important the plan is accurate.

Unless the rent review clause provides to the contrary, the valuer must value the demised premises in the state that they are in at the rent review date. Consequently, if the corporate occupier intends to carry out alterations which would be likely to increase the value of the rent on review, it is important that it secures a specific disregard in the rent review clause requiring the rent review valuation to ignore any tenant alterations (see *Ponsford* v *HMS Aerosols* [1978] 2 EGLR 81).

Corporate occupiers do have a tendency to undertake alterations without first obtaining landlord's consent and without having the works properly documented in a licence for alterations. One common trap is the mezzanine floor that has been installed without obtaining a formal licence for alterations. The landlord will argue (usually successfully) that it should be rentalised. The unwary corporate occupier at this stage will have little option but concede the rent at the higher level thereby providing the landlord with an unwarranted gain. Clearly this is something that needs to be dealt with properly at the time of the signing of the lease, or if not then, in the subsequent licence for alterations. The CREM needs to ensure that proper consents are in place and that the original documents are kept.

Having identified the demised premises, the next step is to determine the floor area and that should be established in accordance with the latest edition of the RICS Code of Measuring Practice. It is usually unsafe to rely upon the plans attached to the lease. There remains scope within the code for certain areas to be treated differently, and there are still a number of surveyors who are not fully conversant with the code. This offers the opportunity for the corporate occupier to make savings on the floor area if the landlord's surveyor is uneducated, but there is also the risk that the landlord's surveyor will start with a gross area and wait to negotiate down the floor areas and the rate per square foot. In the event of referral to a third party expert the floor area will need to be agreed and documented.

7.2.3 The length of term

The length of the term of the hypothetical lease is the factor most likely to affect the new rent. It is perhaps of no surprise that it is an area which gives rise to considerable dispute.

The most common assumptions are that the hypothetical lease has a term equivalent to:

- the term granted by the lease or
- the unexpired residue or
- a fixed period eg five or 10 years or
- a term equal to the rent review cycle.

If a rent review clause does not have an assumed term a court is likely to opt for reality and assume a term equal to the unexpired term. Traditionally landlords favoured a rent review clause which assumed a long term on the basis that tenants desired longer leases which provided adequate security of tenure and allowed tenants the opportunity to write off fit out costs over a long term. This meant that there was greater demand for longer leases and in turn allowed a landlord to argue for a higher rent at rent review. To an extent this analysis still holds good for larger premises. However tenants now tend to crave flexibility and find shorter leases more attractive in the current environment. Consequently, an assumption of a long term in the rent review clause is now increasingly regarded as a disadvantage and will negatively affect the rental value.

The market rent of the hypothetical lease will be increased if the term assumes the existence of a break clause. The impact on the market rent will depend upon whether the break is personal to the original tenant, and how difficult it will be for a hypothetical tenant to operate the break clause. The key issue is whether the break clause which is contained in the actual lease should be taken into account, or disregarded on review. This will always be a question of construing the rent review clause and can be a source of much contention.

7.2.4 Repair

The condition and repair of the demised premises at the rent review date could have a significant impact on their market value. Clearly if they are in a poor state of repair, a tenant is likely to offer less for them or require a rent free period. Most rent review clauses will include an assumption that the tenant has complied with its repairing covenant. If that is the case, the valuer must not value the premises in their actual state of repair at the date of the review, but should assume that the tenant has fully complied with its repairing obligations and ignore any disrepair.

What of the position where the lease does not contain such an assumption? Can the tenant benefit from its own breach of covenant in failing to discharge its repairing obligations and secure a lower rent? The answer to this appears to be no (see for example *Hibernian Property Co Ltd* v *Liverpool Corporation* [1973] 1 WLR 751).

On the other side of the coin, if the repairing clause contained in the lease is deemed to be onerous, as was the case in *Norwich Union Life Insurance Society* v *British Railways Board* [1987] 2 EGLR 137, then this is likely to have a depressing effect on rent. In that case the rent was reduced by 27.5% to reflect an unusually wide repairing obligation (contained in a 150 year lease) which included a requirement on the tenant to rebuild, reconstruct or replace the whole building if necessary.

7.2.5 Restrictions on user

Usually the rent review clause will provide that the hypothetical lease is to be on the same terms as the actual lease. If the lease contains an unduly restricted user clause then this can have a depressing effect on the new rent. This is illustrated by the case of *Plinth Properties Investments Ltd* v *Mott Hay & Anderson* [1979] 1 EGLR 17.

- The lease in this case contains the following restriction:

> not to use the demised premises ... otherwise than as offices ... in connection with the lessee's business of consulting engineers

- The arbitrator made his award on two bases, that:
  - he could take into account the possibility of the landlord being persuaded to relax the restriction in which case the rent would be £130,000 pa or
  - could not in which case the rent would be £89,000 pa.

Held — Court of Appeal

- As a matter of law and construction an arbitrator has to assume that any rights and obligations contained in a lease will be enforced and performed and therefore must assume that no relaxation of the restriction will be possible.
- The landlord was unlikely to relax the restriction without the payment of a considerable premium and consequently the restriction had a depressing effect on the new rent.

However, in *Law Land Co Ltd* v *Consumers Association* [1980] 2 EGLR 109 the user restriction contained in the lease provided that the tenant was not to use the premises "other than as the offices of the Consumer Association". The tenant argued that this meant that the only possible tenant was the actual tenant. The Court of Appeal rejected this on the grounds that it would prevent an open market rent being fixed and was inconsistent with an assumption of vacant possession. The user of the hypothetical tenant would be substituted for the actual tenant.

When negotiating a new lease (whether in an operational or legacy context), a corporate occupier needs to be aware of the ramifications of a restricted user clause on a rent review clause. In an operational context while it may result in a discount of rent at review, it will restrain what the corporate occupier can do with a property. Equally, in a legacy portfolio context, while the corporate occupier may wish to retain control when sub-letting, if the sub-lease contains a rent review clause, an onerous restriction may depress the rent on review.

A restrictive user clause will only have a negative effect on rental value when comparable properties have open user clauses. In the case of a shopping centre, a restricted user clause may have no effect on rental value if all of the comparables are from the same shopping centre and are subject to similar restrictions.

7.2.6 Improvements

In the event that a lease contains an absolute prohibition against alterations, following the reasoning in *Plinth Properties*, the reviewed rent must account for the fact that the tenant will not be able to carry out improvements to the demised premises.

The majority of modern leases will allow alterations subject to the tenant obtaining the landlord's prior written consent (not to be unreasonably withheld), and the rent review clause will instruct the valuer to disregard certain improvements carried out by the tenant. As indicated at 7.2.2 above, this is important for the corporate occupier to avoid the unenviable prospect of having to pay an additional rent for improvements carried out at its own expense (see *Ponsford* v *HMS Aerosols Ltd* [1978] 2 EGLR 81).

This disregard will not usually apply where the improvement has been carried out pursuant to an obligation owed to the landlord. Often draftsmen simply incorporate section 34 of the Landlord and Tenant Act 1954.

An issue which arises all too often is whether improvements should be disregarded where the tenant has failed to obtain landlord's consent. This is likely to depend on the precise wording of the disregard. If the valuer is directed to disregard improvements carried out by tenant "with landlord's consent", then if the tenant carries out improvements without consent, these works will be rentalised on review. However, if the disregard makes no express reference to landlord's consent, then it may be possible for the tenant to argue that the works should not be rentalised. However, the safe course for any corporate occupier is to ensure that it obtains the necessary consent before it embarks upon expensive improvements, and if the rent review clause does not contain the relevant disregard, it should seek to have one inserted in the licence for alterations.

The approach to be taken when applying the disregard is essentially a question of valuation and not law. The valuer must use the valuation method which fairly identifies the increased rental value of premises in their unimproved state and ensures that the tenant is not paying an additional rent referable to the improvement carried out at its own expense (see *GREA Real Property Investments* v *Williams* [1979] 1 EGLR 121 and *Estates Projects Ltd* v *Greenwich London Borough* [1979] 2 EGLR 85).

7.2.7 Restrictions on sub-letting

A restrictive alienation clause can have a depressing effect on rent at review. In practice there tend to be few if any problems with assignment. However, a corporate occupier which finds itself in the reluctant landlord position with surplus premises can have serious problems sub-letting if the lease contains restrictions on sub-letting. In particular, clauses that prevent sub-letting at below passing rent and/or require the sub-lease to be on the same terms as the lease can make the premises virtually inalienable, particularly, as is often the case, where the premises are overrented. This problem is considered in detail at 5.6.1 above. Following the decision in *Allied Dunbar Assurance plc* v *Homebase*

Ltd [2002] 2 EGLR 23 it is now much more difficult to mitigate the effects of such restrictions. The modern view, therefore, is that in a falling market with overrented premises, such restrictions are likely to have a depressing effect on rent. In effect, the passing rent restriction constitutes a total prohibition against subletting in a falling market.

7.2.8 Schedule of condition

A further factor that can impact upon rental value at review is a repairing obligation that is limited by reference to a schedule of condition. The landlord may argue that a corporate occupier should pay more rent in view of the limited repairing obligation. However, the corporate occupier's riposte should be that rent is lower to reflect the true condition of the space. When negotiating lease terms, careful thought must be given the inter-relationship between these competing issues.

7.2.9 Inducements

There is a whole body of complex law which has been generated around this issue. It arose as follows. Most rent review clauses contain an assumption that the demised premises are vacant. This is necessary for the purpose of engineering the letting of the hypothetical lease in an open market. However, this enabled tenants to argue that if the property is vacant, the hypothetical tenant would reduce its bid to account for fit out works and the accompanying dead time which results from the tenant not being able to use premises whilst the works are carried out (see *99 Bishopsgate* v *Prudential Assurance Co Ltd* [1985] 1 EGLR 72). This is usually addressed by a rent free period. Consequently, tenants at review would argue for a discount based on a rent free period. Landlords then sought to deal with this by drafting the rent review clause with a requirement that inducements and fitting out be disregarded, or alternatively an assumption that the premises were fully fitted out at the review date. The cases which have followed on this topic are as complex as they are numerous. A lucid and detail account of these cases can be found in paragraph 6.4 of the *Handbook of Rent Review* by Kirk Reynolds QC and Guy Fetherstonhaugh QC.

7.2.10 Review date and pattern

The standard review term has become five years. Occasionally there will be three or seven year reviews, and more infrequently 21 year patterns. Corporate occupiers need to take account of the possible implications of taking a lease that has rent reviews which are at variance to the norm. The difference between a five year review cycle and other review cycles tends to be around 1% per year. Consequently, a tenant on a seven year review cycle can expect to pay 2% more. A 21 year review pattern could see an additional 16% added to the rent (although this will depend very much on the nature of the property and the market). For a corporate occupier the long review pattern can be an advantage or a major disadvantage depending upon the timing. It becomes more complicated in a legacy portfolio scenario where a unit is sublet, for example, on a five-year review pattern in circumstances where the headlease is on a 21-year pattern. The mismatch between the two can see the rents determined at different times and therefore with different underlying rates. This is not an issue if the headlease is geared at, for example, only 10% of the open market value. However, if both are at full market rent and the headlease is paying 16% more if the peaks align with the headlease review, it is possible to see the head rent substantially above the rent payable under the occupational lease.

7.3 Management issues

7.3.1 Systems

A corporate occupier with many leased properties (both operational and legacy portfolio) will need a management system to record details of the properties in its portfolio. One of the most important data entries will be the rent review information from each property and in particular time critical information about those properties such as time of the essence trigger notices or counter-notices. The level of sophistication required will vary depending upon the scale of the portfolio, but these days most systems are modular and will allow the system to be added to as the portfolio grows.

While the system needs to accommodate the day to day tasks such as paying invoices and the like, of critical importance is the diary entry system. What diary notification system or alarms does it have for forthcoming reviews? Does it indicate to the CREM when there are time of the essence issues? Will the system provide a profile of reviews over a set period of time? These are all points which should be considered.

The CREM should not under estimate the amount of time that it will take to set up a system. Due consideration needs to be given to the various software packages available including:

- the scope for expansion of the system
- the charging structure
- how much adaptation of the system will be needed.

If a significant amount of adaptation is required then it is probably not the right system for that corporate occupier. The CREM should also allow for ongoing management and updating issues. Once a system has been installed, a significant and often time consuming exercise will be inputting data onto the system. When changing systems, while a transfer from an existing system may be quick, there may be problems with the integrity of that data and it is always prudent to carry out a data verification exercise which will involve checking the data in the system against the leases and sub-leases.

Finally a CREM should consider whether the use of a system is widespread or not. Being a user of a minority system can have advantages but generally there are major disadvantages in finding staff that can run the system without training and also when there are portfolio transactions, the transfer of data can be more difficult than it might otherwise be with a system that has a wider use.

7.3.2 Time-limits for notices and counter-notices

Like the operation of a break notice, the timing of the service of a rent review trigger notice (by the landlord), or a counter-notice (by the tenant) can be critical, with rights being lost if the relevant notice is not served in time. It is therefore extremely important that a corporate occupier or its professional advisors are familiar with the detailed machinery of the relevant rent review clause. As discussed above, a responsive and reliable diary system is essential.

Although it will ultimately depend upon the wording of the clause, a rent review is usually initiated in one of two ways.

(a) Automatic review

This type of review does not require a trigger notice to be served by the landlord. Often the lease contains a mandatory requirement for a rent review to take place on a specific date and then periodically thereafter (eg on 1 January 2007 and every five years after that).

The effect of such a requirement is to create a contract between the landlord and the tenant for a revised rent to be ascertained in accordance with the rent review clause and to be substituted for the current rent, and there is an ongoing requirement for the lease to be reviewed at the end of each cycle, eg every five years (see *Addin* v *Secretary of State for the Environment* [1997] 1 EGLR 99).

The clause will invariably allow the parties a period of time to negotiate a new rent, after which (usually) the landlord will have the right to initiate a third party determination procedure (ie expert determination or arbitration) if no agreement is reached.

(b) Trigger notice

The second type, and the one which the parties need to take care with, is a review clause that is initiated by the landlord serving a notice calling for a review. This is often referred to as a trigger notice because it has the effect of triggering the rent review. If the landlord does not serve a notice, no review will take place. In other words, the landlord is given a right to call for a review in contrast to the first type, where there is a mandatory requirement for a review to take place.

Often the lease will not only provide for the service of a trigger notice, but it will also provide for the service of a counter-notice by the tenant if it wishes to contest the rent put forward by the landlord in its trigger notice. As with all notices and counter-notices, great care should be taken to ensure that they contain the information required by the lease, and that they are served in the appropriate way. To that extent, the points made in 12.3.1 in the context of break notices apply equally to rent review notices and counter-notices.

Time-limits in the context of break notices tend to be critical. That is not necessarily the case in relation to rent review notices, but often the rent review clause will stipulate a time-limit for service, and it goes without saying, that where ever possible the time-limits should be complied with to avoid arguments about whether or not those limits were mandatory.

If a time-limit is missed, then the rent review clause and indeed the lease as a whole will have to be considered very carefully. In *United Scientific Holdings Ltd* v *Burnley Borough Council* [1977] 2 EGLR 6 the House of Lords held that in the absence of any contra-indications in the express words of the lease or in the interrelation of the rent review clause and other clauses, or in the surrounding circumstances, the presumption is that the timetable specified in a rent review clause is not of the essence of the contract.

The starting point is, therefore, a presumption that time is not of the essence. However, the presumption will be displaced where:

- the lease states that time is of the essence
- where the wording of the time-limit demonstrates that the parties' intention was that the timetable should be strictly enforced
- there are other indicators in the rent review clause which are only consistent with strict compliance with the time-limit
- where there is an inter-relationship with another clause which demonstrates the time limit was to be strict
- where the lease provides what will happen in the event that the time-limit is not met.

Consequently, if the presumption is displaced, and the landlord fails to serve a trigger notice within the time-limit prescribed by the rent review clause, it will lose its right to review the rent until the next review comes around. Equally, if the tenant fails to serve a counter-notice and the time-limit is of the essence of the contract, it may lose its rights to contest the rent review, depending on the context and the consequences of failure provided for in the lease.

Starmark Enterprises Ltd v *CPL Distribution Ltd* [2000] 3 EGLR 37 demonstrates how a corporate occupier can be caught out. In this case the rent review clause required the landlord to initiate the review process by serving a trigger notice which specified the landlord's proposed new rent. The tenant was entitled to serve a counter-notice within one month calling on the landlord to negotiate the new rent. The rent review clause went on to specify that if no counter-notice was served within the specified time-limit, the tenant "shall be deemed to have agreed to pay the increased rent specified in the [landlord's trigger notice]". The landlord served a trigger notice specifying a new rent of £84,800 pa. The tenant served a counter-notice specifying a rent of £52,725 pa, but it was served late. The Court of Appeal held that the deeming provision ousted the presumption making the time-limit strict. Consequently the new rent was the figure contained in the landlord's trigger notice and the tenant lost the opportunity to challenge the landlord's figure.

Time limits can be very short. In *Starmark* the time-limit was 28 days. Occasionally it can be as short as 14 days. Consequently, a corporate occupier needs to ensure that it has an adequate system in place to ensure the notice is referred to the correct person promptly (particularly if it is sent to a registered office which is distant from the CREM).

7.4 Negotiating a rent review

7.4.1 General points

Assuming that there is a diary system in place, a CREM should be considering the review at least one full financial year before the review date so that the likely impact on the corporate occupier's budget for the following year can be assessed and also to check the requirements of the rent review clause and any time limits. Having a management system is helpful but care is needed to avoid over-dependence on it.

If there is any doubt about the wording of the review clause and its effect then the CREM should seek legal advice. Forming an early view on any issues that are likely to arise will ensure the CREM is in a good position to deal with the rent review when it is initiated. The CREM clearly needs to be in a position to decide whether to accept the landlord's quoted rent or not, and will therefore need to have an awareness of the rental levels for the property.

If the time-limits for service of trigger notices and counter-notices are critical, the drafting and service of such notices are best left to experienced property litigation solicitors, not least because, if they make a mistake, the corporate occupier will have the benefit of their professional indemnity insurance.

In the main, negotiations will continue to the point where there is an impasse and one party, usually the landlord, will initiate the reference to a third party expert or arbitrator in accordance with the lease.

7.4.2 Comparables

As indicated in paragraph 7.2.1 above, the valuation method is largely dependant upon identify comparable property transactions. However, the property market is not totally transparent and obtaining details of suitable transactions can be difficult not least because on occasions, transactions are masked by confidentiality clauses. There is a risk of a corporate occupier operating from imperfect knowledge. Landlords tend to co-operate with each other in trying to improve rental and capital values in ways which corporate occupiers, who are often competing for the same space, do not. In significant rent reviews it is possible to obtain a witness summons to obtain evidence and copies of relevant documents relating to known transactions. Generally speaking, a court is likely to give precedence to the public interest in ensuring rents are fixed at a proper level over the private desire for confidentiality.

7.4.3 In-house v outsourcing

A corporate occupier must determine whether the business is best served by the review being handled in-house, within the corporate real estate team, or whether it should be outsourced. The benefit of it being dealt with in-house is primarily that the CREM will have knowledge of the property and may be in contact with other CREMs of companies that occupy space in the same area. The disadvantage is that the skills of the CREM may not include rent review work. This may not be important where a corporate occupier has a portfolio which comprises a small number of properties with low rental levels, but where there are a significant number of properties and/or involving larger buildings, the lack of rent review experience can have a significant impact on the costs of the business. A corporate occupier can easily outsource its rent review work without having to outsource the entire portfolio management function. The benefit of out sourcing the rent review work is that it will receive a much higher level of expertise and knowledge of the review process and of the specific market place in which the rent is being reviewed. The specialist nature of the rent review process should make the default position that the CREM will outsource the large reviews at the very least, and arguably all reviews.

7.5 Dispute resolution

If the parties are unable to reach agreement on what the new rent should be, then one party will initiate the mechanism contained in the rent review for third party determination. The detailed procedure and time limits will vary from lease to lease. Generally the lease provides that the process is to be started by the landlord although that is not always the case, and it is now clear that even if the presumption applies and time is not of the essence, it is possible for the tenant to serve a notice on the landlord making time of the essence (see *Barclays Bank plc* v *Savile Estates Ltd* [2002] 2 EGLR 16). This may be important to a corporate occupier where there is an automatic rent review clause, and where the landlord delays in initiating the reference to a third party determination. This is likely to be an issue where the review clause allows upwards and downwards reviews and the tenant believes that the market rent has fallen, or where the landlord is delaying until conditions become more favourable for a review. In both of these instances it may be in the corporate occupier's interests to drive the review forward.

Where the lease allows it, in addition to the points raised above there can be other benefits for the tenant in instigating the process, especially if the market is slow and the corporate occupier is able to finalise the review as close to the review date as possible. A tenant can also benefit from a proactive approach when the market is moving ahead quickly as the landlord may see bigger gains if they delay the process to obtain more evidence. The CREM therefore needs to be aware of the movement of the market and what can and cannot be done under the terms of the rent review clause.

Modern leases will usually specify that if the parties are unable to agree upon the choice of a third party expert or arbitrator, the matter is to be referred to the President of the Royal Institution of

Chartered Surveyors. He will then appoint an expert/arbitrator to determine the new rent. The precise procedure after that will either be contained in the lease or the appointed third party will set out a timetable (or occasionally, a combination of the two).

7.5.1 Arbitration v independent expert

A matter which is often debated at some length is whether the third party determination should be by way of arbitration or expert determination. This issue may confront a corporate occupier when negotiating a new lease, or later, when the parties are unable to reach agreement on the new rent.

Both procedures have their advantages and disadvantages. Generally speaking, the larger scale reviews with difficult issues of valuation and lease construction are likely to be better disposed of at a formal arbitration, whereas the smaller lower value reviews are more economically disposed of by way of expert determination. In between, either may suit and will largely depend upon what the lease prescribes, the nature of the dispute, and the amounts at stake.

The traditional view of arbitration was of a formal process that largely mirrored the formality of the court process. While this may still be correct in some cases, the introduction of the Arbitration Act 1996 has made the process a good deal more flexible. The parties can now agree the procedure to be followed, or in default, this will be fixed by the arbitrator. Rent review arbitrations can vary greatly in their value and complexity. One of the benefits of modern arbitration is its procedural flexibility. A substantial case involving a large office building and high rents may require a formal arbitration with an oral hearing before the arbitrator with directions similar to court proceedings. However, at the other end of the scale, a rent review dispute concerning modest shop premises may be determined by an arbitrator without the need for an oral hearing and with a decision being reached on the basis of written representations and counter-representations made by each party's valuer.

The benefits of arbitration include:

- witnesses can be compelled to attend and produce documents
- matters of law can be referred to the court for determination
- the arbitrator can order the parties to disclose relevant documents
- there is a procedure to award costs
- an arbitrator must give reasons for his decision
- an appeal from an arbitrator's award is possible although only in limited circumstances.

In contrast, an expert determination is a distinctly low key affair. It involves the landlord and the tenant jointly instructing an expert with a detailed knowledge of rent reviews in the relevant geographical area to use his own knowledge and experience to fix the rent. The expert derives his authority and powers from the contract which will either be the rent review clause in the lease or a separate agreement referring the matter to expert determination. The expert may (and usually does) invite submissions from the parties but need not do so.

The benefits of an expert determination are:

- the expert will know the market place and therefore will have information that may not be available to either party
- it can be quicker and cheaper
- the decision is final and binding on the parties and there is no appeal against an award of an expert
- the expert can be sued for negligence as he does not have immunity in the way that an arbitrator has.

7.5.2 Without prejudice and subject to contract

Correspondence between the parties involved in a rent review is often sent under the cover of the label without prejudice and sometimes subject to contract and occasionally both. It is important that the parties are familiar with the consequences of these labels or they could result in an unforeseen and unintended result.

The words subject to contract mean that anything done in response to that letter is not intended by its author to produce a legally binding result. The words without prejudice mean that unless a binding result is achieved the letter cannot be referred to in the litigation (or in this case the arbitral) process.

If a landlord serves a trigger notice specifying the new rent or the tenant serves a counter-notice specifying an alternative rent, and those notices are sent under the cover of a letter which is marked subject to contract and without prejudice it is highly likely that the notice or counter-notice will be held to be invalid (see *Shirlcar Properties Ltd* v *Heinitz* [1983] 2 EGLR 120). All letters or notices which a party may wish to show to a judge or an arbitrator during the course of the trial should be open letters (ie with no labels).

The misuse of the term without prejudice works both ways. Parties often take the view that anything said in a letter which bears this label is privileged and cannot be produced in evidence. This is not the case. A letter marked without prejudice will only be privileged where it is written in the context of a genuine offer or negotiations to settle an existing or potential dispute.

7.5.3 Types of offers

The costs of a contested rent review can be substantial and it is therefore important for the corporate occupier to take steps to protect its position on costs which in turn may put pressure on the landlord (or a sub-tenant in a legacy situation) to settle.

There are three types of offer which can be made to settle a rent review arbitration/expert determination.

(a) An open offer

An open offer is not marked without prejudice and may be referred to by both parties at any stage of the third party determination process. A party will make an open offer if it wants the expert/arbitrator to know what a party was prepared to offer (or accept). This can be an effective tactic if a party wants to show its stance has been reasonable. It may also be relevant on the question of costs. Clear reference should be made to the letter being an open letter which can be shown to the expert/arbitrator.

(b) A without prejudice offer

This offer may not be referred to by either party at any stage of the process, even on the question of cost when the expert/arbitrator has made his award. A party who wishes to make an offer in hope that it will be accepted, but does not want the expert/arbitrator to know what it was prepared to offer (or accept) will make a without prejudice offer.

(c) A Calderbank letter/sealed offer

A party which is concerned that it will be prejudiced if the expert/arbitrator is aware of the offer before he makes his award, but nevertheless wishes to draw the offer to the attention of the expert/arbitrator on the issue of costs, can do so in two ways. First, it can do this by making a "sealed" offer which takes the form of an offer being sent to the other party, with a copy being sent to the expert/arbitrator in a sealed envelope. The expert/arbitrator then opens the envelope only after the award has been made. Second, and more usually, a Calderbank offer is sent to the other party which sets out the settlement proposal, including proposals as to costs, but which is headed "without prejudice save as to costs". The offer should either indicate that it remains open indefinitely for acceptance, or should be accepted within a defined period. After the award has been made, Calderbank letters can then be shown to the expert/arbitrator on the question of what costs award is appropriate (if any).

Rent Arrears

8.1 Introduction

Non payment of rent, service charges and insurance rent is a perennial problem facing landlords. Failing to deal promptly and effectively with rent arrears can result in a landlord incurring significant financial losses. A familiarity with the various options available and the factors which can influence the likelihood of a satisfactory outcome is important to avoiding or minimising any financial losses.

A corporate occupier can have a number of perspectives in relation to non payment of rent. It may be:

- a tenant that has failed (or perhaps is unable) to pay rent and is facing/defending action by the landlord or
- a guarantor, former tenant or guarantor of a former tenant which is being pursued following default by the current tenant or
- a landlord in a legacy situation where the premises are surplus and have been sub-let to offset the head rent (and as such is likely to be influenced by substantially the same considerations as a superior landlord) or
- a sub-tenant in circumstances where its immediate landlord is in financial difficulties.

Chapters 8, 9 and 10 are all inextricably linked and should be considered together. This chapter primarily focuses on the various courses of action available to a landlord and the likely factors which will influence its choice of action. It also considers default from the other perspectives which a corporate occupier may have and discusses what options are available in its capacity as tenant or sub-tenant.

Chapter 9 considers the vexed issues that arise when a tenant succumbs to a formal insolvency process and the likely consequences for a landlord. Chapter 10 then goes on to consider the plight of a corporate occupier with contingent liability, and from a different perspective in a legacy situation, the advantages (and in some cases disadvantages) of pursuing a third party with contingent liability in addition to, or instead of, the tenant in arrears.

8.2 The landlord's options — overview

As we have seen, one of the options available to a corporate occupier of surplus premises is to sub-let those premises for the whole or part of the remaining lease term. The sub-rent can be used to offset the corporate occupier's liability to pay rent under the head lease. If the sub-tenant fails to pay rent, swift action is required by the corporate occupier and/or its agents to ensure that rental blockage is unblocked as quickly as possible. In this section, the reference to landlord is to a corporate occupier in a legacy situation, and reference to tenant is a reference to a sub-tenant following a sub-letting by the corporate occupier. However, substantially the same principles would apply to a superior landlord in taking action against a corporate occupier.

When arrears arise, the first step should be to speak to the tenant and try to ascertain why the rent has not been paid. There are a number of possibilities and these may include the following:

- the rent is disputed or the tenant raises a potential counterclaim
- the tenant is a reluctant payer and is playing a cashflow game
- the tenant may be having cash flow difficulties which are unlikely to threaten its business and need time to pay
- the tenant's business may be failing
- the tenant's business may have failed.

The reason for non payment may have a significant bearing on the action which the landlord should take. For example, in the first instance, and in particular where it is thought that the tenant is likely to pay, the landlord may well decide that it is appropriate to allow further time for payment. Most well drafted leases will contain an interest provision which will enable the landlord to recover interest after a certain number of days have elapsed. Modern leases tend to provide 14 days grace and charge interest at 4% above base rate once the 14 days have elapsed until payment is received, with the calculation being from the due date. The tenant should be reminded of this clause and the fact that it will be liable for interest. This may encourage swift payment of the arrears.

In contrast, if the tenant appears to be heading for formal insolvency, quick action may be necessary to avoid a moratorium which is likely to restrict the landlord's remedies, particularly in light of changes to corporate insolvency legislation in recent years (see Chapter 9).

The options available to a landlord where a tenant has failed to pay rent are:

(1) forfeiture of the lease
(2) surrender
(3) service of a notice to elect on a liquidator/trustee in bankruptcy
(4) draw down from a rent deposit deed
(5) service of a section 6 notice on any sub-tenant
(6) distress
(7) sue for the arrears of rent and obtain a judgment
(8) service of a statutory demand and issuing winding up proceedings
(9) pursuing claims against third parties — contingent liability of guarantors, former tenants and guarantors of former tenants.

The most appropriate course of action will depend upon a combination of factors. These factors are likely to include:

- the tenant's reason for non payment and the prospect of the payment being made (albeit late)
- the solvency or otherwise of the tenant's business
- whether the premises are overrented

- whether the premises can be re-let easily
- whether the value of the premises is worth more with vacant possession than with an occupier in place. In the case of a freehold landlord, if the premises are ripe for redevelopment it may precipitate early possession to bring forward the redevelopment
- the likely vacant costs if the tenant is evicted (eg empty rates, service charge and insurance).

The options available to a landlord fall into a number of categories. A landlord will only adopt options (1)–(3) in circumstances where it has resolved to end the lease and go back into possession. If the premises are overrented, not likely to be re-let easily, or there is some other reason to preserve the lease (eg contingent liability of third parties with good covenant strength), then such options are unlikely to be commercially attractive to a landlord.

Options (4)–(6) are direct, swift and relatively cheap remedies which do not necessarily involve the termination of the lease. Indeed, (5) and (6) are inconsistent with the lease coming to an end and will constitute an act of waiver which will prevent a landlord from subsequently forfeiting the lease until a further sum falls due under the lease. In contrast, options (8)–(9) may involve slower, more expensive and to an extent uncertain litigation options. Very often a landlord's initial view is to issue court proceedings for the arrears; however, in most cases this is unlikely to be the most effective action.

It is worth considering each of the possible remedies in turn.

8.3 Forfeiture of lease for arrears of rent

8.3.1 Is it appropriate?

The ultimate sanction available to the landlord is to take steps to terminate the lease. Clearly the landlord needs to be satisfied that it is in its commercial interests to do so. There will be many a tenant locked into an overrented lease that would jump for joy if the landlord were to end its financial misery by forfeiting the lease on the grounds of rent arrears.

There may be other reasons for the landlord not taking steps to forfeit the lease. It may want to preserve the lease because the tenant is a good covenant, or there are former tenants or guarantors who are good for the money.

Forfeiture may also be inappropriate where the tenant is able to pay and is likely to make a court application for relief from forfeiture. In those circumstances there are likely to be cheaper, less complex and quicker remedies eg distress. If a tenant makes an application for relief promptly and pays the arrears, any interest (if claimed) and the landlord's reasonable legal costs, it will obtain relief from forfeiture, and the lease will be re-instated.

8.3.2 Legal requirements

There is no implied right of re-entry for breach of covenant and therefore forfeiture is dependant on an express forfeiture clause being contained in the lease. A typical forfeiture clause can be found at 11.3.3.

A well drafted forfeiture clause will usually stipulate:

- if any rent is outstanding for 21 days (this is often referred to as the grace period and can sometimes be 14 days)
- whether formally demanded or not

- or there is a breach of any other covenant by the tenant
- the landlord can re-enter without prejudice to any rights or remedies that have already accrued to the parties.

There are two regimes in place for forfeiture. In the case of forfeiture for rent arrears no prior written notice need be served on the tenant. However, if the breach of covenant relied upon is not rent arrears, the landlord must serve a notice on the tenant that complies with section 146 of the Law of Property Act 1925 (see 11.3.3).

If the landlord is seeking to recover arrears of other sums due under the lease (eg service charges, insurance rent), it is important to identify whether those sums are defined by the lease as rent or further rent, or the lease provides that such sums should be treated as rent. If insurance and services charges are not treated as rent then it will be necessary to serve a section 146 notice before the landlord can forfeit the lease for non payment of these items. If, however, these sums are to be treated as rent, no section 146 notice is required.

8.3.3 Waiver — has the right been lost?

It is critical that a decision whether or not to forfeit the lease is taken promptly. A landlord can lose its right to forfeit a lease if it is aware of a breach of covenant (in this case a failure to pay rent) and it takes some action which assumes the continuation of the lease. In short, on discovering a breach of covenant a landlord must choose either to:

- forfeit the lease or alternatively
- treat the lease as continuing.

If the landlord takes any action which treats the lease as continuing then it will have waived its forfeiture rights. Below are some examples of actions which are likely to constitute waiver.

- A demand or acceptance of rent which has accrued *after* the right of forfeiture as arisen and *after* the landlord becomes aware of the breach of covenant. This extends to the acceptance of rent on a without prejudice basis. However, rent received inadvertently (eg paid directly into the landlord's account) will not constitute waiver provided it is returned immediately.
- Instructing certificated bailiffs to distrain for rent arrears.
- Service of a notice to quit.
- Service of any notices under the lease eg *Jervis* v *Harris* notice (see 14.4.2).

The principle of waiver is often misunderstood in respect of rent arrears. It is sometimes argued that any demand for rent that is made after the 'grace period' (usually 14 or 21 days) constitutes an act of waiver. This is not necessarily the case and is best illustrated by the facts of *In Re Debtors Nos 13A10 and 14A10 of 1994* [1995] 2 EGLR 33

Facts

- T failed to pay rent due for the quarters commencing 1 September and 1 December 1994.
- The forfeiture clause provided a right for the landlord to re-enter in the event of rent being unpaid for 14 days after becoming payable.

- On 16 December L's solicitors wrote to the T stating that unless the outstanding rent was paid by a given date, "immediate steps will be taken to enforce payment".
- On 5 January 1995 L issued proceedings to recover the outstanding rent (ie *not* forfeiture proceedings).
- Judgment in default of defence was entered by L on 8 February.
- On 13 February L forfeit the lease by peaceable re-entry.
- T argued that the forfeiture was ineffective because L had waived its right to re-enter by:
  - demanding the December rent after the 14 day grace period (ie by the solicitors letter) and
  - issuing proceedings and obtaining a judgment for the arrears.

Held

- L's right to forfeit in respect of the September rent had been waived by the demand for the December rent, which was only consistent with the lease continuing after the September breach of covenant.
- However, L's demand for the December rent, outside the 14 day grace period, and the subsequent issuing of proceedings for the December rent, did not amount to a waiver.

When a quarter's rent falls due and is unpaid within the 14 (or 21) day grace period, a landlord will not, therefore, waive its right to forfeit the lease by demanding that rent. The key distinction is set out below.

- If the rent demanded accrues *on or before* the right to forfeit (eg where it is the breach which founds the right to forfeit), the demand will not amount to a waiver. Demanding rent that has already accrued is not inconsistent with the landlord forfeiting the lease.
- However, if the rent demanded accrues *after* the breach — eg where rent is demanded for the June quarter when the March quarter is still outstanding — the right to forfeit will be waived. The service of the June demand is only consistent with the continuation of the lease. Of course, a new right of forfeiture may arise if the tenant fails to pay the June quarter within the 14 day grace period.

When an act of waiver has occurred, because non payment of rent is a once and for all breach the landlord will have to wait until a further rental instalment falls due before it can forfeit the lease.

It is sometimes suggested that when a tenant is in arrears with a number of different instalments of rent, service charge or insurance rent, the landlord can choose which of the outstanding debts to apply a payment to. This is not necessarily so and ignores the doctrine of appropriation which was applied in the Court of Appeal case of *Thomas* v *Ken Thomas Ltd* [2007] 01 EG 94. The principles can be summarised by the following propositions.

1. A tenant can appropriate (ie allocate) a particular payment to a specific debt by notifying the landlord at, or shortly before, the payment is made.
2. If the tenant exercises this right, the landlord must either accept the appropriation and allocate accordingly, or, return the money promptly indicating that it is not prepared to accept the tenant's appropriation.
3. The landlord cannot accept the tenant's money and purport to allocate the payment to a different debt (eg to an earlier rental instalment so as to avoid the suggestion it has waived its right to forfeit). If the payment is not returned, the landlord will be stuck with the tenant's appropriation.
4. Only if the tenant does not expressly allocate the payment to a specific debt can the landlord choose which debt to apply the payment towards.

In *Thomas* v *Ken Thomas Ltd* the tenant indicated that a payment should be applied to rent due in December 2004. The landlord, contrary to this contention, sought to apply the payment to rent due in November 2004 and then forfeit for non payment of the December 2004 rent. The Court of Appeal held that the landlord could not appropriate the payment in this way and by not returning the payment (intended by the tenant to pay subsequent rent) within a reasonable time; the landlord had waived its right to forfeit for the November rent.

8.3.4 What must the landlord do?

There are two possible ways of forfeiting a lease:

- by the issue and service of court proceedings seeking forfeiture or
- by peaceable re-entry of the premises.

(a) Forfeiture by court proceedings

By far the safest course of action is for the landlord to issue forfeiture proceedings. The service of the court proceedings constitutes the actual re-entry, although there is inevitably a period of uncertainty (often referred to as the twilight period) between the service of the proceedings and the final court hearing. During the twilight period it will be unclear whether the forfeiture will be confirmed at the hearing or not.

Unless there are exceptional circumstances, the proceedings should be issued in the county court for the district in which the premises are located. A claim will only be appropriate in the High Court where there is a complicated dispute on the facts, and/or there are points of law of general importance (CPR PD 55 paragraph 1.3).

The disadvantages of forfeiting by court proceedings are:

- they can be slow (particularly if the tenant defends or applies for relief from forfeiture)
- they can be expensive (and even if the court awards the landlord its costs there is no guarantee that the landlord will be able to enforce a costs order)
- the landlord cannot collect rent for the duration of the proceedings. The landlord will usually include a claim for mesne profits from the service of proceedings until the date when possession is obtained. However, whether the landlord will be able to enforce a money judgment after the tenant has left is another matter. In lengthy cases it may possible to apply to the court for an interim award of damages to cover either rent, or damages for trespass, until the final hearing.

On the question of speed, prior to the Civil Procedure Rules coming into force it used to be possible to obtain a judgment in default after 14 days if the tenant failed to serve an acknowledgement of service. Unfortunately, under the Civil Procedure Rules this is not now possible. When the proceedings are issued the case will be listed for a hearing before a district judge. The delay will vary from court to court depending how busy that particular court is. It is not uncommon for there to be a delay of four to six weeks between the issue of proceedings and the hearing date.

If the proceedings are undefended and the tenant does not apply for relief from forfeiture, an order for possession is likely to be made at the first hearing. If the tenant defends the proceedings or applies for relief, the case is likely to be adjourned on directions for the preparation for trial. In those circumstances it could take many months for the case to be determined. Whilst there is a requirement on the tenant to file a defence before the first court hearing, the court will still permit a tenant to take part in the hearing even if it has failed to serve a defence.

If the landlord considers that a tenant's defence is without merit, to try and avoid expensive and protracted litigation, it can seek to obtain summary judgment under CPR Part 24.

(b) Forfeiture by peaceable re-entry

The alternative to litigation is for the landlord (or more usually its agent) to peaceably re-enter the premises. Peaceable re-entry can be effected by:

- changing the locks and posting a notice on the premises indicating the landlord has re-entered (it is also prudent to send a letter to the tenant or its agent — see 8.15.1 for a suggested precedent letter)
- granting a new lease to a third party (*London & County* v *Wilfred Sportsman* [1971] Ch 764)
- granting a new lease to an existing sub-tenant, but not allowing a sub-tenant to remain in occupation on the basis of the existing sublease (see *Ashton* v *Sobleman* [1987] 1 WLR 177).

The landlord is entitled to use reasonable force in exercising its right but this is restricted to forcing an entry when the premises are empty. It is a criminal offence to use violence to gain entry to a building if there is someone in the premises that is opposed to the entry (section 6 of the Criminal Law Act 1977). It is for this reason that most landlords who re-enter will instruct certificated bailiffs who are well versed in, and insured against, the practical difficulties that may arise.

If the premises are still occupied by a trading tenant, to avoid problems it usually makes sense for the re-entry to take place late at night or early in the morning when the premises are likely to be empty. The tenant does not have to be warned before hand of the landlord's intention to re-enter and indeed in most cases this will be inappropriate. It may lead the tenant to take steps to oppose the re-entry eg by arranging an around the clock vigil at the premises.

There is no doubt that forfeiture by peaceable re-entry is a very potent remedy. If the tenant accepts the forfeiture, it can be a very quick, cheap and effective answer to rent arrears.

However, it can be regarded by the courts as a high handed and aggressive tactic and where the tenant is still trading from the premises, it is not without its risks.

- The landlord is likely to be confronted with the tenant's belongings. If the premises are offices, this is likely to include office furniture, computer hardware, filing cabinets containing documents and the like. The landlord will become the 'involuntary bailee' of the tenant's goods and will immediately become responsible for their safe keeping. The landlord will not be free to dispose of these goods as it wishes, or to sell them with a view to offsetting the sale proceeds against any outstanding sums due under the lease. On the contrary, it will need to make an inventory of the items found and carefully document (and possibly photograph) all items so as to have some protection should the tenant bring a claim for items which it subsequently alleges were missing or damaged. If the goods are not collected by the tenant it may be necessary for the landlord to serve a notice under the Torts (Interference with Goods) Act 1977.
- If the landlord gets it wrong and the tenant has a valid defence, the re-entry will be unlawful and the landlord could face an application for an injunction and/or a claim in damages for trespass.

A claim in damages could include a claim for any losses caused by the interruption of the tenant's business and could therefore be substantial.

- The tenant will still be entitled to make an application to the court for relief from forfeiture. This can lead to a period of uncertainty.

A careful landlord is likely to opt for forfeiture by the issue of court proceedings. However, if the landlord is not risk averse, or alternatively those risks are not judged to be significant, peaceable re-entry can be the quickest and most cost effective solution. A good example is where the tenant has already vacated and the premises are empty. In those circumstances the risks identified above may well be minimal.

8.3.5 What is the effect of forfeiture?

The effect of a lawful forfeiture in circumstances where the tenant either fails to make an application for relief, or alternatively, that application is unsuccessful, is to terminate the lease for all time and for all purposes. Consequently, all future liability under the lease comes to an end on the date of re-entry. This has the consequence of not only releasing the tenant from future liability under the lease, but also discharging any guarantors or former tenants from contingent liability. However, the tenant, former tenant or guarantors may well be liable for any existing breaches of covenant at the date of re-entry.

By way of example:

- If a landlord peaceably re-enters or issues a claim form for possession on 25 June, assuming rent is payable on the usual quarter days in advance, the tenant will be liable for the full June quarter's rent. Any guarantor of the tenant will also be liable for this sum.
- A former tenant or guarantor of a former tenant may also be liable for this sum (see Chapter 10).
- However, the tenant, guarantor, former tenant and its guarantor will not be liable for the September quarter's rent or any sums that would have become payable after 25 June had the lease not been forfeit.

The release of contingent liability on forfeiture is occasionally overlooked. Given that forfeiture of the lease will let any party with contingent liability off the hook, it is clearly an important factor for the landlord to take into consideration when considering what action to take. There will be occasions where a landlord will want to preserve a lease because of the covenant strength of those with contingent liability. Investor landlords will often be prepared to pay significant premiums for freehold premises with tenants backed by substantial plc guarantees.

8.3.6 Relief from forfeiture in rent default cases

The courts have always viewed the landlord's right to re-enter as security for its rent and have granted relief as a matter of course where the landlord is able to be put in the same position as before the breach, and provided that all outstanding rent and legal costs are paid. The rules concerning relief from forfeiture for non payment of rent vary according to which form of re-entry is used by the landlord.

(a) County court proceedings

If the landlord has issued county court proceedings the position is as follows.

- A tenant can obtain automatic relief from forfeiture by paying into court, not less than five clear days before the possession hearing, all arrears together with the landlord's costs (section 138(2) of the County Courts Act 1984). If the payment is made in accordance with section 138(2) the proceedings will cease and the lease will be re-instated without the need for a new lease.
- If no payment is made before the possession hearing, and the judge is satisfied that the landlord is entitled to enforce the right of re-entry, the court *must* make an order for possession. The order for possession must take effect at the end of a period specified by the court (being not less than four weeks from the date of the court order) unless the tenant (or subtenant) pays into court all of the arrears and costs within that period (section 138(3) of the 1984 Act). In other words, the tenant (or subtenant) will have a further opportunity to obtain automatic relief from forfeiture provided all outstanding sums are paid into court within the period set by the court.
- Importantly, when a court makes an order for possession subject to the automatic relief from forfeiture provision, the arrears which are to be paid by the tenant must be the arrears at the date of the hearing and not the date of the service of the proceedings (see *Maryland Estates Ltd* v *Bar-Joseph* [1998] 2 EGLR 47). The point here is that the arrears may have increased from the date of service of the proceedings and it would be very inconvenient to a landlord if relief were granted even though sums were still due under the lease.
- If a possession order is made and the tenant (or subtenant) fails to pay the rent and costs into court within the period specified by the judge then their automatic rights to relief from forfeiture will be lost.
- *However*, the tenant will be entitled to issue a court application for relief from forfeiture at any time within six months from the date on which the landlord actually recovers possession (section 138(9A) of the 1984 Act).
- By the same token a subtenant is entitled to apply for a vesting order under section 138(9)(C) of the 1984 Act at any time within six months of the landlord actually recovering possession.

(b) High Court proceedings

Slightly different rules apply to relief from forfeiture if the landlord's proceedings are issued in the High Court. Under the Civil Procedure Rules most forfeiture cases will now be dealt with in the county court and not the High Court. However, importantly, even if the case is dealt with in the High Court, Section 210 of the Common Law Procedure Act 1852 imposes a limitation period of six months from the date of execution of a possession order on an application for relief from forfeiture.

(c) Peaceable re-entry

If the landlord forfeits by peaceable re-entry, the statutory six month period referred to above does not apply. In theory the tenant can apply under the courts equitable jurisdiction at any time. However, the longer the tenant delays, the more difficult it will be to succeed with an application for relief.

The effect of the various opportunities for a tenant or sub-tenant to apply for relief from forfeiture (or a vesting order in the case of a sub-tenant) means that there is an element of uncertainty with forfeiting a lease for non payment of rent. It can raise difficult practical issues: if forfeiture is implemented by court proceedings, does the landlord have to wait six months before reletting to rule out the possibility of a tenant applying for (and obtaining) relief? How long must the landlord wait if it peaceably re-enters given that there is no six month limitation? The other complicating factor is what might need to be done to the premises to bring them up to a suitable standard. If extensive works are needed then the landlord not only needs to remove the tenant's furniture and equipment but strip out partitions. It is essential therefore to try and get finality on potential re-occupation.

(d) Uncertainty

Although the tenant's rights to apply for relief give rise to uncertainty, in practice the risks involved can be minimised. The obvious concern is that if a landlord retakes possession (either after a court order or following a peaceable re-entry), any re-letting could be unravelled by a court subsequently granting relief from forfeiture to the original tenant. The new tenant would have to vacate the premises and may bring a claim against the landlord for derogation from grant or breach of contract. It is possible to avoid/minimise such a risk by the following.

- Write an open letter to the tenant immediately after re-entry, indicating that it is the landlord's intention to re-let and inviting the tenant to apply for relief from forfeiture promptly (say within 14 or 28 days) if it intends to do so. A precedent letter can be found at see 8.15.1. This precedent can be adapted if a forfeiture judgment has been obtained.
- Fourteen to 28 days later a further letter should be written confirming that as no application for relief has been made by the tenant, the landlord intends to proceed with a new-letting and that this correspondence will be shown to the court if a subsequent application is made.
- If the tenant does subsequently make an application for relief after a new lease has been granted, the previous correspondence can be drawn to the judge's attention. A judge has discretion whether to grant relief or dismiss the application. If a landlord, acting reasonably, re-lets premises after it has re-taken possession, but before an application for relief is made, a court is very unlikely to grant relief (see *Silverman* v *AFCO (UK) Ltd* [1988] 1 EGLR 51). This is particularly the case where the landlord has written to the tenant inviting it to make an application within a specified period and notified the tenant of its intentions to re-let if no application is made.

Despite the points made above, there can be no doubt that one of the principle drawbacks to using forfeiture for non payment of rent is the ease by which the tenant can obtain relief from forfeiture. As can been seen above, if the rent and costs are paid, automatic relief will be granted. The justification is that the landlord will not be out of pocket. However, this overlooks:

- the fact that the landlord is unlikely to recover all of its actual costs (if the costs go to an assessment hearing, the landlord is unlikely to recover more than 50%–70% of its actual costs)
- the management time involved pursuing forfeiture and
- that a tenant is entitled to automatic relief even if it has an appalling payment history.

8.3.7 Restrictions on insolvency

If the tenant's business is in the process of failing, or has failed, there may be restrictions on the landlord's right of forfeiture. These restrictions are imposed by the Insolvency Act 1986 (as amended) and are dealt with in detail in Chapter 9.

The position can be summarised in the following table.

Insolvency procedure	*Peaceable re-entry*	*Forfeiture by proceedings*
IVA — pre approval	If the court has made an interim order, not without the prior permission of the court (s 252 (2) (aa) of the 1986 Act)	If the court has made an interim order, not without the prior permission of the court (s 252 (2) (b) of the 1986 Act)
IVA — after approval	No (bound by IVA)	No (bound by IVA)
IVA — new debt post approval	Yes	Yes
CVA — pre approval	Yes unless a small company and moratorium applied for, in which case not without the court's prior permission (para 12(1)(f) of schedule A1 to the 1986 Act — substituted by the 2000 Act	Yes unless a small company and moratorium applied for, in which case not without the court's prior permission (para 12(1)(h) of schedule A1 to the 1986 Act
CVA — after approval	No — The pre CVA arrears are converted into rights under the CVA	No — the pre CVA arrears are converted into rights under the CVA
CVA — new debt post approval	Yes (see *Thomas* v *Ken Thomas Ltd* [2007] 01 EG 94)	Yes (see *Thomas* v *Ken Thomas Ltd* [2007] 01 EG 94)
Administration	Not without the prior permission of the administrator or the court (para 43(4) of schedule B1 to the 1986 Act)	Not without the prior permission of the administrator or the court (para 43(6) of schedule B1 to the 1986 Act)
LPA receivership	Yes	Yes
Administrative receivership	Yes	Yes
Voluntary liquidation	Yes	Yes but a court can restrain/stay any proceedings if an application is made to the court (s 112 of the 1986 Act)
Compulsory winding up	Probably yes. However, arguable that "no action or proceedings ... against the company or its property" may extend to peaceable re-entry (see *Re Memco Engineering* [1986] Ch 86)	No At any time after a winding up order has been made or a provisional liquidator appointed a party must obtain the prior permission of the court to commence or continue court proceedings (s 130 1986 Act)
Bankruptcy	Yes (*Razzaq* v *Pala* [1997] 2 EGLR 53)	Yes, according to *Ezekiel* v *Orakpo* [1977] 2 EGLR 47, CA. This decision has, however, been criticised in subsequent cases. It may be argued that a forfeiture action should not be issued without the permission of the court.

8.4 Surrender

A further radical option open to a landlord is to accept a surrender of the lease. This is the consensual equivalent to forfeiting the lease, and as such, the points made in 8.3.1 above concerning whether it is in the landlord's interest to terminate the lease apply. The requirements of a valid surrender can be found at 11.3.6. Suffice it to say that a surrender can only arise if both the tenant and the landlord agree that the lease should come to an end by surrender.

This option may be particularly attractive if the tenant has terminal financial problems and is heading towards, or is already in the clutches of, formal insolvency procedures (see Chapter 9). Its great virtue is speed. If both parties agree the lease can be terminated immediately and with the minimum of fuss, allowing the landlord to cut its losses and market the property quickly.

Nowhere is this more useful than when a tenant has gone into administration or liquidation and where the landlord may have technical impediments to simply re-entering. If the administrator/ liquidator can be persuaded to surrender the lease quickly, the landlord can avoid what could otherwise be a costly and protracted procedure to obtain the courts permission to forfeit the lease. It will also remove the uncertainty caused by the possibility of a subsequent application for relief from forfeiture.

If the parties agree that there should be a surrender, it is always prudent to formalise this with a deed of surrender (see 11.3.6 (b)). Furthermore, if the parties are considering entering into an agreement for surrender care must be taken to ensure that the agreement is not struck down, and its obligations rendered unenforceable, by section 38A of the Landlord and Tenant Act 1954 (see 11.3.6(b)).

8.5 Service of a notice to elect

If the tenant has been made bankrupt or is in liquidation, the landlord can serve a notice to elect on the trustee in bankruptcy or the liquidator requiring him to decide whether to disclaim the lease or not. This can be a powerful tactic, which if successful, avoids the need to apply to court for consent to forfeit (see Chapter 9). A copy of a precedent notice to elect can be found at 8.15.2.

Again, as with forfeiture and surrender, care needs to be taken when deciding whether serving a notice to elect is in the landlord's commercial interests. The position in relation to former tenants and guarantors is unaffected by a disclaimer, but the landlord will have to think carefully before deciding to retake possession following a disclaimer (which does have the effect of determining third party liabilities under the disclaimed lease — see 10.7.2 (c)).

8.6 Draw down from rent deposit/bank guarantee/bond

A landlord may attempt to pre-empt a tenant default by insisting on a rent deposit, bank guarantee or a bank bond. If these forms of security are available then one option open to a landlord when a tenant defaults is to drawdown from the deposit or call on the bank guarantee/bond.

In the case of a bank guarantee or bond, the amount secured is likely to be restricted to tenant default up to a certain limit. The specific requirements which must be fulfilled before the landlord can call on a bank guarantee or bond will be contained in the document which created the security and must be considered carefully by the landlord or its advisors. Any trigger requirement must be satisfied before payment becomes due. It is also important to identify precisely what type of tenant default is covered. In some cases this may be restricted to the payment of rent only (and may not extend to service charges or insurance rent).

Rent deposit deeds take many forms and it is unwise to make general statements about how they will operate in practise. Their drafting can be very complex and thorough, dealing with most eventualities (eg assignment by the landlord or formal insolvency of the tenant) but often this is not the case. Difficult issues can therefore arise in circumstances which were not catered for in the rent deposit deed.

A landlord should therefore obtain detailed legal advice on the provisions of a specific rent deposit deed before attempting to draw down. The key points to consider are:

- What types of tenant default are covered by the deed?
- What trigger requirements are required before payment is due? Does the rent deposit deed require a formal notice to be sent by the landlord to the tenant prior to drawdown?
- Can the landlord require the tenant to 'top up' the deed after a valid drawdown?
- If the tenant fails to top up the deed within the required time limit does this give rise to a fresh ground for forfeiture?
- If there has been an assignment of the reversion by the original landlord (and signatory to the rent deposit deed) can the new landlord drawdown from the rent deposit deed or should the deposit have been returned to the tenant on assignment?
- Can the landlord drawdown if there has been a change of tenant?

A common issue that arises is whether a landlord can drawdown from a rent deposit deed on insolvency. This can be an extremely difficult issue to unravel. It will depend upon the wording of the deed and the mechanics of how the deposit is held. It will also depend on the wording of the various moratoriums. Often compelling arguments can be constructed for both sides. If the money is within the control of the landlord it may have the upper hand (eg if it is held by the landlord's agent or solicitor in the landlord's name). One approach is to serve any notice required, drawdown and see what response is received from the administrator, trustee in bankruptcy or liquidator, as the case may be. This will put the onus on the tenant to take action and challenge the legitimacy of the drawdown. However, it would be unwise of the landlord to spend the money until the storm has passed in case it becomes necessary to transfer the money back.

A rent deposit entered into by a company must be registered under section 396 of the Companies Act 1985. Failure to do so will result in the rent deposit deed being void.

Insolvency issues aside, the principal advantages of drawing down from a rent deposit deed are that it is cheap, simple and results in a quick recovery of money. It is also a remedy that may be used in conjunction with other remedies. Care must be taken if used in conjunction with forfeiture. If a drawdown is made from the rent deposit deed this may constitute an act of waiver. However, the rent deposit deed may stipulate that a failure to top up the rent deposit after a drawdown of money and a request for replacement funds, constitutes a fresh right to forfeit. Consequently, a landlord could drawdown for a failure to pay a quarter's rent, require the tenant to top up, and if the tenant fails then forfeit the lease on this ground.

The downside is that once the money has gone from the rent deposit, unless it is replaced by the tenant, it is no longer available to cover future liabilities eg dilapidations.

8.7 Service of a section 6 notice on any sub-tenant

If the tenant is in arrears of rent and there is a sub-tenant, the landlord can require the sub-tenant to pay rent direct to the head landlord unless and until the arrears have been discharged. This is not achieved by contract (because there is no contractual relationship between head landlord and tenant) but by section 6 of the Law of Distress (Amendment) Act 1908.

The head landlord exercises this right by serving a notice on the sub-tenant requiring the sub-tenant to pay all future rent directly to the superior landlord until the tenant's arrears are settled. A precedent notice can be found at 8.15.3.

The key points to note about this procedure are set out below.

- There is no prescribed form for the notice which must be served on the sub-tenant. However this must specify the amount of arrears owed by the tenant to the head landlord, and must also require the sub-tenant to pay rent direct to the head landlord until these arrears have been discharged.
- Section 3 of the 1908 Act provides that immediately after the service of a valid notice, the relationship of landlord and tenant arises between head landlord and sub-tenant. Consequently, if the sub-tenant fails to pay future rent to the head landlord, it can seek to recover rent direct from the sub-tenant using many of the remedies outlined in this chapter.
- The effect of payment by the sub-tenant to the head landlord in response to a valid notice is to reduce (or extinguish) the sub-tenant's liability to pay rent to the tenant (ie its immediate landlord). The tenant has no right of action against the sub-tenant in respect of money paid to the head landlord under section 6.
- If the tenant's failure to pay rent continues, it will be necessary to serve further section 6 notices on the sub-tenant on each occasion when the payment of subrent has discharged the headlease arrears as specified in the original notice. Consequently, while this can be a quick and effective remedy, it is by no means an ideal long term solution and may involve the service of a number of section 6 notices. There will also inevitably be a time lag between head rent arrears becoming due and sub-rent payments being diverted to the landlord.

Many commentators suggest that this remedy remains available notwithstanding a tenant succumbing to a formal insolvency procedure. If that is correct then this is a very potent remedy in the landlord's armoury. However, this proposition has been challenged by some commentators. The summary of the current position set out below is, pending clarification by the courts, necessarily tentative:

Insolvency procedure	*Service of section 6*
IVA — pre approval	Yes
IVA — after approval	No — the landlord will be bound by the terms of, and entitled to the dividend under, the IVA (in substitution for rent)
IVA — new debt post approval	Yes
CVA — pre approval	Yes (unless a small company) If a small company — arguably yes, but query whether the court's prior permission is required. This depends whether a notice to redirect rent constitutes a step to enforce ... security over the company's property (para 12(1)(g) of schedule A1, pt III to 1986 Act as amended)

Insolvency procedure	*Service of section 6*
CVA — after approval	No — the landlord will be bound by the terms of, and entitled to the dividend under, the CVA (in substitution for rent).
CVA — new debt post approval	Yes (see *Thomas* v *Ken Thomas Ltd* [2007] 01 EG 94)
Administration	Arguably yes, but query whether the prior permission of the administrator or the court is required. This depends whether the service of the notice is a "legal process (including ... execution, distress and diligence) ... against the company or property of the company" (para 43(2) and 43(6) of schedule B to the 1986 Act)
LPA receivership	Yes
Administrative receivership	Yes
Voluntary liquidation	Yes
Compulsory winding up	Arguably yes, but query whether the court's prior permission is required. This depends whether service of the notice is "action or proceedings against the company or its property" (s130 of the 1986 Act)
Bankruptcy	Yes

8.8 Distress

Distress is an ancient self help remedy. It entitles a landlord which is owed rent to enter the leased premises, seize goods which it finds there, and secure those goods (impounding). If the tenant fails to pay the rent within five days (or 15 if the tenant requests an extension), the landlord is entitled to sell the goods and allocate the proceeds towards settlement of the arrears.

The remedy of distress can be a very potent remedy in the armoury of a landlord owed rent by its tenant. Unfortunately its antiquity means that the legal rules which have developed are complex. Like most self help remedies it tends to be frowned upon by the courts. There have also been mutterings about whether the remedy complies with the Human Rights Act 1998. It is therefore perhaps of no great surprise that the Tribunals, Courts and Enforcement Act which received royal assent in July 2007 will abolish distress in its current form and replace it with a different system when this 2007 Act comes into force.

8.8.1 Is distress appropriate?

In the right kind of arrears case, levying distress can be quick, effective and cheap. Indeed, in most cases, it is a remedy that can be carried out entirely at the tenant's expense.

However, there are some circumstances where it will not be appropriate.

- The demised premises are empty and there are no goods to seize. A key feature of the law of distress is that only goods found on the premises demised by the lease under which the arrears are due can be seized. A landlord is not entitled to seize goods from other premises of the tenant.
- Arrears are disputed or where the tenant has raised a valid counterclaim.

- There are service charge arrears. Distress should only be used for the arrears of rent in the narrow sense: ie the payment for occupation of premises under a lease. Although it may be possible to recover service charges if they are reserved "as though they were rent", this can be legally complex and in reality it is much more likely that such sums could be disputed and the landlord may find itself facing a complicated and expensive claim for damages for illegal distress.
- The landlord is still considering forfeiture. Levying distress constitutes a waiver of a landlord's right of re-entry.

8.8.2 When can it be used?

A landlord is only entitled to distrain where:

- there is a subsisting lease — distress cannot take place after forfeiture (eg where forfeiture proceedings have been issued but have not yet been determined by the court) and
- there are arrears of rent and
- those arrears are not disputed and
- there are no statutory restrictions preventing it (eg under the Insolvency Act 1986 — see 8.8.5 below). Distress cannot be levied against premises let to the Crown or those that are Crown protected.

8.8.3 Implementation

The landlord should instruct a certificated bailiff to carry out the process. The documents which a bailiff will require will vary according to which bailiff is used. However, it is good practice to provide the bailiff with a copy of the lease and an up to date statement of arrears (including interest) if appropriate.

Generally the process will be as follows:

1. The bailiff must gain access to the demised premises. He must not use force on, or break open, any outer door. However, once inside, he is entitled to use force on any inner door or lock.
2. Once inside, the bailiff will usually make an inventory (and possibly photograph) all goods he wishes to seize up to the estimated level of the arrears plus his fees. In most cases the seized goods will not actually be removed. There will be a constructive seizure. The bailiff will leave a list of the goods seized with the tenant (or at the property) together with a note that the goods should not be removed until the rent is paid. Alternatively the tenant may sign a 'walking possession' agreement used by the bailiff.
3. To enable the landlord to sell the goods, he must serve a notice of distress on the tenant. This notice must identify the goods seized, the amount of the arrears and the time when the goods will be sold. The landlord must wait at least five clear days before selling the goods (this excludes the day of seizure and the day of sale).
4. The tenant can obtain an extension of not more than 15 days by a written request to the landlord provided it gives additional security for any additional costs that might be incurred by the delay.
5. If the tenant fails to pay (i.e. the rent, interest and bailiffs costs), the goods can be sold. The landlord will keep the arrears of rent and costs. The remainder (if any) will be kept for the owner of the goods sold.

8.8.4 Goods not owned by the tenant

A landlord is, at common law, entitled to seize all goods found on the demised premises irrespective of ownership. If the seizure is legal and the goods sold, the actual owner of the goods may be entitled to an indemnity from the tenant. It will not, however, invalidate the seizure. There is a little used procedure under the Law of Distress Amendment Act 1908 which allows a third party to serve a statutory declaration on a bailiff/landlord after seizure but before sale, but this is a complex and expensive procedure and unlikely to be of any real assistance in respect of relatively low value goods.

There are some goods which cannot be the subject of distress. Certificated bailiffs will be familiar with these categories.

8.8.5 Insolvency restrictions

A landlord's right to distrain for rent arrears is subject to the following restrictions if the tenant falls into formal insolvency.

Insolvency procedure	*Can L levy distress?*
IVA — pre approval	No — if an interim order is made, for the duration of the interim order no distress may be levied against the debtor's property (s 252(b) of the 1986 Act inserted by the Insolvency Act 2000)
IVA — after approval	No — the landlord will be bound by the terms of, and entitled to the dividend under, the IVA (in substitution for rent).
IVA — new debt post approval	Yes
CVA — pre approval	Yes (unless a small company). Small company — No — not without the courts prior permission during the moratorium period (para 12(1)(h) of schedule A1, pt III to 1986 Act as amended)
CVA — after approval	No — the landlord will be bound by the terms of, and entitled to the dividend under, the CVA (in substitution for rent).
CVA — new debt post approval	Yes
Administration	No — not without the prior consent of the Administrator or the permission of the court (para 43(6) of schedule B to the 1986 Act)
LPA receivership	Yes
Administrative receivership	Yes
Voluntary liquidation	Yes — however the liquidator may apply to the court to try and prevent any distress (see *Herbert Berry Associates Ltd* v *IRC* [1978] 1 All ER 161)
Compulsory winding up	Depends upon when distress commenced. 1. If distress commenced and then a petition is presented, an application can be made to the court to stay the distress — s126 of the 1986 Act. If no application is made, L can complete the process (provide this is completed before a winding up order is made). See also *Re Memco* [1986] Ch 86. 2. If distress is commenced after a petition has been presented to the court, the distress will be void (ss 128(1) and 129 of the 1986 Act.

Insolvency procedure	Can L levy distress?
	3. Once a winding up order has been made, the landlord can apply to the court for permission to effect distress (s130 of the 1986 Act). If the rent fell due before the winding up, the court is unlikely to grant permission. However, if the rent has fallen due during the liquidation, the courts permission is likely to be granted, or at least if it is refused, it will be on condition that the liquidator pays ongoing rent (see *Re Atlantic Computer Systems plc* [1992] 1 All ER 476).
Bankruptcy	Yes (s 347 of the 1986 Act — but in relation to pre bankruptcy arrears, restricted to six months rent)

It will be clear from most of the statutory restrictions identified above, that the key to achieving a successful outcome is for the landlord to act quickly if insolvency is suspected.

8.9 Sue for arrears of rent and obtain a judgment

8.9.1 When is a monetary claim appropriate?

A landlord that is owed arrears of rent can issue court proceedings with a view to obtaining a monetary judgment for the arrears, interest and legal costs. A judgment can then be enforced against the tenant, usually by instructing a court bailiff to seize goods to the value of the judgment. The key difference between a seizure of goods based upon a judgment debt, and a seizure of goods by way of distress, is that when enforcing a judgment debt the bailiff can seize goods from the tenant's premises anywhere in England and Wales and is not restricted to a seizure at the leased premises.

Alternatively, if seizure of goods is not appropriate (perhaps because there are unlikely to be sufficient goods to satisfy the judgment), a landlord can take steps to obtain a charging order against any freehold property that the tenant owns in England and Wales.

A simple claim for arrears tends to be used when other remedies are unsuitable, for example where:

- forfeiture has been discounted because the landlord wants to preserve the lease eg where there is a guarantor or former tenant, the lease is overrented and it would be difficult to secure a new tenant and/or the tenant is a good covenant
- there is a dispute about rent or the tenant has a counterclaim and is withholding rent
- there are no goods on the leased premises which would make a seizure by way of distress viable
- there is no subtenant on whom a section 6 notice can be served
- there is no rent deposit deed, or if there is, the landlord wishes to reverse this to cover an anticipated dilapidations claim.

A claim for arrears can also be brought either in conjunction with forfeiture proceedings, or alternatively, if the landlord peaceably re-enters, after the premises have been repossessed. In the former case, the landlord will seek a judgment for arrears up to the date of the service of proceedings, and a claim for mesne profits after that date up to the date repossession is secured. In the latter case, a landlord will clearly want to be satisfied that it knows the former tenant's whereabouts and that it is worth pursuing.

The landlord will need to issue a claim form and particulars of claim and then serve these documents on the tenant (this can be done by the court or by the landlord). The tenant then has 14 days to lodge an acknowledgment of service, and provided it does so, a further 14 days to lodge its defence and/or counterclaim.

If the tenant does not serve on the landlord an acknowledgment of service within 14 days, and (if that is done) a defence and/or counterclaim within 28 days, the landlord can enter judgment in default (without the need for a hearing). However, if a defence is served, the court will give directions for this matter to go to full trial, which could be many months after the issue of proceedings. The only way a landlord can shortcut this process is if it makes a successful application to the court for summary judgment where the tenant has filed a questionable or holding defence only. This will require the landlord to satisfy the court that the tenant has no realistic prospects of succeeding in its defence and that there is no other reason why the matter should go to full trial. However, even a summary judgment application can take four to eight weeks to come before a court depending how busy the relevant court is.

8.9.2 Combined with dilapidation claims

If the tenant is worth pursuing for post termination claims, it may be sensible to combine the claim for arrears of rent with a dilapidations claim. The decision whether to do so will turn on the amounts outstanding and the likelihood of the dilapidations claim being settled quickly. If the arrears are modest, it may be sensible to tag the arrears claim on to a much bigger dilapidations claim and avoid the additional irrecoverable costs of two sets of proceedings. However, if the arrears are significant, a landlord may want to secure the outstanding rent quickly and not have it delayed as part of a lengthy dilapidations case. If the landlord brings a separate claim for the arrears, it may be able to obtain a judgment in default or alternatively apply for summary judgment of the arrears if the tenant puts forward a questionable or holding defence. These possibilities are unlikely if the arrears are included with a dilapidations claim.

8.9.3 Drawbacks

If a tenant does not defend a claim, a landlord can obtain a judgment fairly quickly. However, if the reason for the proceedings is that the arrears are disputed or there is a counterclaim which the tenant is entitled, as a matter of law, to offset, then the proceedings could become protracted and expensive with both parties incurring significant legal costs. If a serious dispute does arise, and the landlord does not intend to forfeit the lease, the parties should consider mediating the dispute.

Furthermore, even if a judgment is obtained, further expense will be incurred seeking to enforce a judgment and there is no guarantee that a successful recovery will be achieved. In many cases, issuing court proceedings for a judgment will be a slow and costly procedure and should not be the remedy of first choice for the landlord.

A further drawback is that because defended proceedings can be protracted, circumstances may change the during the litigation process which make a recovery less likely, or impossible. A good example is if the tenant succumbs to a formal insolvency procedure.

8.9.4 Insolvency restrictions

The table below sets out in what circumstances a landlord is entitled to issue proceedings for an arrears judgment where the tenant is subject to a formal insolvency procedure.

Insolvency procedure	*Court proceedings for judgment*
IVA — pre approval	If an interim order is made by the court — not without the prior permission of the court (s 252 of the 1986 Act)
IVA — after approval	No. The landlord will be bound by the terms of, and entitled to the dividend under, the IVA (in substitution for rent arrears covered by the proposal)
IVA — new debt post approval	Yes
CVA — pre approval	Yes unless a small company in which case not without the court's prior permission if moratorium applied for (para 12(1)(h) of schedule A1 to the 1986 Act)
CVA — after approval	No. The pre CVA arrears are converted into rights under the CVA
CVA — new debt post approval	Yes (see *Thomas* v *Ken Thomas Ltd* [2007] 01 EG 94
Administration	Not without the prior permission of the administrator or the court (para 43(6) of schedule B1 to the 1986 Act)
LPA receivership	Yes
Administrative receivership	Yes
Voluntary liquidation	Yes but a court can restrain/stay any proceedings if an application is made to the court.
Compulsory winding up	No. At any time after a winding up order has been made or a provisional liquidator appointed a party must obtain the prior permission of the court to commence or continue court proceedings (s 130 1986 Act)
Bankruptcy	No. After the making of a bankruptcy order no person can commence any legal proceedings against the company without the prior permission of the court (s285(3) of the 1986 Act)

8.10 Service of a statutory demand/winding up proceedings

A remedy which often finds favour with landlords is the service of a statutory demand on a tenant in arrears of rent. The service of a statutory demand is often perceived as a prerequisite to a winding up of a debtor company or an individual debtor being made bankrupt (although in relation to a compulsory winding up procedure, this is not strictly necessary).

8.10.1 Requirements

A landlord can serve a statutory demand where:

- there are undisputed arrears
- which exceed £750.

The sums owed must be liquidated sums and if those sums are disputed the statutory demand procedure is not appropriate. A tenant can issue an application to the court to set aside the statutory demand, and if successful, is likely to obtain an order for its costs to be paid by the landlord. There are many decided cases where judges have frowned upon the practice of creditors using this procedure for debt collection.

The statutory demand is a prescribed form and should be completed carefully. In the case of an individual tenant, the statutory demand must, if possible, be personally served. In the case of a corporate tenant, the demand must be served at its registered office either personally or by post. However, if the statutory demand is served by post, the landlord must prove that it was actually delivered. In practice, therefore, it should be personally served or sent by recorded delivery (although there can be problems with the recorded delivery postal system).

The service of a valid statutory demand gives the tenant 21 days in which to pay the outstanding sums failing which the landlord is entitled to present a bankruptcy or winding up petition to the court.

8.10.2 Advantages

In cases where the tenant is not in significant financial difficulties, the service of a statutory demand can be very effective. It will put immediate pressure on the tenant to pay up by raising the very real possibility that if no payment is made, serious legal consequences could follow which, at the very least, could cause significant disruption to the tenant's business. The service of a statutory demand also has the virtue of being very quick and cheap to implement.

8.10.3 Drawbacks

The procedure can, however, involve a certain amount of poker; ie bluff and counter bluff. If a tenant fails to pay up in response to a statutory demand, the costs involved in issuing a petition are significantly higher than the costs of issuing a claim form for forfeiture and/or a judgment. They will include the need to pay a deposit towards the costs of the Official Receiver. In addition to this, if there is a real concern about the ability of the tenant to pay its debts, taking this further step may in fact be counter-productive and may mean:

- that there is little or no prospect of making a recovery because the landlord will rank behind secured and preferential creditors on distribution and
- if an order is made, this is likely to hinder the landlord's prospects of relying upon other remedies because of the restrictions imposed by the moratorium on proceedings and legal remedies.

Consequently, the service of a statutory demand may be a bluff which the landlord has no intention of carrying through. The tenant may suspect this and do nothing. However, there are certain landlords who will see the process through, if only to ensure that they have a reputation in the market as taking a no nonsense approach to rent collection. It may therefore come down to who blinks first.

8.11 Pursue claims against third parties — contingent liability

The final option available to a landlord is to take steps to recover arrears from third parties who may also be liable in addition to the tenant. Where a lease has been in existence for a number of years there can be a number of potential targets for a landlord, some that may be obvious, and others that may be less obvious. These targets may represent a greater prospect of recovery than the tenant, which may be struggling financially. It is therefore very important for the corporate occupier and its professional advisors in a legacy situation to look carefully through the relevant lease, licences for assignment together with any surety or guarantor covenants. It is not uncommon for third party targets to be overlooked by a landlord.

A list of potential third party targets, depending upon the precise obligations contained in the lease documents and the general law, may include:

- the original tenant
- a guarantor of the original tenant
- subsequent assignees (provided they have entered into a direct covenant with the landlord)
- guarantors of subsequent assignees (again assuming the guarantee was given to the landlord)
- a guarantor of the current tenant.

The rules governing the contingent liability of third parties, and in particular the circumstances in which a third party will be liable for the current tenant's default and the steps which a landlord must take to protect its rights, are dealt with in detail in Chapter 10.

When a tenant falls into arrears, a corporate occupier (and/or its professional team) should investigate promptly whether there are third parties who may be liable. This is important, not only because it may provide a very quick and effective payment of the arrears, but also because it may be necessary to serve a notice under section 17 of the Landlord and Tenant (Covenants) Act 1995 within six months of the arrears falling due to preserve the landlord's right to sue.

The advantages of pursuing a third party are:

- it may result in a quick payment for no more than the cost of drafting a section 17 notice or a written demand for payment
- it may result in a recovery in circumstances where the current tenant is unlikely to be able to pay
- if the third party is a good covenant, the process can be repeated on many occasions
- it places the onus on the third party to pursue the current tenant for the arrears
- the tenant's insolvency in most case will not affect the landlord's rights against the third party.

The disadvantages may include:

- if the third party pays all outstanding arrears demanded by a section 17 notice the third party will be entitled to take control and may require the landlord to grant it an overriding lease (which may or may not be desirable to the landlord)
- the landlord may have to serve demands or section 17 notices every time a sum falls due under the lease.

8.12 Without prejudice marketing

In circumstances where it appears the tenant cannot pay, and is unlikely to be able to pay in the future, it may be prudent for a landlord to take steps to market the premises. A landlord can take steps to remarket premises without prejudicing its legal position provided that it does not re-take possession. In *Oastler* v *Henderson* [1977] 2 QB 575 the landlord showed a prospective tenant around the premises and this was held not to constitute the re-taking of possession. This will only occur when a new lease is actually granted. Nevertheless, if the landlord wants to keep its options open it would be sensible to write to the tenant and confirm that its actions will not constitute a re-entry and are without prejudice to its rights and remedies.

There is no reason why the marketing of premises cannot be run in parallel with other action. It is not uncommon for some institutional landlords to market premises for example as soon as a tenant goes into administration or liquidation, while still keeping their options open. Clearly, this may not be appropriate if the premises are still occupied by the tenant and access is required to show a prospective tenant around.

8.13 Corporate occupier as tenant

A corporate occupier that falls into arrears may find itself in that position for a number of reasons. It may:

1. continue to trade from the premises but dispute the arrears or have a counterclaim (or indeed be withholding rent)
2. continue to trade from the premises but have cash flow problems
3. be insolvent or need to restructure the business
4. have no further need for the premises and wish to dispose of them.

If 1, 2 or 3 are the case, it is important that the corporate occupier tries to maintain a dialogue with the landlord. In particular, it would be prudent to seek written confirmation from the landlord that it will not attempt to peaceably re-enter for a defined period while there is an attempt to resolve the issues that have led to the arrears.

In respect of 2 and 3, the key to a successful outcome will be good communication with the landlord, and putting forward realistic proposals that can be adhered to. If further time for payment is negotiated and it subsequently becomes clear that the timescale cannot be achieved, it made be better to let the landlord know before the agreed period has expired. In the case of serious financial difficulties, Chapter 9 deals with the various formal insolvency procedures which are available to the directors of a tenant company which needs a breathing space to restructure its business or its debts, or both.

If the premises are surplus, one strategy that can be used by a corporate occupier to try and provoke an end to its ongoing rental liability is to simply stop paying rent. If the corporate occupier strikes gold, the landlord will forfeit the lease and the on going rental liability will stop immediately (although there may be a subsequent dilapidations claim). Such a tactic is unlikely to work with sophisticated institutional landlords. It might work, however, with less clued up landlords. Even if it does not result in the lease being forfeit, it might result in surrender negotiations that result in the payment of a premium that is acceptable to the landlord, and is less that the corporate occupier's provision in its accounts for that property.

8.14 Corporate occupier as sub-tenant

It is not uncommon for a corporate occupier to be a sub-tenant. Difficulties can arise if its immediate landlord, the tenant under a superior lease, falls into arrears of rent. The first the corporate occupier might learn of the difficulties is when the superior landlord either takes steps to forfeit the lease, or alternatively serves the corporate occupier with a notice under section 6 of the Law of Distress (Amendment) Act 1908 (see 8.7 above).

If the corporate occupier is served with a section 6 notice, it should pay rent to the superior landlord in accordance with that notice (in so far as the rent has not already been paid to its own landlord). If there is no shortfall between the head rent and the sub rent, this procedure could go on for months, if not years. However, if the sub-tenancy is of part, or there is a substantial discrepancy between the head rent and the sub rent, the superior landlord may take steps to forfeit the headlease, which in turn would terminate the sublease.

A sub-tenant is entitled to apply for relief from forfeiture (strictly speaking a vesting order under section 146(4) of the Law of Property Act 1925). However, a court is only likely to grant such an order on condition that the sub-tenant pays any arrears that exist under the headlease together with the superior landlord's costs. The corporate occupier can find itself significantly out of pocket, particularly if it has paid its rent to the tenant, which in turn has failed to pay its own rent to the superior landlord (including the sub-rent). In these circumstances the corporate occupier sub-tenant may have to, in effect, pay its rent twice. However, if the premises occupied by the sub-tenant are only part of a much larger building leased by the tenant, the sub-tenant will usually only be required to pay a proportionate part of the arrears as a condition for obtaining relief (*Chatham Empire Theatre (1955) Ltd* v *Ultrans Ltd* [1961] 2 All ER 381).

If the tenant becomes insolvent and the lease is disclaimed, the sub-tenant is entitled to apply for a vesting order under section 181 of the Insolvency Act 1986 (see 10.7.4). Alternatively the sub-tenant can remain in the premises for the remainder of the term granted by the sub-lease provided that it complies with the terms of the headlease (see 11.3.4).

8.15 Precedents

8.15.1 Notice of peaceable re-entry

Dear Sirs,

Peaceable re-entry of Unit 12, Twilight Industrial Estate, Sod-in-Chipbury

We act for No Nonsense Limited your former landlord of the above premises.

Our clients have today forfeited your leases by peaceable re-entry on the grounds of rent arrears. As at today's date your rent arrears totalled £40,000 plus interest.

As a result of the forfeiture you are no longer entitled to enter the premises and if you do so you will commit an act of trespass for which you may be liable in damages. Please contact us in respect of any of your belongings that remain on the premises so that we can arrange for them to be collected at a mutually convenient time.

You are entitled to make an application to the court for relief from forfeiture and you may wish to take legal advice on this. Please be aware, that in an attempt to mitigate its losses, our client intends to re-let the premises as quickly as possible. In the circumstances, if it is your intention to apply to the court for relief from forfeiture we would urge you to do so promptly and in any event with the next 28 days. If you fail to make an application to the court within that period, our client will re-let the premises and we will show this letter to the court in the event that you subsequently apply for relief from forfeiture.

Yours faithfully,

8.15.2 Notice to elect

To: [the Liquidator]
Address

Notice to Elect

I, [insert name] landlord of the premises known as [address] under a lease dated [ ] between [ ] and [ ] require the liquidator to decide within 28 days of receiving this notice whether he will disclaim the property or not and to notify me of his decision.

Signed .... Dated ....
For and on behalf of the Landlord

8.15.3 Section 6 notice

To: [Name and address of sub-tenant] "the Sub-tenant"

From: [Name and address of superior landlord] "the Superior Landlord"

Notice under section 6 of the Law of Distress Amendment Act 1908

The Superior Landlord of [description of premises] hereby gives the Sub-tenant notice that its immediate tenant [name and address of mesne landlord] is in arrears with it rent in the sum of [amount] and the Sub-tenant is required to pay all rent accrued due or to become due from you as sub-tenant direct to the Superior Landlord until such arrears due to the Superior Landlord have been paid.

Signed .... Dated ....
For and on behalf of the Superior Landlord

Tenant Insolvency

When a tenant finds itself in financial difficulties it may be able to solve these difficulties by refinancing and restructuring. However, if this is not possible, or does not succeed, then there are a number of formal insolvency procedures which might come into play. These formal procedures may ensure that a company's business is wound up in a way that maximises recovery for the creditors. Alternatively these procedures may actually assist a company to recover and become solvent again. This is often referred to as the rescue culture and underpins much of the UK's current insolvency legislation. While they may assist a struggling company by providing a breathing space in which to trade out of difficulties, these procedures can dramatically affect a landlord's rights and remedies under the lease.

Once again, a corporate occupier can have different perspectives. It could be a tenant in financial difficulties where its directors are exploring various solutions to restore the financial health of the company. It may need protection from claims by creditors to allow one of those solutions to be implemented. By way of example, the directors may decide to put the company into administration to explore the possibility of entering into a company voluntary arrangement with creditors. The automatic moratorium from being in administration would protect the company from court proceedings by its creditors while this possibility was properly investigated. However, if the boot is on the other foot, and the premises have been sub-let by the corporate occupier to a tenant which gets into financial difficulties, there will be a desire to know what action can lawfully be taken to ensure that the much needed income flow is restored as quickly as possible to offset against the head rent (which it will still be obliged to pay to the superior landlord).

There are a number of formal insolvency procedures that may be encountered by a landlord seeking to recover outstanding rent:

1. voluntary arrangements
2. administration
3. receiverships
4. liquidation
5. bankruptcy.

It is worth considering each of these procedures and how they can affect a landlord's rights and remedies.

9.1 Voluntary arrangements

If a tenant is in financial difficulties, in addition to being in arrears with its rent, it is also likely to have failed to pay other creditors eg suppliers. In an attempt to avoid bankruptcy/liquidation, a debtor can enter into a formal arrangement with its creditors to allow for the gradual or partial discharge of its outstanding liabilities and to provide him/it with an opportunity to trade out of difficulties. These arrangements are governed by the Insolvency Act 1986 (as amended) and allow the debtor to make a proposal to its creditors in satisfaction of the debts. This usually involves the creditors accepting lesser sums than they are legally entitled to.

Such an arrangement can be made by an individual leading to an individual voluntary arrangement (IVA) or by a company leading to a company voluntary arrangement (CVA). The primary purpose of a voluntary arrangement is to give the debtor an opportunity to be rescued and trade out of financial difficulties.

9.1.1 Individual voluntary arrangement

(a) Procedure

The first step is for the debtor to apply to court nominating a person to act as supervisor of the arrangement (the nominee). The nominee must be either an insolvency practitioner, or a person with specific authorisation to act as a nominee under the 1986 Act.

A court may make an interim order which has the effect of imposing a moratorium for a short period to allow the feasibility of the proposed arrangement to be investigated.

During the period of the interim order the nominee will prepare a report containing the debtor's proposal and indicating whether it has reasonable prospects of success. If the court accepts that it does, the nominee will convene a meeting of the creditors to consider the proposal.

The creditors are entitled to vote at the meeting on whether to accept or reject the proposal. Generally speaking the weight of a creditor's vote will be determined by the amount of debt it is owed. In the case of a landlord this will usually be the extent of the arrears at the date of the meeting. It may also be appropriate to have regard to the rent for the period of the proposed arrangement.

For the proposal to be accepted there must be a majority in excess of three-quarters of the value of the creditors voting.

(b) Moratorium

If an interim order is made under section 252 of the 1986 Act, no bankruptcy petition can be issued, and no other proceedings or other legal processes can be taken, against the debtor or the debtor's property for the duration of the order. If, for example, a landlord has already issued court proceedings for rent arrears, they will be stayed for the duration of the interim order.

The moratorium also has the effect of preventing a landlord from exercising its rights of distress or forfeiture (whether by peaceable re-entry or the issuing of proceedings — see section 252 (aa) and (b) of the 1986 Act) except after obtaining permission from the court.

(c) Who is bound by IVA?

Importantly, an approved arrangement binds every person who was entitled to vote at the creditors meeting, or would have been entitled if it had had notice of the meeting (section 260 of the 1986 Act). A landlord will be bound by an IVA even if it votes against the proposal and loses. The effect of the IVA is to transform the landlord's right to recover arrears into rights under the IVA (usually a amount reduced to so many pence in the pound).

Any existing court proceedings for arrears which are covered by the IVA will be deemed to be dismissed on the date the interim order expires. If a landlord subsequently issues court proceedings based upon arrears covered by an IVA, these are bound to fail.

(d) Challenge of IVA

A landlord can challenge a decision to approve an IVA by making an application to the court within 28 days of that decision.

Section 262 of the 1986 Act provides two grounds for challenge, namely that:

(1) the arrangement unfairly prejudices the interests of a particular creditor or
(2) there was a material irregularity in relation to the meeting.

For a useful summary of the principles which a court will apply when considering such an application see *Prudential Assurance Company Ltd* v *PRG Powerhouse Ltd* [2007] EWHC 1002 Ch (9.1.2 (d) below).

9.1.2 Company voluntary arrangement

(a) Procedure

The procedure leading to a CVA has some differences but is broadly comparable with those set out above in respect of an IVA. The key practical difference is that two meetings are necessary to consider the proposal; a creditors meeting and a company meeting. Both must vote to accept the proposal.

There is no statutory moratorium while a court application is being pursued except where a small company proposes a voluntary arrangement. This is because CVAs often arise in the course of formal administrations in which an automatic moratorium will arise.

(b) Moratorium for 'small companies'

A concession to small companies (for which administration is unlikely) was brought in by the Insolvency Act 2000. This provides that where directors of a small company intend to make a proposal for a voluntary arrangement, they may take steps to obtain a short moratorium for that company during which a proposal for a CVA can be put to its creditors (see schedule 1 of the Insolvency Act 2000 which inserts a new section 1A and schedule A1 to the 1986 Act).

To obtain a moratorium, the directors of a small company must file with the court:

- the terms of the proposed CVA
- a statement of the company's affairs
- a statement that the company is eligible for a moratorium

- a statement from a nominee that he has given his consent to act and that in his opinion the CVA has a reasonable prospect of success, the company has sufficient funds to carry on business during the moratorium and a company and a creditors meeting should be convened to consider the proposed CVA.

A moratorium will usually be for 28 days and during that period will prevent:

- the company being wound up
- an administrator being appointed
- a landlord from peaceably re-entering without permission of the court
- the issuing of court proceedings, enforcement of security or distress without the court's permission.

(c) Who is bound — landlord's proprietary rights

The way in which CVAs affect a landlord's rights, obligations and remedies was comprehensively considered by the Court of Appeal in *Thomas* v *Ken Thomas Ltd* [2007] 01 EG 94.

The position can be summarised as follows:

1. A CVA which is approved in accordance with the 1986 Act binds every person entitled to vote at the creditors meeting or who would have been entitled to vote if they had had notice of it (section 5(2) of the 1986 Act). On approval, therefore, it becomes binding on all creditors including a landlord to whom rent is due at the date of the approval. Consequently, prior debts such as rent arrears are converted into rights under the CVA.
2. Rent falling due *after* the approval of the CVA should not be included. It is wrong in principle that a tenant should be able to trade under a CVA for the benefit of its past creditors at the present and future expense of its landlord. If the CVA does purport to apply to future rent, an application can be made to the court objecting to the proposal on the grounds that it unfairly prejudices the landlord's rights.
3. Prior to *Thomas* there was a belief in some quarters that although a landlord was prevented from suing a tenant for pre CVA rent arrears, it could nevertheless rely upon its proprietary right to forfeit for non payment of this rent. If the tenant then applied for relief, it should, the argument continued, only be granted on the payment of all outstanding arrears including the rent covered by the CVA. The landlord could, therefore achieve via the back door what it could not achieve via the front door. This was rejected by the Court of Appeal in *Thomas*. It was inconsistent with the rescue culture embodied in the 1986 Act. Further, the landlord was bound to accept payments in accordance with the CVA in substitution for rent. In short, his right to sue had gone because there was no rent owing, merely a substituted debt due under the CVA.
4. A landlord with a right of forfeiture is not treated for these purposes as a secure creditor. Suggestions in earlier cases to the contrary are wrong.
5. However, if the forfeiture clause in the lease is well drafted, and provides the landlord with a right to forfeit if the tenant proposes a CVA, the landlord is not prevented from forfeiting the lease on this ground.
6. The date for considering the position is the date of the forfeiture hearing. This means that a landlord cannot trump a tenant by issuing forfeiture proceedings before the CVA is approved.

Consequently, a CVA can have dramatic financial consequences for a landlord. It may find itself bound by an arrangement under which it is paid a dividend that is a fraction of what it was entitled to receive as rent. All too often landlords are outvoted at creditors meetings. If served with a proposal, a landlord should obtain advice quickly and ensure that a realistic value is put on the debt owed to it. It goes without saying that the landlord should make its voice heard and vote.

(d) Challenge of CVA

A landlord can apply to the court to challenge a CVA if it unfairly prejudices its interests or there has been a material irregularity at or in relation to the company or creditors meeting (section 6 of the 1986 Act).

Prudential Assurance Company Ltd v *PRG Powerhouse Ltd* [2007] EWHC 1002 provides a useful illustration of how a court approached an application by a group of landlords alleging unfair prejudice.

Facts

- T was the third largest electrical retailer in the UK and had over 88 stores in UK.
- G (a parent company) had stood as guarantor for a number of leases granted to T.
- T got into financial difficulties and its director proposed a CVA which would close 35 underperforming stores, and the remaining stores, it was hoped, would trade profitably.
- G put forward funds of £1.5m to be applied in full and final settlement of claims covered by the CVA.
- The CVA affected landlord's rights as follows:
  - landlords of the profit making premises that would remain open would be paid in full
  - landlords of the closed loss making premises with five to eight years remaining under the lease were to be paid the equivalent of eight months rent
  - landlords of the closed premises with over eight years remaining under the lease were to be paid 12 months rent
  - the landlords of the closed premises would lose their rights to sue G under the guarantees, and in effect would lose their rights to future rent on being paid the above sums (which equated to 28p in the pound).
- Landlords of the closed premises made applications to the court under section 6 of the 1985 Act arguing that the CVA would unfairly prejudice them.

Held

- Unfairness is to be assessed by a comparative analysis from a number of different angles. These include a vertical comparison with the position on winding up, and the horizontal comparison with other creditors or classes of creditors.
- The treatment of the landlords of the closed premises when compared with other creditors (in particular the landlords of the profit making premises that remained open) showed that they had been unfairly prejudiced. Their present, future and contingent claims were to be discharged at a fraction of their value so that the other creditors could be paid in full.
- If the company had been wound up, the landlords would still have had the benefit of the guarantee claims against G, and therefore would have been better off whereas the other creditors would have received nothing.
- The landlords of the closed premises would have suffered least on a winding up of the company, but were prejudiced most by the CVA. The dividend placed no value on the guarantees which

improved their position over all other unsecured creditors, and which were intended to benefit the landlords if T became insolvent.

The wider implications of this case, and the obvious attempt at what has become known as guarantee stripping are considered at 10.4.5.

9.2 Administration

9.2.1 Purpose

A limited company in financial difficulties has the option of going into administration. The rules governing administration were changed by the Enterprise Act 2002 and are now contained in a new schedule B1 to the Insolvency Act 1986 which was inserted by the 2002 Act. The 2002 Act introduced a new administration regime which among other things allows the company in financial difficulties or its directors to appoint an administrator.

The purpose of administration is to create a breathing space which allows the company to be rescued through continued trading, or if that is not possible, to allow the business and assets to be sold as a going-concern which is likely to result in a greater recovery than if the company was immediately wound up. This procedure has been a popular choice for insolvent corporate tenants and is now the most popular insolvency procedure.

9.2.2 Appointment and duties of an administrator

An administrator can be appointed by:

- the court
- the holder of a floating charge or
- the company in difficulties or its directors.

Once appointed, the administrator must perform his function with the objective of:

- rescuing the company as a going concern or
- achieving a better result for the company's creditors as a whole than would be likely if the company were wound up or
- realising property in order to make a distribution to one or more secured or preferential creditors.

9.2.3 Moratorium

To provide a breathing space which enables an administrator to achieve one of the desired objectives, schedule B1 of the 1986 imposes a moratorium on court proceedings or other legal processes against the company for the duration of the administration. The moratorium operates as follows:

- paragraph 42 of schedule B1 provides that no resolution may be passed, or order of the court made, for winding up the company while it is in administration (with certain limited exceptions)

- paragraph 43 provides that for the duration of the administration and except with the consent of the administrator, or with the permission of the court:
  - no step may be taken to enforce security over the company's property
  - a landlord may not exercise a right of forfeiture by peaceable re-entry in relation to premises let to the company
  - no legal process (including legal proceedings, execution, distress) may be instituted or continued against the company or the property of the company.
- if the landlord applies to the court for permission, the court may impose conditions when granting or refusing consent.

The consequences to a landlord of a corporate tenant going into administration can therefore be both dramatic and immediate. While that tenant is in administration the landlord will not be able to levy distress, issue court proceedings for rent arrears and/or forfeiture, or peaceably re-enter without first obtaining either the consent of the administrator or without making a court application and persuading the judge to grant permission.

9.2.4 Application for consent to distrain or forfeit

Once a landlord is notified that its tenant has gone into administration, its immediate concern will be to try and recover existing arrears and continuing rents as they fall due. The operation of the moratorium does not wipe the slate clean. The tenant's liability for accrued rent arrears and future rent remains, but the landlord is temporarily prevented from taking steps to recover those arrears from the company or repossess the premises.

The landlord's best prospects of recovering rent arrears in most case will be by making an early strike (eg by levying distress on the tenant's goods). There is no guarantee that when the moratorium is lifted there will be sufficient funds to pay past or future rent. There is great tension between the aims of a landlord who will want to take immediate steps to recover rent and a tenant that craves protection to provide it with an opportunity to trade out of difficulties. It requires the court to strike a careful balance when considering a landlord's application for consent to distrain or, alternatively if the landlord considers it a lost cause, to forfeit the lease. To allow the landlord to distrain may enable it to recover rent lawfully due. Alternatively, giving consent to the forfeiture of the lease will enable the landlord to re-let quickly. However, granting consent for the landlord to exercise either of these remedies may prevent a company from trading out of difficulties or result in its immediate demise thereby substantially disadvantaging other creditors.

The courts when considering this conundrum have held that the 'greater good' argument can prevail to defeat a landlord that is seeking to exercise its proprietary rights (distress or forfeiture), but only if the circumstances dictate it is unavoidable and even then, it will only usually be acceptable to "a strictly limited extent" (see *Re Atlantic Computer Systems plc* [1992] 1 All ER 476 at p500).

When a tenant company enters administration a landlord will want to speak to the administrator quickly to ascertain what his intentions are in relation to the premises. There are a number of possibilities.

1. If he considers that the company can be rescued he may be keen to ensure the company continues trading from the premises. Future rent may therefore be paid as an expense of the administration and have priority over the ordinary unsecured debts of the tenant company. In these circumstances the landlord may be content to do nothing.

2. He may consider whether the company can be sold as a going concern and therefore wish to keep the premises pending the sale of the business to a third party. If a purchaser is found, the administrator is likely to seek landlord's consent to assign the lease to the purchaser in the event that the purchaser buys the business and not the company. If the purchaser buys the company then there will be no need to assign the lease as the tenant will remain the same — all that will change is the ownership of the company's shares.
3. He may decide that the company is unlikely to be sold or survive but nevertheless wish to continue trading from the premises for a limited period to ensure an orderly winding up of the company's affairs and to maximise a recovery for creditors.
4. The premises may be surplus to the administrator's plans. In such circumstances he may seek to surrender the lease or, if approached, consent to the landlord peaceably re-entering the premises.

In relation to 2–4 above, if the administrator fails to pay rent as it falls due as an expense or necessary disbursement of the administration and is not prepared to hand the premises back, a landlord will need to consider carefully what to do.

There can be confusion as to whether future rent must be paid by an administrator as an expense of the administration when the company remains in possession of the premises. If the administrator decides that the premises are required for the reasons set out in 1–3 above, there is a respectable argument for saying that the administrator should pay rent as it falls due as "expenses properly incurred by the administrator in performing his function ..." (rule 2.67 (1)(a) of the Insolvency Rules). This category of expense ranks at the top of the priority order of payment. Alternatively, it is arguable that it could fall within "necessary disbursements" (rule 2.67 (1)(f) of the Insolvency Rules) which falls further down the priority list but still a good deal higher up the pecking order than unsecured creditors.

There are, however, some unresolved legal points surrounding the question of whether post administration rent should be treated as an expense. It is clearly important, therefore, that this is clarified with the administrator quickly. Prompt action is the key to minimising any losses.

9.2.5 Strategy for obtaining possession

If the tenant in administration is in arrears of rent, and there is no reason for a landlord to preserve the lease (eg the existence of former tenants or guarantors the landlord can pursue), a strategy along the following lines may be effective.

(a) Written request for consent

The landlord should write an open letter to the administrator requesting consent to forfeit the lease or levy distress. He should be reminded of the guidelines set out by the Court of Appeal in *Re Atlantic Computer Systems plc* [1992] 1 All ER 476 (see also *Metro Nominees (Wandsworth)* v *Krayment* 2006 WL 3933204 for an application of these guidelines in a landlord and tenant forfeiture case). The letter should also provide that if the administrator refuses consent, he is required to provide the following information:

- the end result which is sought to be achieved by the administrator
- the prospects of that result being achieved
- the period for which the administration order is expected to remain in force

- the consequences which the administrator anticipates and the effect on the company in administration if consent were to be granted.

A precedent letter can be found at 9.6.1 below.

(b) Letter before action

If the administrator requires the continued use of the premises he may refuse permission but agree to pay continuing rent (as opposed to the arrears) as expenses incurred in the administration. In most cases this will be as much as the landlord can achieve at this stage. However, if the administrator refuses consent and does not agree to pay continuing rent, or fails to respond, then the landlord should send a seven day letter before action to the administrator threatening to make an application to the court for permission. The landlord should point out (if appropriate) that:

- the administrator is an officer of the court and is expected to make his decision speedily
- if he refuses consent he must state why he has refused permission and why he has not paid rent arrears or at least the current rent as it falls due
- the Court of Appeal held in *Re Atlantic* that a landlord should be granted permission to forfeit the lease unless the premises are required to facilitate the objects of the administration. However, if the premises are necessary to the administration plan, normally the court will require the administrator to pay current rent as a condition of refusal of permission.

(c) Court application

If the letter before action is not successful, the landlord should consider carefully whether to apply to the court for permission to forfeit. As with all court proceedings there will be a cost benefit analysis to carry out. However, some administrations can continue for many months (occasionally years) and a landlord cannot allow its tenant to continue in possession indefinitely without payment of rent. Furthermore, a court may make an order for costs against an administrator if it considers the failure to pay (or to offer to pay) ongoing rent was unreasonable.

Any delay in taking action is likely to result in the landlord suffering further losses. It may provide the administrator with time to fulfil the purpose for which continued occupation is required thus removing any opportunity of securing rental payments. The key to maximising a landlord's prospects of recovering rent is to take prompt action.

It is important that if the landlord is considering forfeiting the lease no step is taken that could constitute a waiver of the landlord's right to forfeit (see 8.3.3 above).

9.2.6 Strategy for securing payment of rent

There will be occasions where the landlord does not wish to forfeit the lease. This may be because the market is weak (and the prospects of finding a replacement tenant at the same rent are poor) and it wants to see if the tenant in administration can recover.

In these circumstances a landlord should write in open correspondence to the administrator seeking an assurance that the rents will be paid as an expense of the administration. If that assurance is not forthcoming, and the administrator continues to use the premises (either for the purposes of the company's business, or as part and parcel of the sale of that business), the landlord can make, or threaten to make, an application to the court challenging the administrator's conduct of the company.

Under paragraph 74(1)(a) and (b) of schedule B1 to the Insolvency Act 1986 a creditor (which a landlord that is owed rent clearly is) may apply to the court if the administrator is acting, has acted or proposes to act so as unfairly to harm the interests of the applicant. Alternatively, the landlord can make an application under paragraph 74(2) of schedule B1 to the 1986 Act if the administrator is not performing his functions as quickly or efficiently as is reasonably practicable.

The court has very wide powers to order relief and has the power to order that rent be paid by the administrator as expenses incurred by him in performing his function. It is unlikely that this order will be open ended. It may be time limited. The administrator is unlikely to be required to pay rent as an expense for any period after the premises are needed for the purposes of the company in administration or the sale of its business. The appropriate order may be that rent be paid as an expense of the administration until the premises cease to be required by the administrator and thereafter the landlord have consent to forfeit the lease should it so wish. This would balance the needs of the landlord against the needs of the administrator.

9.3 Receiverships

If the tenant has borrowed money and the lease is a lengthy term, the lease may be mortgaged. If the tenant defaults on the mortgage payments, the mortgagee may appoint a receiver under the Law of Property Act 1925. Strictly, the LPA receiver acts as agent on behalf of the company in receivership notwithstanding that the receiver has been appointed by a mortgagee.

The main function of a receiver is to manage the company's business and use and/or dispose of the assets subject to the charge in order to repay the secured creditor and discharge the charge. The primary duty of the receiver is to the secured creditor.

The important point from a landlord's perspective is that if a LPA receiver is appointed, all the landlord's rights and remedies remain enforceable. There is no moratorium on enforcement of rights and remedies contained in the Law of Property Act 1925. However, from a landlord's perspective, notification that a receiver has been appointed should put it on enquiry. The signs are not good and careful consideration should be given as to whether any action is possible (the appointment of a receiver may give rise to a forfeiture ground), and if so, should be taken. If the landlord does not take quick action, it may miss the opportunity to forfeit or distrain for rent arrears if the tenant succumbs to another more severe insolvency procedure and is prevented from taking action by a moratorium.

A similar remedy was available to a creditor which had a floating charge over the company's assets and involved the appointment of an administrative receiver. However, this is being phased out and an administrative receiver can only now be appointed under a debenture created before 15 September 2003.

9.4 Liquidation of tenant (company)

The most extreme insolvency procedure that a tenant company can encounter is liquidation. If a tenant company is wound up its assets will be liquidated and distributed among its creditors and (if anything remains) its shareholders.

There are two types of liquidation:

- voluntary liquidation
- compulsory liquidation.

9.4.1 Voluntary liquidation

A voluntary liquidation occurs where the company in financial difficulties initiates the winding up procedure itself. A compulsory winding up usually occurs following the issuing of a winding up petition by a third party creditor.

The key points in relation to a voluntary liquidation are set out below.

- A solvent liquidation can occur where the directors pass a resolution to wind up the company (perhaps to restructure the business or to terminate a dormant company) and the company is able to pay all of its debts within a specified period not exceeding 12 months. The process is within the control of the company and is also known as a "members' voluntary liquidation".
- If the company is not able to pay all of its debts within 12 months, but the directors nevertheless pass a resolution for the company to be wound up, a joint meeting of both the creditors and the company will be held to vote on the appointment of a liquidator. This is known as a "creditors' voluntary liquidation".
- Once a resolution to wind up the company has been passed the company must cease to carry on its business.
- There is no automatic moratorium. A landlord is not prevented from levying distress or forfeiting the lease by peaceable re-entry. However an application can be made to the court to stay or restrain any action or court proceedings (eg forfeiture proceedings as opposed to peaceable re-entry). However the onus is on the liquidator/other creditors to make such an application.

9.4.2 Compulsory winding up

A corporate occupier seeking to recover rent from a sub-tenant in arrears is more likely to encounter compulsory winding up. It will invariably arise where another creditor has issued and served a winding up petition on the grounds that the tenant company is unable to pay its debts. Occasionally it may be action which the landlord considers taking against a company tenant that has failed to pay rent (see 8.10 above).

A company can be wound up if it is unable to pay its debts. Section 123 of the Insolvency Act 1986 deems that a company is unable to pay its debts if:

- the company owes more than £750 to a creditor and fails to pay the full amount within 21 days of being served with a statutory demand or
- if a judgment is unsatisfied (in whole or in part) or
- the company is unable to pay its debts as they fall due or
- the value of its assets is less than the amount of its liabilities taking into account its contingent and prospective liabilities.

9.4.3 Moratorium in compulsory winding up

There is no automatic moratorium when a winding up petition is presented to the court. However, under section 126(1) of the 1986 Act, at any time after issue of the petition and the making of a winding-up order, an application can be made to the court seeking an order staying any proceedings which are on-going against the company.

However, an automatic moratorium does apply following the making of a winding-up order. Section 130(2) provides:

> When a winding-up order has been made or a provisional liquidator has been appointed, no action or proceedings shall be proceeded with or commenced against the company or its property, except by leave of the court and subject to such terms as the court may impose.

Consequently, if a landlord wishes to issue forfeiture proceedings on the grounds of rent arrears after a winding up order has been made against its tenant, it will need to obtain the permission of the court.

The position in respect of peaceable re-entry is less clear. In the bankruptcy context, and on different statutory wording, it has been held that a landlord can peaceably re-enter without prior permission from the court (see 9.5 and in particular *Razzaq* v *Pala* [1997] 2 EGLR 53). It is often said that the position is the same in respect of winding up. However, the wording of section 130(2) differs from section 285(3) — the moratorium imposed after a bankruptcy order.

In *Re Memco Engineering* [1986] Ch 86 on similar statutory wording, the court held that distress was "an action or proceeding" within the meaning of section 231 of the Companies Act 1948. It is, therefore, at the very least arguable, that section 130(2) could extend to forfeiture by peaceable re-entry and the landlord does therefore need the prior permission of the court. The point is open until decided by the courts.

9.4.4 Permission of the court and the Blue Jeans order

If it is necessary for a landlord to apply for permission to forfeit, unless the liquidator has a defence, the court is very likely to grant consent and make a possession order. Importantly, it is not necessary for the landlord to issue fresh proceedings in another court. The insolvency court has jurisdiction to make an order for possession within the liquidation (see *Re Blue Jeans Sales Ltd* [1979] 1 All ER 641).

9.4.5 Forcing the liquidator's hand — disclaimer and the notice to elect

If a provisional liquidator is appointed, or a winding up order made, a landlord will be prevented from levying distress or forfeiting the lease without leave of the court. In most cases a liquidator will serve a notice disclaiming the lease under section 178 of the 1986 Act. The lease will be an onerous contract and the liquidator will want to put a stop to liability as quickly as possible.

The effect of a validly served notice of disclaimer is to terminate the lease from the date of the notice, and more particularly, the rights and liabilities of the insolvent tenant in the disclaimed lease. To be effective, a formal notice of disclaimer must be sealed by and filed with the court, and served on all parties affected by the disclaimer. Importantly, it does not affect the rights or liabilities of any other person (eg guarantor or former tenant). This is dealt with in more detail below (see 10.7 below).

If the liquidator does not serve a notice of disclaimer, one option available to the landlord is to serve a notice to elect on a liquidator requiring him to decide whether to disclaim the lease or not. The liquidator has 28 days in which to disclaim. If he fails to do so, he will be prevented from disclaiming in the future.

If the liquidator does not disclaim, but continues to use the premises, in most (if not all) cases ongoing rent will rank either as:

- "expenses or costs which... are properly chargeable ... in preserving ... any of the assets of the company"(r 4.218 (1)(a)(i) of the Insolvency Rules) — which rank above all other expenses or
- "other expenses incurred or disbursements made by the official receiver in carrying on the business of the company" (r 4.218 of the Insolvency Rules) which rank above many expenses including the liquidator's remuneration.

Consequently, there may be a real disadvantage to the liquidator in losing the right to disclaim, and this may result in an election to disclaim. In practical terms, therefore, this may prove to be a much quicker and cheaper route to securing possession than issuing a court application for permission to forfeit. A draft precedent of a notice to elect can be found at 8.15.2.

9.4.6 Prove in the liquidation for rent arrears

In the case of pre-liquidation rent arrears a landlord is entitled to lodge a claim in the liquidation for any outstanding sums due under the lease and/or for damages for any breaches of covenant. In a compulsory liquidation, the claim must be lodged with the liquidator by completing a prescribed form signed by the landlord or its agent. In the case of a voluntary liquidation, the liquidator can require the landlord to lodge a written proof of debt, although this is not contained in a prescribed form.

A landlord is only likely to make a recovery if the tenant company has significant assets. A landlord ranks as an unsecured creditor in respect of claims which have arisen pre-liquidation. Such claims are not at the top of the payment list.

Unless the lease is disclaimed, a landlord will only be entitled to make a claim in the liquidation for any sums outstanding at the date of the issuing of the winding up petition (compulsory), or the passing of a resolution to issue a winding up petition (voluntary). It is not entitled to claim in the liquidation for any sums accruing after these dates. The possibilities in respect of rent (or service/insurance rent) after this date are as follows.

- The liquidator takes time to consider whether to disclaim or not and/or he continues to use the premises with a view to assigning or surrendering the lease. In each of these cases the rent is likely to be paid by the liquidator as an expense or necessary disbursement in the liquidation. The landlord should, therefore, recover these sums.
- If the landlord forfeits the lease (with permission of the court), or accepts a surrender, any future liability under the lease will come to an end on the date that the lease is forfeit or surrendered.
- If the liquidator assigns the lease, the primary claim for arrears will be against the assignee and the landlord's future loss (which in theory it is entitled to prove for in the debt) is restricted to the rather nebulous claim based on the loss of opportunity to sue the company for former tenant liability or under an AGA, as the case may be.

Usually a claim in the liquidation will be relatively modest (eg one or two quarters rent and/or service charge plus interest). Occasionally it may include a dilapidations claim.

The position is different if the liquidator disclaims the lease. Here, a landlord is entitled to claim in the winding up for any losses suffered as a result of the disclaimer. If the lease term has a number of years to run, a landlord's claim could be very substantial, particularly if it is likely to be difficult for the landlord to find a new tenant at the current rent. These losses will extend to sums which the landlord would have recovered had the lease continued for the remainder of the lease term, subject to taking account for early receipt and likely re-letting (see *Re Park Air Services* [1999] 2 WLR 396). The landlord may therefore be able to establish a substantial claim based on loss of rent (including any service and insurance rent) and dilapidations. Such a claim will only usually be worth pursuing (and a worthwhile recovery likely) in a solvent liquidation.

9.5 Bankruptcy of tenant (individual)

An individual who is a tenant may be made bankrupt. Bankruptcy is the equivalent to liquidation in the corporate context. The processes are similar in many respects, but not identical.

A bankruptcy petition is usually issued by a creditor (although can be issued by others including the debtor him/herself). A creditor can issue a petition if the debtor has liquidated debts of £750 or more, and is unable to pay or has no real prospects of paying. This is proven by evidence that:

- the debtor has failed to pay a sum (of not less than £750) within 21 days of being served with a statutory demand or
- a judgment debt of not less than £750 is unsatisfied.

Once a petition is issued the Official Receiver supervises the bankruptcy procedure and may be appointed as interim receiver pending a bankruptcy order. If appropriate, an application can be made to the court at any time after a bankruptcy petition has been issued to obtain a stay of any action, execution or other legal process against the property or person of the debtor. It has been held in *Re Debtors Nos 13A10 and 14A10 of 1994* [1995] 2 EGLR 33 (albeit in the context of another section in the 1986 Act) that the words "other legal process" do not apply to peaceable re-entry. Consequently, the court does not have power to restrain a peaceable re-entry by a landlord after a petition is issued.

The moratorium which applies after a bankruptcy order is made is contained in section 285 (3) of the 1986 Act which provides that:

- a creditor has no remedy against the property or person of the bankrupt in respect of a debt provable in the bankruptcy and
- a creditor must not commence any action or other legal proceedings against the bankrupt except with the leave of the court and on such terms as the court may impose.

This wording is different to the moratorium imposed on the granting of a winding up order in the liquidation context. There have been numerous cases considering, in particular, whether forfeiture proceedings or peaceable re-entry after a bankruptcy order required the court's prior consent. The position is now clear:

1. A landlord is entitled to exercise its right of peaceable re-entry without obtaining the court's permission.
2. According to the Court of Appeal decision in *Ezekiel* v *Orakpo* [1977] 2 EGLR 47, the court's prior

permission is not required before issuing forfeiture proceedings. However, this decision has been criticised in subsequent cases. It is possible, therefore, that case may be challenged in the future.

A landlord is entitled to lodge a claim in the bankruptcy in the same way that it can make a claim in the liquidation of a company (see 9.4.6 above). An important difference from liquidation is that the landlord is entitled to claim all outstanding sums at the date of the bankruptcy order and not the date when the petition was issued. Again, however, a landlord's claim ranks down the pecking order as an ordinary debt and a recovery in most cases will be unlikely.

A further important difference is the position of the trustee in bankruptcy. The trustee becomes the assignee of the lease and unless and until he disclaims the lease, he will be personally liable for all rent and obligations under the lease from the date of his appointment. A trustee in bankruptcy will therefore be acutely aware of his need to consider disclaiming the lease promptly.

9.6 Precedents

9.6.1 Letter to administrator where tenant in arrears

Dear Sirs,

Re: Pennyless Ltd in Administration — Unit 9 Bure Valley Estate

We act for Grabittall Ltd the landlord of the above premises and note that you have been recently appointed administrators of the tenant.

The tenant occupies the premises under an underlease dated 6 April 1995 between our client and Pennyless Ltd (a copy of which we enclose for ease of reference). The current rent is £500,000 pa.

The tenant is currently in arrears of rent in the sum of £125,000 and we enclose a rental statement. Clause 5 of the underlease allows the landlord to re-enter the premises if the tenant is in arrears with its rent 14 days after it falls due. In the circumstances the landlord intends to forfeit the lease and we formally request your consent to the landlord forfeiting the lease by peaceable re-entry pursuant to paragraph 43(4)(a) of Schedule B1 to the Insolvency Act 1986.

Alternatively, in the event that the tenant is continuing to trade from the premises and the landlord decides to forfeit by court proceedings we also formally request your consent to the issue of forfeiture proceedings pursuant to 43(6)(a) of Schedule B1.

In considering the landlord's application we would draw your attention to the guidelines set down by the Court of Appeal in *Re Atlantic Computer Systems plc* [1992] 1 All ER 476. In that case Nicholls L.J. indicated that an administrator is expected to make his decision speedily and when he withholds consent he should state the reason why.

In relation to a landlord's request for consent to forfeit, Nicholls LJ stated at p500:

> "In carrying out the balancing exercise great importance, or weight, is normally given to the proprietary interests of the lessor... The underlying principle here is that an administration for the benefit of unsecured creditors should not be conducted at the expense of those who have proprietary rights which they are seeking to exercise, save to the extent that this may be unavoidable and even then this will usually be acceptable only to a strictly limited extent".

If you do withhold consent:

1. Please give your reasons and state what loss and consequences you anticipate will result from the grant of consent.

2. What are the administrator's proposals for the discharge:
   - of the arrears
   - future rent.

3. If the administrator does not propose to pay future rent please explain the reason why.

4. Please state the period for which the administration is expected to continue.

5. Please specify the end result which is sought to be achieved by the administrator and the prospects of that result being achieved.

6. Please indicate when the administrator anticipates that he will be in a position to allow the landlord to re take possession.

We reserve the right to show this letter to the court should the need arise.

We look forward to hearing from you within seven days of the date of this letter.

Yours faithfully,

10 Contingent Liability

10.1 Introduction

10.1.1 *What is contingent liability?*

A contingent liability is a potential claim that may arise in the future but which has not yet crystallised and is dependant on an event that has not yet occurred (usually the default of a third party). A corporate occupier who no longer needs premises may assign the lease of those premises to a new tenant (ie an assignee). However the assignment of a lease does not necessarily mean that a corporate occupier's potential liability under that lease comes to an end. On the contrary, the corporate occupier may remain potentially liable for breaches of the tenant covenants under that lease until it expires many years later. This is not an immediate liability because unless and until the assignee (or any subsequent assignee) defaults, the corporate occupier's liability will not crystallise.

From a landlord's perspective, where a current tenant is in default, the existence of former tenants/ guarantors with contingent liability will be a reassuring sight, particularly where they are juicy targets. Often their existence will provide a very strong reason for the landlord to preserve the lease and not to forfeit it. Instead the landlord is likely to pursue those with contingent liability.

On the other side of the coin, contingent liability can be a very serious problem for corporate occupiers. It arises from the inflexible nature of most commercial leases (particularly those granted in the 1980s and 1990s) with lease durations that extended way beyond the actual requirements of most corporate occupiers. For example, in the 1980s and early 1990s corporate occupiers were often granted lease durations of 25 years. Research has shown that the average period that corporate occupiers actually occupied leased premises is between six and seven years. Consequently, a corporate occupier that was granted a 25 year lease in 1985 and which was assigned after seven years would be potentially exposed to contingent liability in respect of that lease for a further 18 years. Put another way, it will be at risk of a substantial claim until 2010 from premises it occupied for merely seven years and which it disposed of over a decade ago.

10.1.2 *The Landlord and Tenant (Covenants) Act 1995*

Prior to the reforms brought in by the Landlord and Tenant (Covenants) Act 1995, a corporate occupier could be pursued by a landlord for arrears of rent many years after it had assigned the lease where the arrears were substantial and covered long periods of non payment by the current tenant. While the corporate occupier usually had a legal indemnity from the party to whom it assigned the lease (ie the assignee), this was of little practical use if the assignee was insolvent or had no assets. This led to some very harsh cases with corporate occupiers having to find significant sums of money for historic arrears of which they had no prior knowledge or warning.

The 1995 Act was passed to redress this problem in three broad ways:

1. It abolished former tenant/guarantor liability in the context of leases which were granted on or after 1 January 1996 (new tenancies) although contingent liability remains in respect of tenancies granted before this date (old tenancies). The only exception to the statutory automatic release on the assignment of a new tenancy is where the landlord requires (and is entitled to require) the former tenant to enter into an Authorised Guarantee Agreement (AGA). Under an AGA a former tenant guarantees the performance of the tenant obligations by the assignee, but only for such period as that assignee remains tenant. Once it assigns, the AGA comes to an end.
2. In the context of both old and new tenancies, if a landlord wishes to recover a fixed sum from a former tenant or its guarantor, it must serve a notice on that party under section 17 (in prescribed form) within six months of the sum becoming due. If it fails to serve the notice, it will be statute barred from seeking to recover that sum. Gone are the days when a landlord could seek to recover significant sums where it had failed to notify the former tenant/guarantor that arrears were accumulating.
3. A former tenant/guarantor that has made full payment in response to a section 17 notice can now require the landlord to grant it an overriding lease, therefore enabling the paying party to gain control, forfeit the lease of the defaulting tenant, and either go back into occupation or assign or sublet the premises.

There can be no doubt that the 1995 Act has eased the burden of a corporate occupier with contingent liabilities, however, it has by no means removed the problem. There are still a significant number of 25 year leases that were granted before 1996 and that have been assigned by corporate occupiers. They could and do come back to haunt corporate occupiers on a regular basis. Moreover, the nature of contingent liability means that the larger the corporate occupier and the stronger its covenant, the more likely the prospect it has of facing a claim.

From the landlord's perspective, a corporate occupier which has sub-let premises may be able to bring a claim against a party with contingent liability, although if the sub-letting has taken place since 1 January 1996 it will be subject to the restrictions brought in by the 1995 Act.

10.1.3 Contingent liability under old tenancies

Where a lease has been assigned on a number of occasions there are likely to be four potential candidates who may have contingent liability.

1. an original tenant
2. an intermediate/former assignee
3. a guarantor of a former tenant (original or intermediate)
4. a guarantor of the existing tenant.

Whether contingent liability remains following assignment of a lease will depend upon a number of factors. It will depend upon the principles of contract (and in particular the terms of the agreements entered into by the various parties involved), common law principles and statute. In particular, and as indicated above, since 1 January 1996 the principle of former tenant/guarantor liability has been restricted in its operation by the 1995 Act.

10.1.4 Working example

The principles of this chapter are best illustrated by reference to a simple (but not uncommon) example which will be use throughout.

- In 1990 the landlord (L) granted a lease to the original tenant (OT) for a term of 25 years.
- A guarantor (G1) entered into a covenant with L to guarantee the performance of OT's tenant obligations under the lease.
- In 1995 OT assigned the lease to the first assignee (A1).
- In 1998 A1 assigned the lease to the second assignee (A2).
- In 2000 A2 assigned the lease to the current tenant (A3).
- As A3 was not a good covenant, L insisted on a guarantor (G2) to guarantee the performance of A3's obligations under the lease.

The example can be illustrated by a simple diagram:

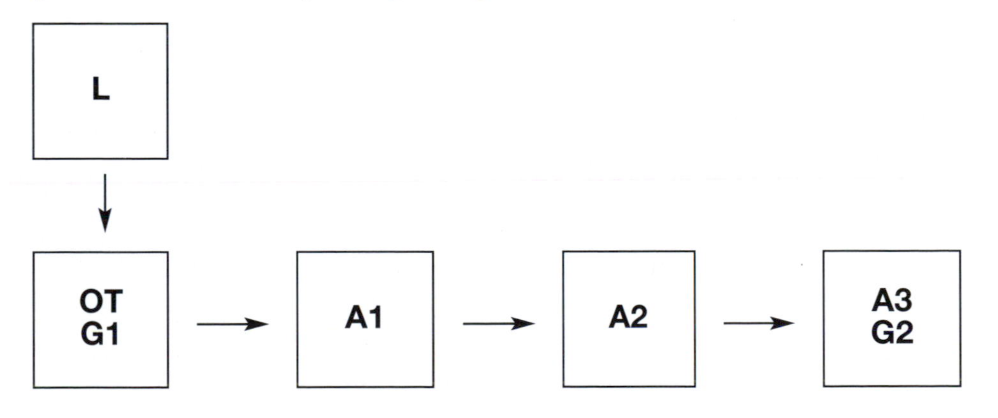

Using the above example, we shall examine the liability of each party with contingent liability in circumstances where A3 becomes insolvent and fails to pay rent.

10.2 Original tenant — OT's liabilities and rights

10.2.1 Legal basis for liability

Original tenant liability arises in the context of old tenancies ie those created before 1 January 1996. It arises by virtue of a lease being a contract between a landlord and a tenant. Under that contract both parties agree to perform the obligations imposed upon the landlord and the tenant by the lease for the duration of that contract. If one of the original parties chooses to assign the lease, the obligations between the original landlord and the original tenant will continue notwithstanding the assignment. This principle is often referred to as privity of contract. Consequently, if a tenant signs up to a 20 year lease it will be liable to pay the rents due under that lease (and perform the other tenant covenants) for the entire 20 year period, unless the lease is terminated before that date or it obtains a deed release from the landlord.

It is important, therefore, that a corporate occupier is fully aware of its commitments under any old tenancies that it may have signed up to in the past. Using the example above, if the premises become surplus to OT's needs, it will remain potentially liable to pay rents (and comply with the other covenants) long after it has ceased to occupy those premises. It may assign the lease to an A1, but if that assignee defaults then L can, subject to the points set out below, recover any arrears from OT.

Faced with a potential claim, there are a number of legal issues which an original tenant should investigate. However, it may have been many years since the original tenant had any dealings with the property and it will almost certainly need to obtain copies of the relevant information and documents from the landlord to enable it to evaluate the claim. A copy of a precedent letter to the landlord making such a request can be found at 10.8.1.

10.2.2 Release of covenant

If a corporate occupier assigned a lease many years ago, there may be nobody who remains within the company with knowledge of the property or the basis on which it was disposed. There are some rare cases where tenants have obtained a deed of release from the landlord at the time of the assignment. This can happen where the assignee, at the time of the assignment, is a very good covenant and/or the original tenant paid a premium to secure a release from future liability.

It is therefore worth scrutinising the original lease and all subsequent deeds carefully to see if a release was obtained. This may be contained in the licence to assign (if it was completed simultaneously with the assignment), or alternatively, in a separate deed.

To be effective, a release must be contained in a deed, or granted by an agreement for valuable consideration between the tenant and the original landlord or its successor in title.

If no deed or agreement of release exists, the contingent liability for an old lease will continue until the contractual term of the lease expires, or until the lease is determined by:

- forfeiture
- the exercise of a break right
- surrender.

Contingent liability does not come to end on the lease being disclaimed by a liquidator or a trustee in bankruptcy (see 10.7 below).

10.2.3 Subsequent variations in the lease

Unlike the position with a guarantor, and subject to one exception, an original tenant is not able to get off the hook by pointing to a subsequent variation in the lease to which it was not a party.

The only occasion where an original tenant might escape liability is where the attempted variation of the lease, as a matter of law, constituted a surrender of the existing lease followed by the re-grant of a new lease. In *Friends Provident Life Office* v *British Railways Board* [1995] 2 EGLR 55 it was held that a surrender and re-grant by operation of law will only occur where the landlord and tenant attempt to vary the existing lease by:

- adding to the existing premises or
- altering the length of the contractual term granted by the existing lease.

In *Friends Provident* an original tenant was sued by the landlord's successor in title for rent arrears. The original tenant argued that it had been released from its obligation by a variation of the lease which took place after the original tenant had assigned the lease. The variation comprised a relaxation in the alienation clause in return for a higher rent. The Court of Appeal held that the variations did not amount to a surrender and re-grant and that the original tenant therefore remained on the hook. However, the original tenant was only liable for the old (unvaried) rent.

Sir Christopher Slade in his judgment stated that the current law was correctly summarised by the following propositions which were contained in an article by Patrick McLoughlin published in *The Conveyencer,* 1994 p443.

1. The assignment of the lease does not destroy the privity of contract between the landlord and the original tenant and consequently the original tenant remains liable on all its covenants in the original lease notwithstanding its assignment.
2. If the contract embodied in the original lease itself provides for some variation in the future of the obligations to be performed by the tenant (for example by a rent review clause), the original tenant may be bound to perform the obligations as varied, even though the variations occur after the assignment of the lease — this will depend upon the construction of the relevant covenants in the original lease.
3. If, on the other hand, an assignee by arrangement with the landlord agrees to undertake some obligation not contemplated by the contract contained in the original lease, the estate may be altered, but the variation does not affect the obligations of the original tenant.

The effect on a former tenant of a variation that has taken place following an assignment of a lease is now also governed by section 18(2) of the 1995 Act. Where section 18 applies, this prevents a landlord from recovering any additional sums over and above the amount that would have been recoverable if the lease had not been varied. Consequently, if a lease is varied, the former tenant will remain liable under the lease covenants, but only to the extent that they have not been varied. If for example an assignee varies the lease to extend the user clause in return for a higher rent, the assignee then defaults and the landlord sues the former tenant for rent arrears, it will only be able to recover the original rent and not the increased rent.

Consequently, an original tenant will be liable for increased rent provided that increase arises in line with the rent review provisions contained in the lease. However, from a landlord's perspective, it is important that the rent review process is followed to the letter of the provisions contained in the lease. In Beegas Nominees Ltd v *BHP Petroleum Ltd* [1998] 2 EGLR 57 the landlord and the assignee had agreed a stepped rent as part of the rent review process. However, the Court of Appeal held that a stepped rent was not permitted by the rent review clause and therefore amounted to a variation that would not bind a former tenant. A different result on the basis of a different rent review clause was reached in *GUS Property Management Ltd* v *Texas Homecare Ltd* [1993] 2 EGLR 63.

10.2.4 Section 17 notice

Section 17 imposes a requirement on a landlord to serve a notice within six months of a fixed charge falling due if it wishes to recover that amount from a former tenant, or guarantor of a former tenant. This provision applies to both old and new tenancies alike.

(a) Fixed charge

The term fixed charge is defined by section 17(6) of the 1995 Act as:

- rent
- service charge
- liquidated sum provided for by the lease if the tenant breaches a covenant.

An example of the third category is a *Jervis* v *Harris* clause which allows a landlord to recover the cost of carrying out works to the premises following the failure by the tenant to comply with a notice to carry out works (see 14.4.2).

(b) Form of notice

The 1995 Act requires the notice to inform the former tenant/guarantor that a fixed charge is due and that the landlord intends to recover the amount specified, and where payable, interest, calculated on the basis set out in the notice. The notice must be in prescribed form. A precedent notice can be found at 10.8.1 below. A notice that is not in the prescribed form, or in a form "substantially to the same effect", will be invalid and the landlord will be unable to recover the fixed charges specified in that notice. Consequently, a landlord should take care to use the correct notice and to complete all the relevant parts.

If a landlord wrongly includes an item which is not properly due, this will not invalidate the notice provided it includes other items which are due and owing. In *Commercial Union Life Assurance Co Ltd* v *Moustafa* [1999] 2 EGLR 44 the original tenant argued that the section 17 notice served by the landlord was defective because it contained an error as to the way in which interest on rent arrears was calculated under the lease. It was held that a notice was not invalidated by the inclusion of items which the landlord was found not to be entitled to recover.

(c) Service of the notice

To satisfy section 17, a landlord must serve the notice on the former tenant/guarantor within the six month deadline. Ensuring that a notice is validly served is, therefore, as important as using the correct form and taking care to complete the form correctly.

Service of notices can be a complex and worrying business (see 12.3.1 (c) in the context of serving break notices). However, the landlord's plight is eased by section 27(5) of the 1995 Act which provides that section 23 of the Landlord and Tenant Act 1927 applies. The effect of this provision is to place the risk of the notice going astray and not being received, on the intended recipient. The notice is deemed to have been served provided the landlord can show that is was:

- served personally
- left for the recipient at its last known place of abode in England or Wales

- sent by recorded delivery addressed to the recipient at its last known place of abode in England or Wales or
- in the case of a local or public authority or a statutory or public utility company, sent by recorded delivery addressed to the secretary or other proper officer at its principal office.

The deeming provision was tested in *Commercial Union Life Assurance Co Ltd* v *Moustafa* [1999] 2 EGLR 44 where the notice was sent by recorded delivery to the last known address of the former tenant but was returned "undelivered". It was held that section 23 of the Landlord and Tenant Act 1927 applied and service was valid even though the notice was never actually received by the intended recipient.

(d) Additional amounts — pending rent review

Section 17(4) provides that a landlord cannot recover an amount in excess of the sum claimed in the notice unless:

- the fixed charge claimed is subsequently determined to be higher and
- the notice informed the recipient of the possibility of a higher charge and
- a further notice informing the recipient that a higher charge will be claimed is served within three months of the date the increased charge is determined.

It follows that if there is a possibility that the charge may be higher, this should be stated in the section 17 notice; and a second notice must be served after determination of the higher charge. Again, this notice must be in the prescribed form. A copy of a precedent notice can be found at 10.8.2. It is important to note that the second notice served under section 17(4) is in a different form to the initial notice. Failure to use the correct notice will be fatal to a landlord's claim.

The ability to serve a second notice under section 17(4) was aimed at liabilities which had not been finally determined at the date the notice was served such as an outstanding rent review or service charges involving on account payments with balancing charges or credits at the end of the financial year. However the draftsmen of section 17 did not consider carefully how such notices would operate where a tenant defaults during the course of an outstanding rent review process.

In *Scottish & Newcastle plc* v *Raguz* [2007] 15 EG 148 the rent reviews of two leases were outstanding for a period of over five years. Although the current tenant did not fall into arrears of rent until the end of this period, it was argued that the landlord could not recover the additional back rent (due as a result of the rent review determination) from an original tenant because the landlord had failed to serve the initial notices every six months, followed by a section 17(4) notice within three months of the rent determination. Surprisingly, this argument was upheld by the Court of Appeal. Lloyd LJ stated:

> ... if a landlord wishes to preserve the possibility of claiming against an original tenant when rent is subject to review, he must serve section 17(2) notices within six months after each rent day in turn, specifying in the Schedule that the sum intended to be recovered is then nil, but subject to paragraph 4 of the notice and the possibility of the rent being determined to be a greater sum...

The effect of this decision is to impose a considerable burden on landlords who may wish to recover back rent from former tenants or their guarantors following the initiation of a rent review process. Even though there are no arrears at each rent payment date, to protect its position a landlord must:

- serve a section 17(2) notice (former tenant) and/or 17(3) notice (guarantor of former tenant) on all relevant potential targets within six months of each rent payment date and
- the notice must state that the sum to be recovered is nil but could be determined to be higher and
- if the current tenant defaults, serve a section 17(4) on the former tenant and/or its guarantor within three months of the rent being determined.

Although it is clear that the draftsmen of section 17 could not have intended such an obscure result, the alternative could lead to a situation where the original tenant is faced with a very substantial claim for back rent following a protracted rent review process. This was precisely the ill which the policy underlying section 17 sought to remedy. Consequently, if a landlord wishes to protect its position during a rent review, it may have to serve many notices on a six monthly basis unless and until the review is completed. This will impose a substantial burden on investor landlords of large portfolios and is a point for the corporate occupier to watch out for when pursued by the landlord in these circumstances. The position is infinitely worse, if, as is suggested by some commentators, this principle applies to service charges which operate by payments on account with balancing charges at the end of the year. At the time of writing, *Raguz* is on appeal to the House of Lords. It remains to be seen whether the House of Lords will uphold the Court of Appeal ruling, and if it does, whether the courts will extend the principle to service charges.

(e) Effect of failure to serve notice

Failure to serve a notice in accordance with section 17 will provide a complete defence to a claim by the landlord. If a landlord issues proceedings to recover a fixed charge after failing to serve a section 17 notice the proceedings can be brought to an end swiftly by an application to strike out the claim and/or an application for summary judgment on the basis that the landlord has no reasonable prospects of success. The only evidence required will be a short witness statement confirming that no notice was served within the six month deadline.

10.2.5 Is the debt disputed by the tenant?

The original tenant is not likely to be in contact with the current tenant of the property. When it receives a section 17 notice from the landlord it will only have the landlord's word that the amounts specified in the notice are in fact outstanding. These amounts could be disputed by the current tenant. The original tenant must, therefore, try and make contact with the current tenant and ascertain whether the sums are due but it simply cannot pay, or alternatively, whether the current tenant disputes the arrears or has a counterclaim.

10.2.6 Remedies following payment

If an original tenant, having carried out all of the investigations identified above, decides that it has to pay sums to the landlord in response to a section 17 notice, it will wish to take immediate action to try to prevent or minimise further claims arising in the future as fresh sums fall due under the lease. Alternatively, or in addition to these steps, it may wish to try and recover its outlay from other parties in the chain.

(a) Overriding lease

Section 19 of the 1995 Act was passed with the intention of mitigating the injustice of former tenants/ guarantors being required to pay rent (or other sums due) for premises they neither occupied nor were entitled to deal with. It provides that a former tenant (or its guarantor) that has made a full payment in response to a section 17 notice can require the landlord to grant it an overriding lease.

An overriding lease is a lease granted by the landlord out of its reversion but subject to the existing lease which has given rise to the default. It will usually be for a term three days longer than the existing lease, and contain identical terms (except in relation to personal terms or spent obligations which are not to be included).

The grant of an overriding lease enables the former tenant or its guarantor to take control of the unfortunate situation it finds itself in. On being granted an overriding lease, it will become the immediate landlord of the defaulting tenant and thus have the full panoply of remedies that are available to landlords when a tenant is in default (see Chapter 8 above). In particular, in the right circumstances, it will allow the lease to be forfeit thus providing the former tenant or its guarantor with an opportunity of re-letting the premises which will generate much needed income to offset against rent due under the overriding lease.

To exercise a right a party must make a written request to the landlord identifying the payment which has given rise to the right to request an overriding lease. This request must be made at the time of making the qualifying payment, or within 12 months beginning with the date of payment (section 19(5)).

On receipt of a request, the landlord is under an obligation to grant an overriding lease within "a reasonable time" of the request and on completion the party requesting the overriding lease must pay the landlord's reasonable costs of and incidental to the granting of the lease (section 19(6)). In practice however, the amount of time it can take to put an overriding lease in place is still generally too long.

There is no doubt that section 19 has improved the lot of a former tenant or its guarantor. However, there are a number of practical and legal issues to keep in mind from the perspective of a former tenant or its guarantor.

1. Stamp duty land tax will be payable on the overriding lease. This may be substantial where a lease has many years to run and/or has a high rent.
2. Entitlement to an overriding lease arises only where full payment, together with interest, has been made in response to a section 17 notice. Part payment (for example where an item is challenged) is likely to cause problems and delays.
3. The terms of the overriding lease should be the same as the current lease (with minor consequential amendment). However, section 19(2)(b) does provide for agreed modifications, and in practice, negotiations can become protracted and costly.
4. It will be required to pay the landlord's legal costs of preparing the overriding lease.

From the landlord's perspective, once a former tenant or its guarantor has paid up in response to a section 17 notice, it *must* grant an overriding lease to the payee if requested. A landlord will want to pick its target very carefully and it should think twice before adopting a scattergun approach. This can be illustrated by the chain set out below. Assume A3 (the current tenant) defaults and the landlord serves a section 17 notice on OT (the original tenant). If OT then pays up, takes an overriding lease and forfeits the original lease, the chain of privity will be destroyed and G1, A1, A2, A3 and G2 will all be off the hook so far as future liability is concerned. Consequently, if a weaker covenant pays up and obtains an overriding lease, stronger covenants (including the current tenant) may be lost.

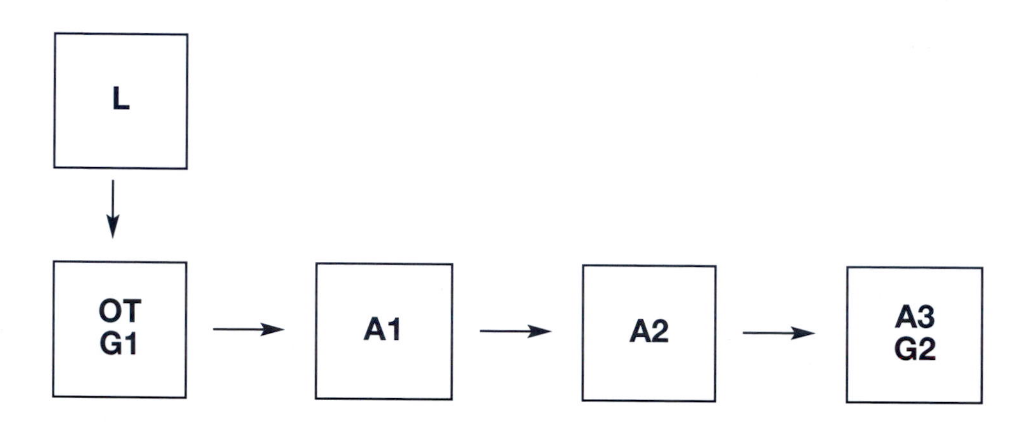

(b) Claims against other parties in chain

If there are other juicy targets in the chain, the original tenant may decide not to seek an overriding lease, but rather, to pass on the claim to another party in the chain. Alternatively, it may request an overriding lease and at the same time seek to recover any payments which it has made to the landlord from other parties in the chain.

(i) *A1 — the immediate assignee of OT*

The most obvious choice is the person to whom the original tenant assigned the lease (ie A1). In the example given above, if A3 defaults and OT pays up in response to a demand from L, OT may have a claim against A1 for the money it has handed over to L based on either an express or implied indemnity.

When a tenant (ie assignor) assigns a lease to an assignee, the deed of assignment may contain an express indemnity by the assignee in favour of the assignor. This will usually take the form of a covenant by the assignee (A1) to pay the rents and observe the tenant covenants for the remainder of the lease term, and also to indemnify the assignor (OT) for any consequences arising out of a breach or non payment by the assignee.

If the deed of assignment of an old tenancy does not expressly contain an indemnity, then subject to an express intention to the contrary contained in the deed of assignment, one will be implied either by section 24(1)(b) of the Land Registration Act 1925 in the case of a registered lease, or alternatively Part IX of schedule 2 to the Law of Property Act 1925. Both provisions give rise to a similar implied indemnity on assignment of a lease.

By way of example, section 24(1)(b) provides:

> ... during the residue of the term the [assignee] and the persons deriving title under him will pay, perform and observe the rent, covenants ... and will keep the [former tenant] indemnified against all actions, expenses, and claims on account of the non-payment of the said rent... or the breach of the said covenants ...

The extent of the indemnity contained in section 24(1)(b) of the Land Registration Act 1925 was considered in *Scottish & Newcastle plc* v *Raguz* [2007] 15 EG 148.

Facts

- L's predecessor granted two leases to OT.
- OT assigned the lease to A1. The assignment was subject to the implied indemnity contained in section 24(1)(b) of the Land Registration Act 1925.
- A1 assigned on to A2. There were further assignments.
- The current tenant defaulted and L claimed rent from OT which it paid totalling over £600,000.
- OT claimed an indemnity from A1.
- A1 argued that OT had a complete defence to L's claim for a substantial part of the rent arrears (which comprised back rent following rent reviews) because L had failed to serve the relevant section 17 notices.
- As OT was under no legal obligation to pay L, A1 argued it equally was under no obligation to pay OT

Court of Appeal — Held

- The amounts claimed were "expenses" or "claims" which arose because the current tenant did not pay the rent due.
- Although at first sight it seems odd that a party should be entitled to be indemnified against payments to a third party which he is not bound to make, the scope of the indemnity was not limited to payments which OT was legally liable to make.
- The appropriate test is whether the expenditure was fairly and reasonably incurred. The section 17 point was not a clear point, and the commercial context meant that OT's action in not antagonising the L to ensure a quick assignment of the lease was reasonable in the circumstances.
- OT was therefore entitled to an indemnity for the monies it had paid L even though L had no legal right to recover those monies from OT.

At the time of writing, *Raguz* is on appeal to the House of Lords.

(ii) *A2/A3 — other assignees*

If the original tenant's immediate assignee cannot be traced or is impecunious, it may have a claim against a later assignee.

At common law each subsequent assignee must indemnify the original tenant against any breaches of covenant while the lease is vested in him (see *Baynton* v *Morgan* (1888) 22 QBD 74 and *Moule* v *Garrett* (1872) LR 7 Ex 101). This is based on a restitutionary right (or common law indemnity) by which OT is entitled to recover from A3 (ie the assignee in possession) any rent which it has had to pay to the landlord because A3 has enjoyed the benefit of occupying the land. Consequently, if the current tenant (A3) is still worth powder and shot, OT can seek reimbursement from A3.

The law becomes more opaque in relation to whether OT can recover from A2 (a remote assignee down the chain). There is no direct or implied covenant by A2 to OT as there was no legal contract between A2 and OT. A2 is not the current tenant and does not benefit from the payment which OT makes to L in the same way that the current tenant in possession of the premises does and so restitutionary principles do not apply. The current thinking is that OT is unlikely to be able to recover a complete indemnity from A2. OT may, however be able to seek a contribution from A2 because they are severally liable in respect of the same debt. The amount OT can recover will depend upon how many other potential targets exist. If the only other potential target for the landlord was A2, OT will be entitled to recover half sum handed over to L. If, however, the current tenant — A3 has not been wound up or made bankrupt, OT will be entitled to recover one third from A2.

(iii) *G2 — guarantors further down the chain*
Can OT seek to recover its outlay to L from G2? The first instances decisions of *Selous Street Properties Ltd* v *Oronel Fabrics Ltd* [1984] 1 EGLR 50 and *Becton Dickinson UK Ltd* v *Zwebner* [1989] 1 EGLR 71 suggest that it can. However, this has been seriously doubted by some leading commentators (see paragraph 7.26 of *Enforceability of Landlord and Tenant Covenants* 2nd ed by TM Fancourt QC). At present, therefore, this issue is open until it is decided by the Court of Appeal or House of Lords.

10.3 Former Assignee — A1 and A2's liabilities and rights

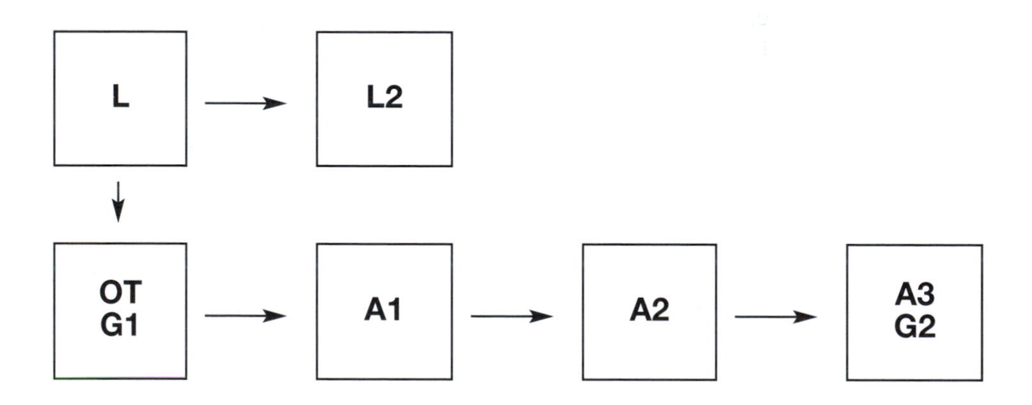

10.3.1 Legal basis for liability

The landlord's ability to sue an intermediate assignee for rent not paid by the current tenant may arise in two ways:

- by 'privity of estate' or
- by a direct covenant.

(a) Privity of estate

When OT assigns the lease to A1, as a matter of contract, although OT remains liable on the covenants for the remainder of the term, unless A1 enters into a direct covenant with L (see below) there is no contractual relationship between A1 and L. The same is true when the lease is subsequently assigned to A2.

In the absence of a direct covenant with L, no contractual relationship with the landlord exists and the position is governed by section 141 of the Law of Property Act 1925. A1 and A2 are liable to pay the rents to the current landlord and comply with the tenant covenants so far as they touch and concern the demised premises for so long as they remain the tenant. The liability of A1 (or subsequently A2) to pay to L the rent and comply with the tenant covenants whilst it is the tenant is often referred to as privity of estate. The liability only continues for so long as A1 (or A2) is the owner of the leasehold estate. In the above example, therefore, once A1 assigns to A2, A1 ceases to be liable for rent and the tenant covenants on assignment. A2 will then become liable until it assigns to A3, following which it too will be off the hook. If L1 were to assign the reversion to L2, then the entitlement of L1 to receive rents would end and pass to L2 on the date of the assignment.

Consequently, if A1 assigns the lease to A2, and A2 falls into arrears, A1 will not be liable for the arrears that accrue under A2's subsequent watch. Similarly, if A2 assigns the lease to A3 which then defaults, then L will not be able to sue A1 or A2 for A3's arrears (although L will be able to pursue OT and/or G1 by virtue of privity of contract).

(b) Direct covenant

In most cases a landlord will not be limited to relying upon privity of estate. It is very common for a landlord to take a direct covenant from each assignee of the term. This will usually be contained in the licence to assign. Although such covenants are usually a matter of negotiation between the assignee and the landlord, in most cases it will involve the assignee covenanting that the rent will be paid and the tenant covenants complied with for the remainder of the contractual term. A corporate occupier which enters into such a direct covenant should therefore be aware of its potentially onerous nature. Unless the wording expressly limits its duration to the period of the assignee's tenure as tenant, it will potentially leave the assignee liable to pay rent and comply with the tenant covenants for the remainder of the lease term, even if the lease is assigned many years before lease expiry.

Importantly, a direct covenant given by an intermediate assignee to a landlord will be enforceable by the landlord's successors in title. Therefore, in the example given above, if A1 gives a direct covenant to L, and L subsequently assigns the reversion to L2, L2 will be entitled to enforce the direct covenant against A1 (see *P & A Swift Investments Ltd* v *Combined English Stores Group Ltd* [1988] 2 EGLR 67).

10.3.2 Liability

In our example, If A1 and A2 have given direct covenants and A3 defaults, L can potentially pursue A1 and A2 for A3's default. However, when faced with such a claim, the issues explored in 10.2 above in the context of original tenant liability need to be considered:

- Has there been a release of the direct covenant?
- Have there been any subsequent variations in the lease?
- Has a valid section 17 notice been served within the six month period?
- Is the debt disputed by the current tenant?

10.3.3 Remedies following payment

If A1, or alternatively A2, pay sums due from A3 following the service of a valid section 17 notice, then an application can be made to the landlord for an overriding lease (see 10.2.6 (a) above).

Alternatively, or in addition to obtaining an overriding lease, A1 or A2 may seek to recover any payments from other parties in the chain.

If A2 is required to pay A3's rent, A2 may seek to recover:

- the full amount from A3 relying upon the express or implied indemnity given on assignment (although commercially if L has pursued A2, A3 may be insolvent or without funds)
- the full amount from G2 (although this is doubtful — see 10.2.6 (b) (iii) above)
- a contribution from OT and A1 who are severally liable for the same debt. This principle may extend to G1 and G2.

If A1 is required to pay A3's rent, A1 may seek to recover:

- the full amount from A2 relying upon the express or implied indemnity given on assignment
- the full amount from A3 (see *Moule* v *Garrett* (1872) LR 7 Ex 101) although commercially it may be a barren target
- the full amount from G2 (although this is doubtful — see above)
- contributions from OT and A2. This principle may extend to G1 and G2.

10.4 Guarantor of former tenant — G1's liabilities and rights

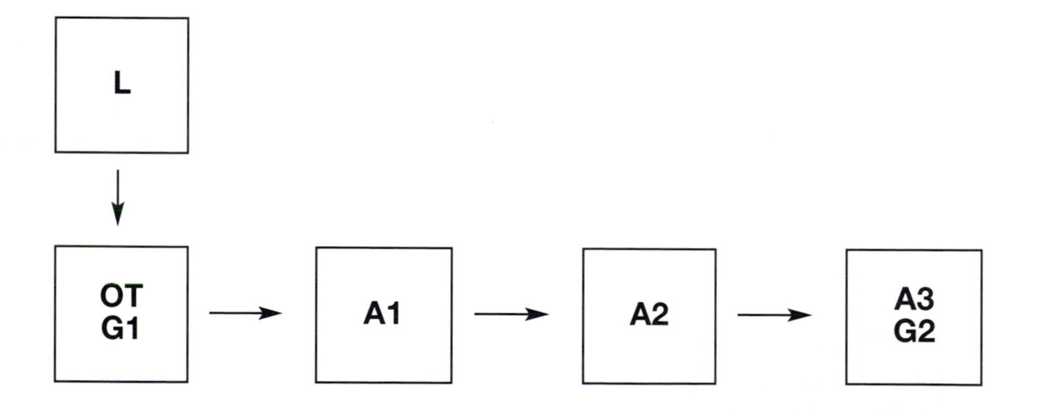

10.4.1 Nature of obligation

A landlord that is unhappy about the covenant strength of a proposed tenant may seek to bolster its position by insisting that an individual and/or a financially stronger company enter into a direct covenant with the landlord to guarantee the performance of the tenant covenants under the lease. If the tenant then defaults, the landlord will have a separate claim against the guarantor (also known as surety) in addition to, or instead of, any claim against the tenant.

The covenant of guarantee is between the guarantor and the landlord and will usually be contained in the lease itself. The precise legal nature of the guarantee obligation will depend upon the terms of the actual covenant. The starting point for either a landlord considering action against a guarantor, or a guarantor considering its potential liability, will be to look carefully at the obligations imposed.

Most guarantee covenants comprise three separate obligations:

- the guarantor covenants that the tenant will perform all its obligations under the lease and
- if the tenant is in breach the guarantor will perform those obligations and pay any sums due and outstanding under the lease and
- if the lease is disclaimed, the guarantor will enter into a new lease on the same terms as the disclaimed lease for the remainder of the contractual lease term.

It is important to note that although the covenant of guarantee is between the guarantor and the landlord, the benefit of that guarantor covenant will pass to the landlord's successor in title if it sells the reversion (see *P& A Swift Investment Ltd* v *Combined English Stores Group Ltd* [1988] 2 EGLR 67).

10.4.2 Operation of guarantee

The formal steps required to activate an obligation of a guarantor will depend upon the wording of the specific guarantee. Some require the service of a demand or notice. If a demand or notice is required to trigger the guarantee, no obligation will arise until this requirement has been satisfied.

However even if no formal notice or demand is required, from a landlord's perspective, once a tenant has defaulted it is sensible to write to the guarantor promptly, referring to the covenant of guarantee, and demanding any outstanding amounts. The landlord may also wish to point out that any delay in payment is likely to result in the guarantor ultimately paying out more by way of interest payments and legal costs.

If payment is not forthcoming and the landlord's other remedies are unattractive, the landlord may wish to issue proceedings for the outstanding amounts and immediately apply for summary judgment. In most cases this will put substantial pressure on the guarantor and is likely to result in swift payment. Alternatively, if the landlord wants to take a more aggressive stance, it may wish to serve a statutory demand (see 8.10 above).

10.4.3 Potential defences

A guarantor of an existing tenant (ie G2 above) should investigate a potential claim carefully. It should consider broadly the same issues identified in the section on original tenant liability.

- Is the debt due and/or disputed (enquiries should be made of the current tenant whose obligations have been guaranteed)?
- The wording of the specific guarantor obligation should be considered carefully — does it cover items which L is seeking to recover? Is a demand required? Was a time-limit imposed?
- Has the guarantor covenant been expressly released?
- Has a valid section 17 notice been served in respect of the amount claimed?

There is one potential defence available to a current tenant that is not available to former tenants. It is an important line of attack and should be investigated carefully before payment is made. A guarantor will normally be discharged from its obligations under a guarantee if the lease has been modified or varied without the guarantor being a party to the variation (see *Holme* v *Brunskill* (1877) 3 QBD 495). Consequently, if L and OT agreed to relax the alienation clause in exchange for an increased rent, and G1 was not a party to that variation, G1 would be released from all liability under the guarantee from the date of the variation. This defence may fail if the wording of the guarantee provides that G1 will not be discharged by variations to the lease to which it is not a party. The court will however scrutinise the wording carefully to ensure that the guarantee was intended to impose such an onerous obligation. It appears, therefore, that the rule in *Holme* v *Brunskill* can be defeated by careful wording of the guarantee.

As part of its investigations therefore, a guarantor should seek to obtain all subsequent deeds and side letters concerning the lease from the landlord (or the current tenant) to see if it has a potential defence. A precedent letter can be found at 10.8.1.This point is sometimes overlooked by guarantors and their advisors.

10.4.4 Onerous obligation

From a guarantor's perspective, such a covenant can be very onerous. It requires the guarantor to have a close relationship with the tenant, and be in a position to influence (if not control) the tenant's actions in relation to the leased premises. At the very least it needs to know whether rent is being paid regularly. However, guarantees are often provided by companies in the same group as the tenant. If there is a subsequent sale of the company or part of it, a situation may arise where the guarantor is no longer in the same group as the tenant. Such legacy problems are often overlooked, particularly where the guarantee covenant is contained in the lease and not a separate document. In those circumstances, a copy of the lease is unlikely to be kept by the guarantor and the obligation forgotten. A claim many years later can come as a very nasty surprise and leave the guarantor in difficulties if no provision has been made for such an event. This again reinforces the need for a company to keep good records of all significant contingent liabilities.

If the tenant succumbs to a formal insolvency procedure the guarantor may be unable to gain control of the situation for some considerable time. If the lease is disclaimed, and the landlord decides not to rely upon a covenant requiring the guarantor to take a fresh lease for the remainder of the term, the guarantor's only option will be to apply to the court for an order vesting the disclaimed lease in it (see 10.7.4 and 10.7.5 below). However, it may take many months before a lease is disclaimed (particularly if the company is in administration before a liquidator is appointed) and during that time the guarantor will be liable for the rents without having an opportunity to gain control of the premises and remarket them with a view to assignment or sub-letting. This opportunity will only arise after the landlord has served a section 17 notice and the guarantor has satisfied the demand in full.

A corporate occupier that stands as surety for another company should also be conscious that depending upon the wording of the guarantee, unless there is an express release of its covenant on assignment, it will potentially remain liable for the remainder of the lease term.

10.4.5 "Guarantee stripping"

From a landlord's perspective, the existence of a strong guarantor covenant (particularly when the tenant company's covenant is much weaker) has been an important safeguard, and the provision of a parent company guarantor has become a common feature of the commercial property market. However, in an insolvency situation the existence of a potential claim by a landlord against a parent company can undermine a CVA. While the landlord's claims for existing rent can be included in the arrangement (see 9.1.2 above), the landlord is nevertheless free to sue the parent company under the guarantee, which in turn, has an indemnity against the tenant company for the sums paid out. A potentially serious challenge to the value of a parent company guarantee arose in *Prudential Assurance Company Ltd* v *PRG Powerhouse Ltd* [2007] EWHC 1002 Ch. The facts of *Powerhouse* are set out in detail above at 9.1.2 (d).

In *Powerhouse* the purpose of the CVA was to enable loss making stores to be closed down and the profit making stores to remain open. The CVA was challenged by the landlords of the loss making stores. Its implementation would have led to the payment of a dividend of 28p in the pound in lieu of future rent. A further sting in the tail for these landlords was that the CVA provided for the parent company (which was putting up the money for the scheme) to be released from its guarantees in respect of the leases of the loss making stores. In contrast, the landlords of the profit making stores were to be paid in full.

If this device had succeeded, it would have had dramatic consequences for the commercial property market by significantly undermining the value of parent company guarantees. However, the High Court held that the proposed CVA unfairly prejudiced the interests of a group of creditors (ie the landlords of the loss making stores) under section 6 of the Insolvency 1986 Act. The court found that the effect of the CVA was to discharge the present, future and contingent claims of these landlords at a fraction of their value so that other creditors (including landlords of the profit making stores) could be paid in full. If the tenant company had been wound up, the landlords of the loss making stores would have been better off. The same could not be said of the other creditors. They stood to gain from the CVA and inevitably swamped the votes of the landlords who were significantly disadvantaged by it.

The coach and horses have, for the time being at least, been diverted around parent company guarantees. While this judgment may not stop future attempts to incorporate contingent guarantor liability into a CVA (indeed it was held that as a matter of law such a provision was possible), it illustrates the difficulty in doing so and that future attempts are unlikely to succeed.

10.4.6 Remedies following payment

If a guarantor of a former tenant makes a payment in full in response to a section 17 notice served by the landlord, it is entitled to apply for an overriding lease under section 19 of the 1995 Act (see 10.2.6 (a) above).

If the guarantor pays up in response to a demand from the landlord, it can seek to recover that amount from the tenant under an express, or implied, indemnity. However, this will be little comfort if the tenant has no assets. If it is worth powder and shot, the guarantor can join the tenant as a third party into any court proceedings brought by the landlord to recover the arrears.

If A3 is in administration, G1 in certain circumstances may be able to bring a subrogated claim against A3 (in administration) seeking the payment of rent as an expense of the administration (see 10.5 below).

In *Selous Street Properties Ltd* v *Oronel Fabrics Ltd* [1984] 1 EGLR 50 it was suggested that the guarantor with closest proximity to the defaulting tenant may have primary liability. If this is correct, G1 may be able to seek re-imbursement from G2 (ie the guarantor of the defaulting tenant (A3). However, as stated above, this decision has been much criticised (see 10.2.6.(b)(iii)).

It may be possible (although the law on this point is far from clear) for G1 to seek a contribution from G2 on the basis that they are both liable to contribute equally to the debt. It is also possible that this principle will extend to OT, A1 and A2.

10.5 Guarantor of current tenant — G2's liabilities and rights

10.5.1 Differences from position of guarantor of former tenant

In the example given above, if A3 defaults in payment of rent, L's first potential target is likely to be G2.

A guarantor of a current tenant (ie G2) is in the same position as a guarantor of a former tenant (ie G1) except in two material respects.

First, the protection brought in by the 1995 Act for the benefit of former tenants/guarantors does not extend to a guarantor of a current tenant. In the example given above, L does not need to serve a section 17 notice within the six month period to enable a claim to be brought, and importantly, G2 is not entitled to an overriding lease if it satisfies a demand made under the guarantee. These protections only apply to guarantors of a former tenant (eg G1 in the example given). Consequently, there is no requirement for a landlord to demand arrears from G2 within six months of the rent falling due.

Second, with the exception of the current tenant which has defaulted (ie A3) and which may not be good for the money, there are no obvious targets to seek an indemnity from. However, G2 may be entitled to seek a contribution at common law from previous tenants and guarantors (ie OT, G1, A1 and A2).

10.5.2 Guarantor of current tenant and administration

The lot of G2 is not a happy one. The following example demonstrates an all too common scenario:

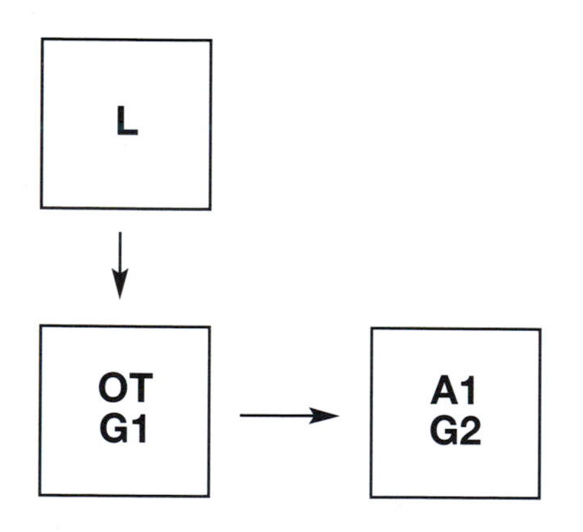

- L grants a 25 year lease to OT. G1 stands as guarantor.
- T1 assigns the lease to A1. G2 (a company in the same group) stands as guarantor.
- A1 and G2 cease to be in the same group of companies when part of the business is sold off to a third party.
- A1 goes into administration.
- The administrators of A1 pay no rent and make no attempt to assign or surrender the lease. They believe that the premises may be required by the purchasers of the business (or part of the business) and are necessary to ensure a greater recovery for the creditors if the company is wound up.
- L demands a quarter's rent from G2.

G2 is in a most unenviable position. It is liable to pay rent (and comply with other tenant covenants) for the remainder of the lease. However, it is not able to obtain an overriding lease and take steps to forfeit and remarket the lease because section 19 of the 1995 Act only applies to guarantors of *former* and not *existing* tenants.

G2 may have the benefit of an indemnity (either express or implied) from A1. However there are two problems with this:

- the permission of the court or of the administrators is required before a claim can be brought against the company in administration and
- even if a judgment is obtained, G2 would merely rank as an unsecured creditor and will be something of a pyrrhic victory.

On the face of it therefore, G2 is at risk of being liable to pay the rent for an indefinite period whilst the administrators attempt to sell A1's business. G2 has no benefit from the premises for which it is paying rent and it has no obvious legal route to obtaining control of the situation.

While there is no simple or clear immediate legal solution, there are a number of steps its advisors can take to try to ameliorate its predicament:

1. Although there may be no legal right to an overriding lease, there may be good commercial reasons why the landlord would be prepared to grant one. Most investor landlords will want to tidy the situation up quickly and at the minimum of cost. They will want to avoid the premises being empty for significant periods and will not want to risk void rental periods. For these reasons it is always worthwhile G2 requesting an overriding lease, and possibly as an opening gambit, offering to make payment conditional on an overriding lease being granted (although there is no legal basis for imposing such a condition).
2. In parallel with a request for an overriding lease, G2's advisors should put pressure on the administrators of A1. There are essentially two points that can be made to the administrator.

    - If the administrators have made a decision to keep the premises for the purposes of selling the company's business, it is arguable that the rent should be paid by the administrators as an expense of the administration properly incurred (see rule 2.67 (1)(a) of the Insolvency Rules 1986) or alternatively, as a necessary disbursement (see (f)). If this is accepted by the administrators, and they agree to reimburse G2 for any payments already made to the landlord, it will recover its outlay quickly. Alternatively, if no payment has been made by G2, it may be possible to persuade the administrators to pay the rent demanded of G2 under the indemnity, direct to the landlord.
    - If the administrators refuse to reimburse G2 for the rent it has had to pay to L under its guarantee, and they fail to agree to pay future rent as it falls due, it is arguable that G2 can take over by subrogation any rights which the landlord had against the administrators, including the right to be paid in full for rent as an expense of the administration (by a parity of reasoning with the position on liquidation: see *Re Downer Enterprises Ltd* [1974] 2 All ER 1074). The advantage of pursuing a subrogated claim for rent as an expense of the administration is that if successful, it will enable G2 to gain priority over unsecured creditors.

3. If L refuses to grant an overriding lease while continuing to demand rent under the guarantee, and the administrators continue to use the premises as an incentive for the sale of A1's business

without paying rent, G2 can make an application to the court to challenge the administrator's conduct of the company. Under paragraph 74(1) of Schedule B1 to the Insolvency Act 1986 a creditor (which G2 will be by virtue of the subrogated claim or a claim based on an express or implied indemnity from A1) may apply to the court if the administrator is acting or has acted so as unfairly to harm the interests of the applicant. Alternatively, a creditor can make an application if the administrator is not performing his functions as quickly or efficiently as is reasonably practicable. The court has very wide powers to order relief. G2 can argue that it is being forced to pay rent wholly for the benefit of other creditors by continuing to keep the premises available for the purposes of a possible sale of the business. G2 will be seeking an order that the administrator pays rent as an expense of the administration for such period as the premises are required by A1 and until it takes reasonable steps to surrender the lease or assign it to G2.

In most cases the sting can be drawn out of the above predicament by a combination of the steps suggested above. However there are no guarantees and the moral of the story is very clear: think very carefully before standing as guarantor for a tenant. If a guarantor covenant must be given then adequate provision should be made for this potentially onerous contingent liability.

10.6 New tenancies — the 1995 Act

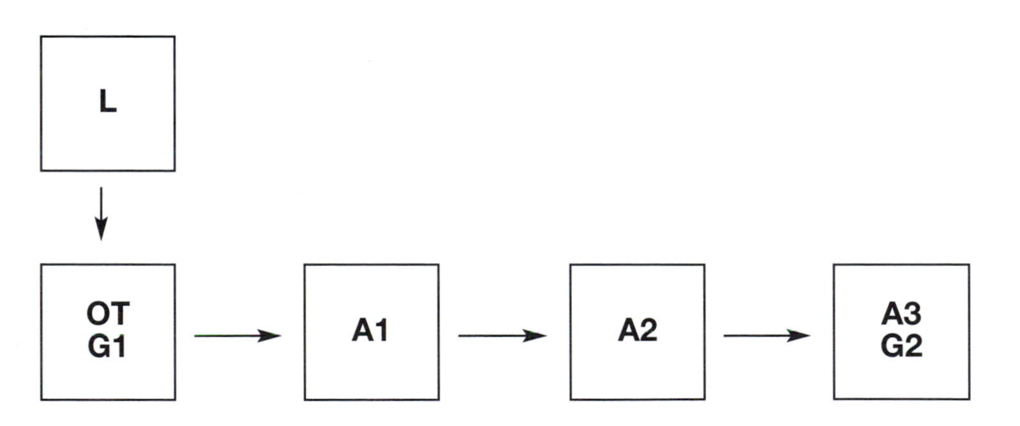

10.6.1 Former tenant/guarantor liability abolished

In all but a limited number of minor exceptions, section 5 of the 1995 Act provides a tenant of a new lease (ie granted after 1 January 1996) with an automatic statutory release from its covenants from the date that the lease is assigned. A guarantor of a former tenant is also automatically released from its liability under section 24 (2) of the 1995 Act.

Consequently, assuming that there is no authorised guarantee agreement (see below):

- OT and G1 will not be liable to pay rent to L for any arrears that arise following the assignment of the lease from OT to A1

- A1 will not be liable for any arrears of rent arising after assigning the lease to A2
- If A3 defaults, L will not be entitled to sue OT, G1,A1 and A2 in the same way as it would be entitled to do if the lease in question was an old lease. G2 (ie the guarantor of the existing tenant) would be the only alternative target.

The automatic release provisions are subject to robust anti-avoidance provisions contained in section 25 of the 1995 Act. Any agreement which attempts to exclude, modify or frustrate the operation of the Act will be void.

It is important to appreciate that the 1995 Act has not done away with all contingent liability for new leases. It will remain in two principal circumstances:

- G2 — a guarantor of an existing tenant and
- a former tenant which has entered into an authorised guarantee agreement (see 10.6.2 below).

The plight of a guarantor of an existing tenant is dealt with in detail above in respect of old leases. Its position is unaltered by the 1995 Act (except to the extent that it will be discharged from its obligations when the lease is assigned). However, the scenario set out at 10.5.2 is now more likely. It has become a practice among some investor landlords to require the surety of an outgoing tenant to also stand as surety for the incoming tenant ie

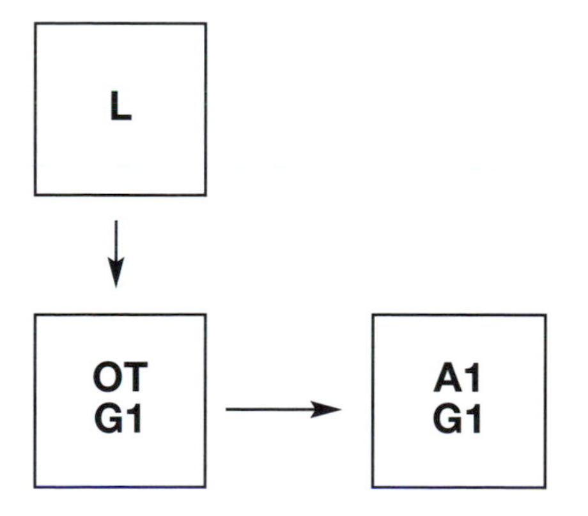

From the landlord's perspective this makes commercial sense in circumstances where the assignee is a weak covenant and is unable to put forward a sufficiently strong guarantor. Following the assignment of the lease from OT to A1, G1 will be release from its guarantee by section 24(2) of the 1995 Act. L will therefore lose its guarantor. However, there is nothing in the anti-avoidance provisions that prevents a guarantor of a former tenant standing as guarantor again, this time for the assignee.

From G1's position this is likely to be very risky. In most commercial situations it is unlikely that there will be a sufficiently close nexus between principal and guarantor to justify such an onerous covenant. In the majority of cases A1 and G1 will be unrelated companies. If A1 defaults, G1 will *not* be legally entitled to an overriding lease on payment of rent to L because section 19 of the 1995 Act only applies to guarantors of former tenants and not to those of existing tenants.

L may indicate that because A1's covenant is weak it will not give consent to assignment unless G1 stands as guarantor. If the market is weak and the premises are surplus, OT and G1 may be under pressure to mitigate the rental loss. However, and depending on the circumstances, a better solution may be for the proposed transaction to be restructured as a subletting to enable OT and G1 to maintain control. In some cases (perhaps where OT is itself insolvent) this may not be possible. If G1 does guarantee A1's obligations for commercial reasons, it should at the very least make realistic provision for the contingent liability.

10.6.2 Authorised Guarantee Agreement (AGA)

The quid pro quo for the automatic release of tenant covenants which arises on assignment is a right given to landlords, in certain circumstances, to require the outgoing tenant, as a condition for granting consent to assign the lease, to stand as guarantor for the incoming tenant (section 16).

An authorised guarantee agreement is a creature of statute and to be fully binding on an assignee must comply with the requirements of section 16 of the 1995 Act.

An AGA can impose three types of obligation on a former tenant:

- a liability as sole or principal debtor for any obligation owed by the assignee and/or
- a liability as guarantor for the assignee's performance of the tenant covenants and/or
- if the tenancy is disclaimed, a requirement to enter into a new tenancy for the remainder of the contractual term provided that the lease covenants are not more onerous than the assigned lease.

Importantly, a former tenant can only be required to stand as guarantor for the period that the assignee remains tenant. In the given example, if OT enters into an AGA with L and stands as guarantor in respect of A1's leasehold covenants, from the moment A1 assigns the lease to A2, OT will be off the hook (section 16(4) of the 1995 Act). Consequently, and in contrast to the position in relation to an old tenancy, OT cannot be required to guarantee A2 and A3's obligations under a new tenancy.

There are three situations where a landlord can require an outgoing tenant (ie assignee) of a new lease to enter into an AGA:

- where the alienation clause in the lease entitles the landlord to require the outgoing tenant to enter into an AGA as a condition of requiring consent (see section 19 (1A) of the Landlord and Tenant Act 1927) or
- where the landlord's consent which is not to be unreasonably refused is required and it is reasonable for the landlord to require the outgoing tenant to enter into an AGA as a condition of granting consent or
- where the lease contains an absolute prohibition against assignment and the landlord is prepared to consent to the assignment but only if the outgoing tenant enters into an AGA.

Section 16(8) of the 1995 Act provides that the ordinary rules of law relating to guarantees apply to AGAs. Consequently, if OT enters into an AGA on the assignment of the lease to A1 and L seeks to recover arrears under the AGA, OT would be released from liability if the lease had been modified or varied without OT being a party to the variation (see 10.4.3 above).

An issue which has attracted much academic debate is whether G1 could be required to stand as guarantor to OT's AGA when the lease is assigned to A1. Those in favour of this sub-guarantee argue that if OT is a fledgling company with weak covenant strength, the loss of the supporting parent guarantee on assignment will effectively make the AGA worthless and allow the parent company off the hook. The riposte, by those against the principle, is that a sub-guarantee would fall foul of the anti-avoidance provisions and be rendered void. Until decided by the courts, there is no definitive answer to this conundrum.

10.7 Disclaimer and vesting orders

A liquidator of a company in liquidation has the power to disclaim onerous property under section 178 of the Insolvency Act 1986 (see 11.3.4). A similar power is conferred on a trustee in bankruptcy of an individual (section 315 of the 1986 Act). A lease contains many potentially onerous contractual obligations and falls within the definition of contracts that are capable of being disclaimed.

10.7.1 The implications of disclaimer

The implications of a valid disclaimer of a lease are not as straightforward as one might immediately think. Indeed the courts have for a number of years been wrestling with what Lord Nicholls referred to as the "artificiality" and "awkwardness" of section 178(4) of the 1986 Act which provides that a disclaimer:

> ... operates so as to determine, as from the date of the disclaimer, the rights, interests and liabilities of the company in or in respect of the property disclaimed ... *but does not, except so far as is necessary for the purpose of releasing the company from any liability, affect the rights or liabilities of any other person.*

Lord Nicholls made these comments in the leading case of *Hindcastle Ltd* v *Barbara Attenborough Associates Ltd* [1996] 1 All ER 737.

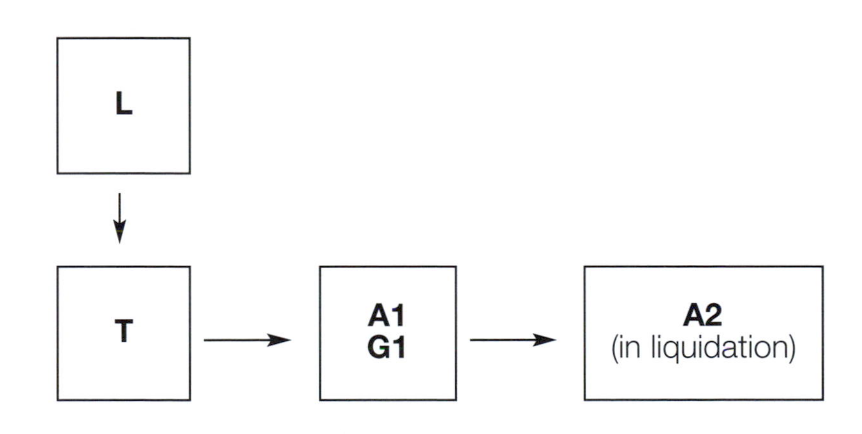

Facts

- L granted a 20 year lease to T.
- T assigned the lease to A1 and G1 guaranteed A1's obligations under the lease.
- A1 assigned the lease to A2.
- A2 went into liquidation and its liquidator disclaimed the lease.

- L obtained a judgment against T, A1 and G1 for arrears of rent. They appealed arguing that their obligations had also been terminated by the disclaimer.

Held by the House of Lords

- The disclaimer did not terminate the rights of former tenants and their guarantors. Rent was lawfully recoverable from T, A1 and G1.
- Section 178(4) operated as a deeming provision. While the lease was determined by the disclaimer, it was deemed to continue for the purposes of any third party liabilities.

10.7.2 Effect of disclaimer on contingent liability (old tenancy)

The current rules in respect of an old tenancy (ie pre 1 January 1996) can be summarised as follows.

(a) Where only landlord and tenant are involved

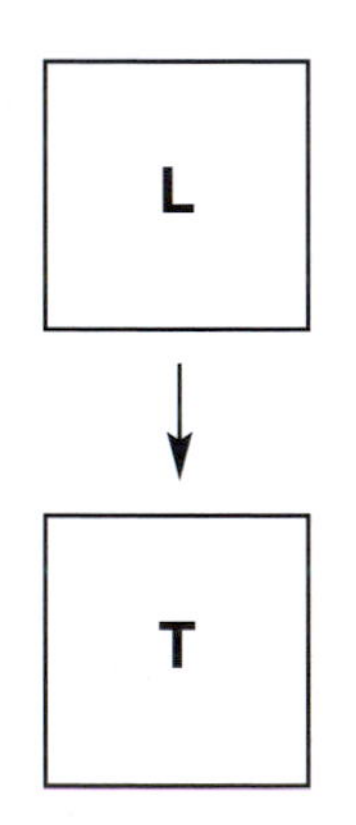

Where no third parties are involved, the effect of a disclaimer is to terminate the landlord and tenant obligations under the lease and also to determine the leasehold interest in land for all time.

(b) Where third parties have contingent or current liability

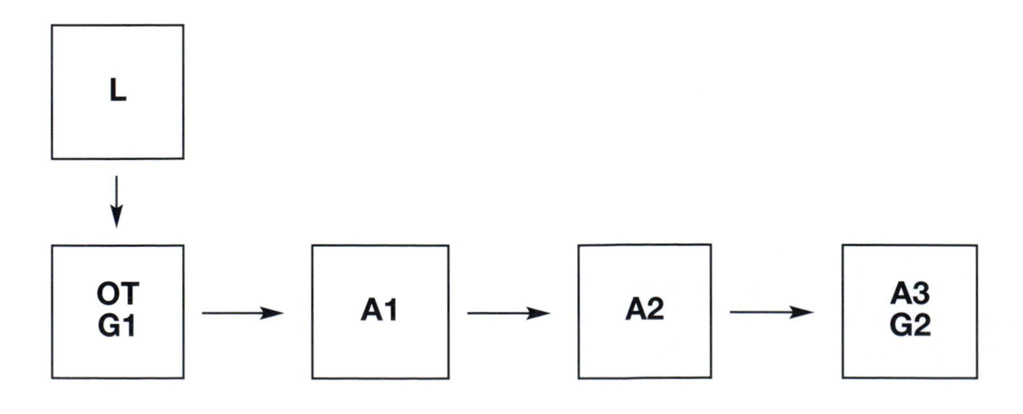

In the above example, if A3's liquidator serves a valid notice of disclaimer, although the lease will come to an end, L will be entitled to pursue G2, A1 and A2 (provided they entered into a direct covenant with L) OT and G1 "as though the lease had continued and not been determined". A further consequence of the disclaimer is that any claims based on indemnity, contribution or restitutionary principles by guarantors and/or previous tenants against A3 will be lost, although each may have a right to claim in the winding up or bankruptcy of A3 (ie the disclaiming tenant). While this is unlikely to bear fruit in a compulsory liquidation; such a claim may be worthwhile if the tenant company is in voluntary liquidation.

(c) Re-taking of possession by the landlord

Disclaimer will only result in those with contingent liability getting off the hook if the disclaimer prompts the landlord to actively re-take possession of the premises. By taking possession the landlord demonstrates that it regards the lease as "ended for all purposes", and the contingent or actual liabilities of guarantors, former tenants and their guarantors will come to an end on the date of re-entry. L cannot be in possession of the premises and claim rent from others.

From a landlord's perspective therefore, following disclaimer it should take great care not to take any steps in respect of the premises which could be construed as re-entry.

From the perspective of those with contingent liability, it is obviously of great importance to establish whether the landlord has re-taken possession. Often, where the premises are empty, this can be difficult to ascertain, particularly if the former tenant/guarantor has not had any recent involvement with the premises. The landlord should in these circumstances be asked to disclose all relevant e-mails, internal memorandum and correspondence with third parties which demonstrate the landlord's attitude toward the premises. If, for example, the landlord carries out extensive works to the premises which go beyond keeping them wind and water tight, perhaps as a prerequisite to re-letting, this may be sufficient to establish that it has re-taken possession. Merely marketing the premises and showing prospective tenants around is not sufficient to constitute re-taking possession (see *Oastler* v *Henderson* [1977] 2 QB 575). If the landlord refuses to disclose documents voluntarily, it is always open to a party to make an application to the court for pre-action disclosure.

10.7.3 Effect of disclaimer on contingent liability (new tenancy)

The same principles apply to contingent liability under a new tenancy. Consequently:

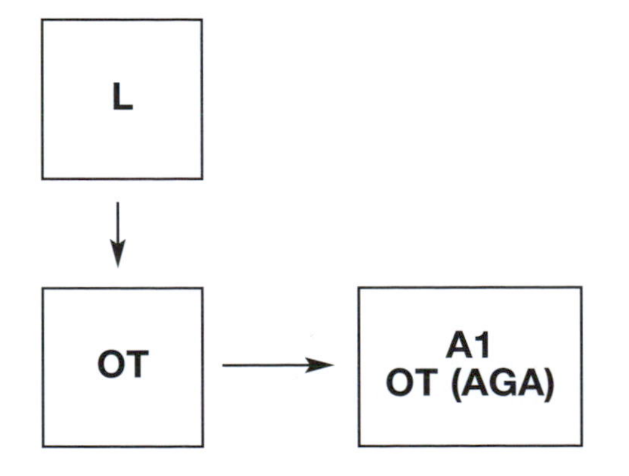

In the above example, if A1's liquidator disclaims the lease, OT will continue to be liable for rent and the tenant covenants notwithstanding the determination of the lease.

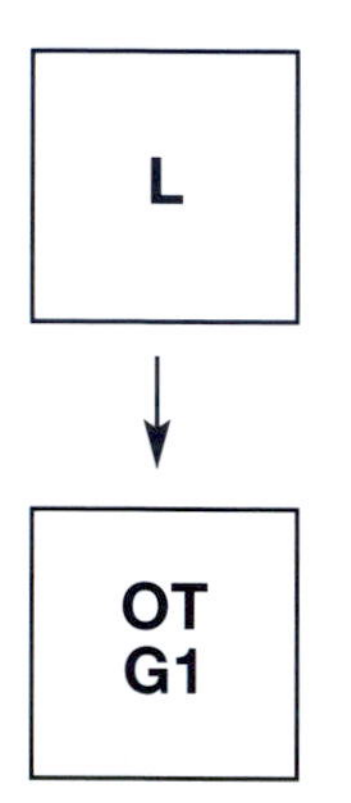

In the above example, G1 as guarantor of OT's obligations will remain liable following disclaimer by OT's liquidator.

10.7.4 *Vesting orders*

The remedy available to a party with contingent liability pursued by a landlord after disclaimer is to seek a vesting order under section 181 (company in liquidation) and section 320 (individual in bankruptcy) of the Insolvency Act 1986. This may be unnecessary in the case of a guarantor which is under an obligation to take a new lease for the remainder of the lease term. However, these obligations tend to be triggered at the option of the landlord, and if the landlord does not exercise the option, the guarantor may remain on the hook to pay rent without being able to use or market the premises. In those circumstances an application for a vesting order would be necessary if the guarantor wishes to take control of the premises.

The court can make an order vesting a lease in a party that either has an interest in the property (eg a sub-tenant) or a party that is under any liability in respect of the disclaimed property. Clearly those with contingent liability would fall within the latter category. An order will not be made by the court unless it appears to the court that it would be just to do so for the purposes of compensating the party subject to a liability. If a landlord has made or has indicated that it intends to make a claim against a party, a vesting order is likely to be made by the court so as to give that party an opportunity to use the premises or to re-let them.

The principal downside of this remedy is that the Insolvency Rules require a party to make an application to the court within three months of becoming aware of the disclaimer, or receiving a copy of the notice of disclaimer, which ever is the earlier: r 4 194(2). The difficulties caused by this time-limit are considered in more detail below.

10.7.5 *The contingent liability dilemma*

A dilemma that can arise as a result of the three month time-limit for applying for a vesting order can be illustrated by a further example:

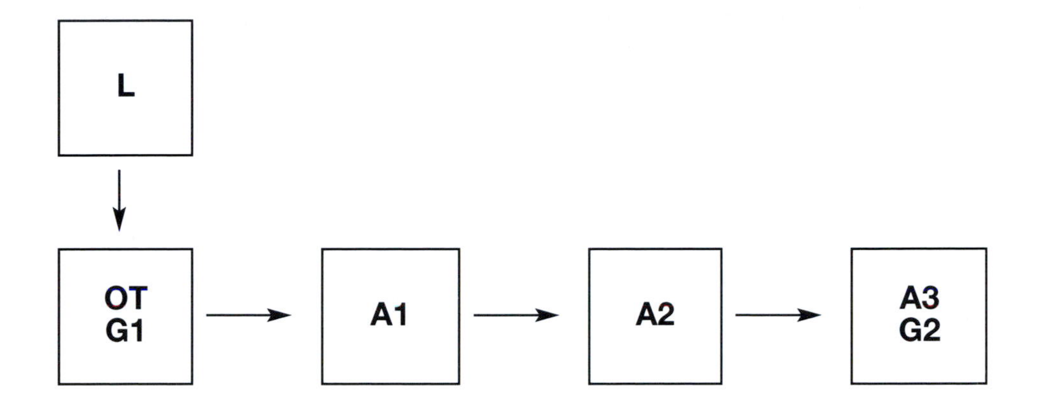

If A3 goes into liquidation and its liquidator serves a notice of disclaimer, those with contingent liability may face a dilemma (ie OT, G1, A1 and A2 (assuming a direct covenant was given) and G2). Each party is likely to be served with a copy of the notice of disclaimer and therefore the three month period will start to run from that date.

If L has made a claim against anyone in the chain, that party can make an application to the court for a vesting order. However, if no claim has been made within the three month period what should they do? If they delay and L makes a claim after the three month deadline has expired, they may have lost their opportunity to gain control of the premises and derive some benefit from their liability. This is likely to be most acute with parties who are remote from the existing tenant. In the above example, L may pursue G2 first before pursuing more remote parties such as G1 and OT. Indeed it could be many years after the service of the notice of disclaimer before L needs to call on OT and G1.

(a) Are former tenants/guarantors entitled to an overriding lease?

With the exception of G2, the former tenants and their guarantors could wait to see if a section 17 notice is served by L, and at that stage request an overriding lease from L. However, there is at least some doubt about whether a former tenant or its guarantor is legally entitled to an overriding lease following disclaimer. The argument runs as follows.

1. Section 19(7) of the Landlord and Tenant (Covenants) Act 1995 stipulates that a landlord is not under an obligation to grant an overriding lease at any time when the relevant tenancy has been determined.
2. The decision in *Hindcastle* was that the disclaimed lease is determined following disclaimer and the reversion accelerated. The language of Lord Nicholls permits no other conclusion.
3. If the lease is determined by disclaimer, section 19(7) bites and the landlord is under no obligation to grant an overriding lease. Indeed, as a matter of logic, it is difficult to see how a former tenant/guarantor could be granted an overriding lease in circumstances where the lease, as between insolvent tenant and landlord, has come to an end.
4. Arguably, their remedy was to apply for a vesting order within three months of disclaimer.

This conundrum is a consequence of treating the lease as being dead as between the landlord and the insolvent tenant, but as being alive as between the landlord and former tenants/guarantors. Despite being cogent, this argument is not attractive. If the argument is correct, it has the curious consequence of the lease being deemed to exist for the purposes of a landlord serving a section 17 notice, but not existing for the purposes of a request for an overriding lease under section 19. It also defeats the rationale underlying the 1995 Act of allowing a party to gain control if they are required to pay arrears in response to a section 17 notice.

The better view is that a former tenant/guarantor is entitled to call for an overriding lease after disclaimer. *Scottish Widows plc* v *Tripipatkul* [2003] EWHC 1874 may contain the key to unlocking the problem. In this case, the landlord, following a disclaimer, sold its freehold interest to L2. Following the transfer L2 sought to enforce surety obligations under the disclaimed lease. The surety argued that the disclaimer had determined the lease and accelerated the reversion and therefore no reversionary interest existed at the date of the freehold transfer to L2 to which the surety's covenant could run. This argument was rejected. The court held that a notional reversion had existed which passed to L2. Pomfrey J stated:

> ... the dead lease is deemed to subsist so far as the landlord and surety are concerned, and if that is the case, I cannot see why the "notional reversion" cannot be assigned with the benefit of the covenant annexed to it as if the lease had not been disclaimed.

By a parity of reasoning, it could be argued that since a dead lease is deemed to exist between a landlord and a former tenant/guarantor, it can be held not to have been determined for the purposes of section 19(7) of the 1995 Act.

(b) Possible strategy

Consequently, on being served with a notice of disclaimer, former tenants or their guarantors face a dilemma; they can take no action and hope they are not pursued by the landlord, or they can apply for a vesting order. If they do nothing they risk being served with a section 17 notice demanding rent arrears outside the three month time-limit for applying for a vesting order. But if they take the initiative they will incur a liability which they might otherwise have avoided. In most cases the commercial temptation will be for a party with contingent liability to keep its head down and hope that the landlord pursues another party from the chain, or alternatively, re-enters the premises thereby terminating all future contingent liability under the lease. If it does not, there may be difficulties obtaining an overriding lease for the reasons discussed above.

There is no simple solution. A possible tactic is to take no action until served with a section 17 notice and then apply for a vesting order if the liability cannot be passed on to another party in the chain. The court has discretion to extend time for making an application for a vesting order: see section 376 of the 1986 Act and r 4.3 of the Insolvency Rules. A court is unlikely to refuse to exercise this discretion where, for instance, the applicant has had no dealings with the property for many years and did not apply for a vesting order because it was unsure as to whether the landlord would look to it for future rent, but where that applicant has promptly made an application for a vesting order following the discharge of a section 17 notice. While this may be a sensible commercial solution, there is no guarantee that a court will exercise its discretion at a later date and to that extent this strategy will always carry an element of risk.

(c) Vesting orders v overriding leases

A party that has paid up in response to a section 17 notice and is unable to recover from another party in the chain will have to decide whether to seek a vesting order or an overriding lease.

The principal advantages of a vesting order are:

- if the landlord has no objection, it can be dealt with by a consent order and is likely to be a speedier and cheaper exercise (although, if opposed, the opposite could be the case)
- stamp duty land tax will not be payable.

In addition to the *Hindcastle* problem, further disincentives for choosing the overriding lease option can be:

- stamp duty land tax will be payable on the overriding lease. This may be substantial where a lease has many years to run and/or is at a high rent
- there could be delays if there is an issue turning on part payment or the landlord insists on negotiating the terms of the overriding lease (see 10.2.6 (a)).

10.8 Precedents

10.8.1 Section 17(2) notice — former tenant

NOTICE TO FORMER TENANT OR GUARANTOR OF INTENTION TO RECOVER FIXED CHARGE[1]
(Landlord and Tenant (Covenants) Act 1995, section 17)

To [name and address]: **Fat Cat Limited, of Invisible House, Lands End, Cornwall**.

IMPORTANT — THE PERSON GIVING THIS NOTICE IS PROTECTING THE RIGHT TO RECOVER THE AMOUNT(S) SPECIFIED FROM YOU NOW OR AT SOME TIME IN THE FUTURE. THERE MAY BE ACTION WHICH YOU CAN TAKE TO PROTECT YOUR POSITION. READ THE NOTICE AND ALL THE NOTES OVERLEAF CAREFULLY. IF YOU ARE IN ANY DOUBT ABOUT THE ACTION YOU SHOULD TAKE, SEEK ADVICE IMMEDIATELY, FOR INSTANCE FROM A SOLICITOR OR CITIZENS ADVICE BUREAU.

1. This notice is given under section 17 of the Landlord and Tenant (Covenants) Act 1995. {*see Note 1 overleaf*}

2. It relates to (address and description of property) **Unit 1, Forget Me Not House, Dark Road N1**

   let under a lease dated **1 February 1987** and made between **Out of the Blue Limited (1) and Fat Cat Limited (2)**

   [of which you were formerly tenant] [~~in relation to which you are liable as guarantor of a person who was formerly tenant~~].[2]

3. ~~I~~/we as landlord[3] hereby give you notice that the fixed charge(s) of which details are set out in the attached Schedule[4] is/are now due and unpaid, and that ~~I~~/we intend to recover from you the amount(s) specified in the Schedule [and interest from the date and calculated on the basis specified in the Schedule].[5]{*see Notes 2 and 3 overleaf*}

4. [6] There is a possibility that your liability in respect of the fixed charge(s) detailed in the Schedule will subsequently be determined to be for a greater amount. {see Note 4 below}

5. All correspondence about this notice should be sent to the landlord/landlord's agent at the address given below.

Date **1 April 2007** Signature of landlord/landlord's agent . . . .
Name and address of landlord

Out of the Blue Limited
Unit 1
Grabbers Paradise
London

Name and address of agent

Duckin And Divein Managing Agents,
1 Plushsville Tower,
London

[1] The Act defines a fixed charge as (a) rent, (b) any service charge (as defined by section 18 of the Landlord and Tenant Act 1985, disregarding the words "of a dwelling") and (c) any amount payable under a tenant covenant of the tenancy providing for payment of a liquidated sum in the event of failure to comply with the covenant.
[2] Delete alternative as appropriate.
[3] "Landlord" for these purposes includes any person who has the right to enforce the charge.
[4] The schedule must be in writing, and must indicate in relation to each item the date on which it became payable, the amount payable and whether it is rent, service charge or a fixed charge of some other kind (in which case particulars of the nature of the charge should be given). Charges due before 1 January 1996 are deemed to have become due on that date, but the actual date on which they became due should also be stated.
[5] Delete words in brackets if not applicable. If applicable, the schedule must state the basis on which interest is calculated (for example, rate of interest, date from which it is payable and provision of lease or other document under which it is payable).
[6] Delete this paragraph if not applicable. If applicable (for example, where there is an outstanding rent review or service charge collected on account) a further notice must be served on the former tenant or guarantor within three (3) months beginning with the date on which the greater amount is determined. If only applicable to one or more charge of several, the schedule should specify which.

NOTES
1. The person giving you this notice alleges that you are still liable for the performance of the tenant's obligations under the tenancy to which this notice relates, either as a previous tenant bound by privity of contract or an authorised guarantee agreement, or because you are the guarantor of a previous tenant. By giving you this notice, the landlord (or other person entitled to enforce payment, such as a management company) is protecting his right to require you to pay the amount specified in the notice. There may be other sums not covered by the notice which the landlord can also recover because they are not fixed charges (for example in respect of repairs or costs if legal proceedings have to be brought). If you pay the amount specified in this notice in full you will have the right to call on the landlord to grant you an overriding lease, which puts you in the position of landlord to the present tenant. There are both advantages and drawbacks to doing this, and you should take advice before coming to a decision.

Validity of notice
2. The landlord is required to give this notice within six months of the date on which the charge or charges in question became due (or, if it became due before 1 January 1996, within six months of that date). If the notice has

been given late, it is not valid and the amount in the notice cannot be recovered from you. The date of the giving of the notice may not be the date written on the notice or the date on which you actually saw it. It may, for instance, be the date on which the notice was delivered through the post to your last address known to the landlord. If you are in any doubt, you should seek advice immediately.

Interest

3. If interest is payable on the amount due, the landlord does not have to state the precise amount of interest, but he must state the basis on which the interest is calculated to enable you to work out the likely amount, or he will not be able to claim interest at all. This does not include interest which may be payable under rules of court if legal proceedings are brought.

Change in amount due

4. Apart from interest, the landlord is not entitled to recover an amount which is more than he has specified in the notice, with one exception. This is where the amount cannot be finally determined within six months after it is due (for example, if there is dispute concerning an outstanding rent review or if the charge is a service charge collected on account and adjusted following final determination). In such a case, if the amount due is eventually determined to be more than originally notified, the landlord may claim the larger amount if and only if he completes the paragraph giving notice of the possibility that the amount may change, and gives a further notice specifying the larger amount within three months of the final determination.

Schedule of Charges

[insert charges]

10.8.2 Section 17(4) — notice of revised charges

FURTHER NOTICE TO FORMER TENANT OR GUARANTOR OF REVISED AMOUNT DUE IN RESPECT OF A FIXED CHARGE[1]
(Landlord and Tenant (Covenants) Act 1995, section 17) . . . . . .

To **Fat Cat Limited, of Invisible Houses, Lands End, Cornwall.**

IMPORTANT — THE PERSON GIVING THIS NOTICE IS PROTECTING THE RIGHT TO RECOVER THE AMOUNT(S) SPECIFIED FROM YOU NOW OR AT SOME TIME IN THE FUTURE. THERE MAY BE ACTION WHICH YOU CAN TAKE TO PROTECT YOUR POSITION. READ THE NOTICE AND ALL THE NOTES OVERLEAF CAREFULLY. IF YOU ARE IN ANY DOUBT ABOUT THE ACTION YOU SHOULD TAKE, SEEK ADVICE IMMEDIATELY, FOR INSTANCE FROM A SOLICITOR OR CITIZENS ADVICE BUREAU.

1. This notice is given under section 17 of the Landlord and Tenant (Covenants) Act 1995. {*see Note 1 overleaf*}

2. It relates to (address and description of property) **Unit 1, Forget Me Not House, Dark Road N1**

let under a lease dated **1 February 1987** and made between **Out of the Blue Limited (1) and Fat Cat Limited (2)**

[of which you were formerly tenant] [~~in relation to which you are liable as guarantor of a person who was formerly tenant~~].[2]

3. You were informed on **1 April 2007** (date of original notice) of the amount due in respect of a fixed charge or charges, and of the possibility that your liability in respect of the charge(s) might subsequently be determined to be for a greater amount.

4. ~~I~~/we as landlord[3] hereby give you notice that the fixed charge(s) of which details are set out in the attached Schedule[4] has/have now been determined to be for a greater amount than specified in the original notice, and that ~~I~~/we intend to recover from you the amount(s) specified in the Schedule [and interest from the date and calculated on the basis specified in the Schedule][5]. {*see Notes 2 and 3 overleaf*}

5. All correspondence about this notice should be sent to the landlord/landlord's agent at the address given below.
Date **1 September 2007** Signature of landlord/landlord's agent . . . .

Name and address of landlord

Out of the Blue Limited
Unit 1
Grabbers Paradise
London

Name and address of agent

Duckin And Divein Managing Agents,
1 Plushsville Tower,
London

[1] The Act defines a fixed charge as (a) rent, (b) any service charge (as defined by section 18 of the Landlord and Tenant Act 1985, disregarding the words "of a dwelling") and (c) any amount payable under a tenant covenant of the tenancy providing for payment of a liquidated sum in the event of failure to comply with the covenant.
[2] Delete alternative as appropriate.
[3] "Landlord" for these purposes includes any person who has the right to enforce the charge.
[4] The Schedule can be in any form, but must indicate in relation to each item the date on which it was revised, the revised amount payable and whether it is rent, service charge or a fixed charge or some other kind (in which case particulars of the nature of the charge should be given).
[5] Delete words in brackets if not applicable. If applicable, the Schedule must state the basis on which interest is calculated (for example, rate of interest, date from which it is payable and provision of Lease or other document under which it is payable).

NOTES

1. The person giving you this notice alleges that you are still liable for the performance of the tenant's obligations under the tenancy to which this notices relates, either as a previous tenant bound by privity of contract or an authorised guarantee agreement, or because you are the guarantor of a previous tenant. You should already have been given a notice by which the landlord (or other person entitled to enforce payment, such as a management company) protected his right to require you to pay the amount specified in that notice. The purpose of this notice is to protect the landlord's right to require you to pay a larger amount, because the amount specified in the original notice could not be finally determined at the time of the original notice (for example, because there was a dispute concerning an outstanding rent review or if the charge was a service charge collected on account and adjusted following final determination).

Validity of notice

2. The notice is not valid unless the original notice contained a warning that the amount in question might subsequently be determined to be greater. In addition, the landlord is required to give this notice within three months of the date on which the amount was finally determined. If the original notice did not include that warning, or if this notice has been given late, then this notice is not valid and the landlord cannot recover the greater amount, but only the smaller amount specified in the original notice. The date of the giving of this notice may not be the date written on the notice or the date on which you actually saw it. It may, for instance, be the date on which the notice was delivered through the post to your last address known to the person giving notice. If you are in any doubt, you should seek advice immediately.

Interest
3. If interest is chargeable on the amount due, the landlord does not have to state the precise amount of interest, but he must have stated the basis on which the interest is calculated, or he will not be able to claim interest at all.

Schedule of Revised Charges

[insert revised charges]

10.8.3 Letter to landlord requesting documents

"Dear Sirs

Re: Claim by Out of the Blue Ltd — Unit 1, Forget Me Not House, Dark Road N1

We have been instructed to advise Fat Cat Ltd in connection with a section 17 notice dated 1 April 2007 recently served on our client.

Our client's involvement with these premises ended on 14 May 1985 when it assigned the premises [the former tenant which our client stood as guarantor for] to Longsince Gone Ltd. We understand that there may have been a number of assignments since.

In order that we can properly advise our client, please provide the following information and copy documents:

1. A copy of the lease. Our client has not retained a copy.

2. A copy of all deeds relating to the lease and its assignments to date, including but not limited to copies of:

    - deeds of assignment
    - licences to assign or sublet
    - subleases
    - licences to alter the building
    - deeds of variation
    - rent review memorandum
    - deeds of rectification

3. Copies of any side letters or agreements relating to the property.

4. All relevant demands sent to, and all relevant correspondence with, the current tenant relating to the sums claimed.

5. Details of the current tenant.

6. An up to date statement of arrears.

7. Confirmation that you have not re-entered the property and that the lease has not been surrendered.

If you do not disclose any deeds of variation, please confirm for the avoidance of doubt that there have been no variations of the lease.

We are grateful for your assistance in this matter and we would request that you do not issue proceedings until we have had an opportunity to consider the above documents and advise our client.

Yours faithfully

Termination

11.1 Introduction

This chapter considers the different ways in which an agreement to occupy commercial premises can come to an end. It deals with the temporary forms of occupation such as licences and tenancies at will; and then considers the most common ways in which a lease terminates.

The legal nature of the agreement will affect its termination and it is important to identify whether the agreement is a lease, licence or tenancy at will (which is dealt with in 3.2), and if it is a lease, it may also be important to identify whether or not it is protected under the Landlord and Tenant Act 1954. If a lease is protected, it will only be able to come to an end by the termination procedures provided for by the 1954 Act. This is dealt with in detail in 11.3.7 and Chapter 13.

Finally the chapter considers practical issues that arise on termination including issuing court proceedings.

11.2 Termination of licences and tenancies at will

11.2.1 Simple permission

A bare licence (ie one for no consideration) can be ended at once by oral or written notice. However, the licensee is entitled to a period of grace to pack up its belongings and move out. This period must allow the licensee reasonable time to make and implement the practical arrangements to go. If a licensor jumps the gun and issues court proceedings for trespass before the packing up period has expired, the proceedings will be premature and will fail (see *Minister of Health* v *Bellotti* [1944] KB 298).

11.2.2 Contractual licence

The duration or termination of a contractual licence will usually be governed by the terms of that licence. A contractual licence will come to an end by one of the following events.

- The expiry of the contractual term (if one is expressly provided for). Once the expiry date has passed, the permission to occupy comes to an end and the licensor is entitled to possession and can issue proceedings immediately without the need to serve a notice to quit. The security of tenure provisions in the Landlord and Tenant Act 1954 do not apply to licences.

- The giving of notice of the length required by the contract. If the contract specifies that the notice must be in writing; oral notice will not do.
- The giving of reasonable notice. Where the contract is silent as to its termination, a court will imply a term that it can be brought to an end by either party giving the other reasonable notice.
- By the assignment of the licensor's interest in the premises.
- By the death (or insolvency) of the licensor or the licensee.

A corporate occupier may grant or be granted a free standing licence. However, it may also grant or be granted a licence to occupy under a conditional agreement for lease. Typically this will occur when the potential tenant wants to go into occupation before the grant of the lease for fitting out purposes or where certain issues have yet to be resolved, such as where superior landlord's consent is still awaited. The termination provisions will be a matter of negotiation and will usually be expressly provided for in the agreement. It is common for the parties to provide for a long stop date after which, if the conditions have not been fulfilled (eg superior landlord's consent has not been given), one or both of the parties are entitled to serve a rescission notice. The service of a valid notice will not only terminate future obligations under the agreement for lease, but it will also terminate the licence to occupy.

Where a licence contains a determination clause, it can only be terminated in accordance with the terms of that clause. If the licensor purports to terminate the licence in breach of contract, the purported termination will be ineffective and in an appropriate case a court will grant an injunction restraining eviction of the licensee. Equally, possession proceedings brought by a licensor will be premature and fail unless the licensee is in serious breach of the licence agreement, in which case a court is likely to imply a term that the licensor can terminate the licence immediately.

Where a licence agreement is periodic and does not contain a termination clause, a court will imply a term that it can be determined on reasonable notice. What constitutes reasonable notice will depend upon the circumstances. In the commercial context, a period of 14 to 28 days will usually be sufficient. However, this period could be significantly longer if there are special circumstances (see *Henrietta Barnett School* v *Hampstead Garden Suburb Institute* (1995) 93 LGR 470).

11.2.3 Other types of licences

There are other forms of licences which a corporate occupier is unlikely to come across such as residential licences; agricultural licences and licences which are not freely terminable. These licences are governed by different rules and principles and go beyond the scope of this book.

11.2.4 Tenancies at will

A landlord or tenant can determine a tenancy at will instantly, without a period of notice. The usual method of terminating a tenancy at will is for the landlord to make a demand for possession of the premises. This can take the form of:

- a demand for the keys
- a letter before action demanding possession
- the service of possession proceedings.

Where the landlord terminates the arrangement with immediate effect, the tenant has a reasonable time to enter the property to remove its belongings. A tenant at will is not entitled to protection under the Landlord and Tenant Act 1954 and consequently the landlord can issue and serve court proceedings without delay.

If a tenant wants to terminate a tenancy at will it must notify the landlord of its intention. It is not sufficient to abandon the property without telling the landlord.

11.2.5 Written notice

It is always good practice to give notice to terminate a licence or tenancy at will in writing even if it is not strictly required. A copy should be kept and the details of how the notice was served should be carefully recorded for evidential purposes.

Unless the agreement specifically stipulates (some agreements for lease attach draft recission notices) there is no prescribed or formal notice required. A letter will suffice. Precedent notices drafted on behalf of the licensee and licensor; and tenant at will and landlord can be found at 11.7.1 to 11.7.4.

11.3 Termination of leases

The most common ways in which a lease may be terminated are by:

1. notice to quit
2. exercise of a break clause
3. forfeiture
4. disclaimer
5. frustration/recission
6. surrender
7. effluxion of time.

11.3.1 Notice to quit

(a) General points

A periodic tenancy can be terminated unilaterally by either party serving a notice to quit. Strictly speaking, the notice to quit prevents a new period of the tenancy from starting.

The circumstances in which the corporate occupier may serve or receive a notice to quit are limited; primarily because most leases that it will encounter are likely to be fixed term leases. However, occasionally periodic tenancies are granted and can come into existence unwittingly, by implication of law, on the payment and acceptance of rent. A corporate occupier may wish to terminate a periodic tenancy that it occupies as a tenant.

It is important to note, that once a valid notice to quit has been served it cannot be revoked or waived, even if both parties agree. The tenancy will automatically end. If the tenant remains in occupation after the expiry of the notice to quit it will be as:

- a trespasser (if the landlord does not consent to the tenant's continued occupation) or
- a tenant at will (if the landlord does consent but does not grant a fresh tenancy) or
- a tenant under a new tenancy on the terms of the old tenancy except as to the period (for example,

on the payment and acceptance of rent in circumstances where the parties' intention, or inferred intention, was to create a new tenancy).

A notice to quit served by a landlord or tenant will also determine any subtenancies.

(b) Formalities

A notice to quit should always be given in writing and a copy kept for evidential purposes. It must be served by the person with the legal estate in land (a beneficial owner cannot serve a valid notice) and care must be taken to ensure it is served on the correct person.

A valid notice must:

- Give notice equal to the period required by the tenancy agreement (if provided for) and/or at least one complete period of the tenancy (except a yearly tenancy in which only six months notice is required). For example, if the tenancy is monthly, one months notice is required; if the tenancy is quarterly, one quarters notice is required.
- State a definite date when the tenant must quit or the tenant intends to leave.
- The date must coincide with the end of a period of the tenancy or the first day of any subsequent period. If there is any doubt about the last day of a period, the notice, after specifying the date should add the following saving words:

> or other the day on which this tenancy expires next after the end of [one week][one month][one quarter][half year] — *delete as appropriate*] from the date of service of this notice

A tenant is entitled to terminate a periodic tenancy protected by the Landlord and Tenant Act 1954 by the service of an ordinary common law notice to quit provided it has occupied the premises for a least one month (section 24(2)). A landlord cannot serve an ordinary notice to quit; it must serve a section 25 notice and establish a ground of opposition under section 30(1) to obtain possession (see chapter 13).

Examples of a precedent notice to quit for use in the commercial context (ie non-residential cases) can be found at 11.7.5 and 11.7.6 below. The landlord notice to quit can only be use in respect of a periodic tenancy which does not have security of tenure under the 1954 Act (eg because the tenant is no longer using the premises for the purposes of its business).

11.3.2 Exercise of a break clause

A lease can be terminated before the contractual expiry date by the exercise of a break right. The existence of a break clause in an operational or surplus lease can be of great importance to a corporate occupier and yet the complexity and draconian nature of some of the legal principles that apply are often overlooked.

Given the importance of this topic to the corporate occupier the whole of Chapter 12 has been dedicated to the operation of break clauses.

11.3.3 Forfeiture

(a) General points

In appropriate circumstances a landlord can unilaterally terminate a fixed term or periodic tenancy by exercising its right of forfeiture or 're-entry'. The right to forfeit arises where the tenant is in default (ie in breach of covenant) and the tenancy contains a forfeiture clause.

If a lease is lawfully forfeit, it will terminate on the forfeiture date and any sublease will also terminate, even though the subtenant may have been up to date with its rent and not in breach of any of its covenants in the sublease. However, both the tenant and sub-tenant may apply for relief from forfeiture.

(b) A typical forfeiture clause

An example of a simple forfeiture clause is:

> If the rents (or any part of them) are outstanding for 21 days after becoming due whether formally demanded or not or there is a breach by the tenant of any covenant or other term of this lease then the landlord may re-enter the premises (or any part of them in the name of the whole) at any time and upon such re-entry the term shall absolutely cease but without prejudice to any rights or remedies which may have accrued to either party against the other

In the absence of a forfeiture clause, a landlord can only forfeit a lease if the covenant is a condition ie that compliance with it is fundamental to the lease.

(c) Forfeiture for arrears of rent

The relevant principles and procedures which govern forfeiture for arrears of rent and relief from forfeiture are dealt with in detail at 8.3.

(d) Forfeiture for a breach other than rent arrears

Almost all of the legal principles and practical issues discussed in 8.3 in the context of rent arrears apply in relation to a forfeiture claim based on a different breach of covenant. The principal difference is that a landlord proposing to forfeit a lease must, except where the breach is rent arrears, first serve a notice on the tenant which complies with section 146 of the Law of Property Act 1925. In broad terms such a notice must specify the breach complained of, require the tenant to remedy the breach (if it is a breach which is capable of remedy) and require the tenant to pay compensation to the landlord (if the landlord requires compensation). Failure to serve a valid section 146 notice will be fatal to the landlord's re-entry. If the tenant fails to remedy the breach within a reasonable time of being served with a valid section 146 notice, or the breach is irremediable, the landlord can forfeit by peaceable re-entry or the issue and service of court proceedings (see 8.3.4).

11.3.4 Disclaimer

(a) General points

The Insolvency Act 1986 provides powers for a liquidator of an insolvent corporate tenant (section 178), and the trustee in bankruptcy of an individual (section 315), to disclaim a lease if it is an onerous contract. A liquidator or trustee can disclaim a lease by the service of a formal notice of disclaimer, which must be sealed by and filed with the court, and served on all parties affected by the disclaimer.

The rules relating to disclaimer are detailed and complex. However, in general terms, the effect of the service of a valid notice of disclaimer is to determine, from the date of the notice, the rights interests and liabilities of the insolvent tenant in the lease disclaimed. It does not, however, affect the rights or liabilities of any other person (section 178(4) of the 1986 Act).

There are three main ways in which disclaimer may be relevant to a corporate occupier:

- where the corporate occupier is a sub-tenant and its immediate landlord (ie the tenant) has become insolvent
- where a corporate occupier with a legacy portfolio is the landlord of a sub-tenant which has become insolvent
- where the corporate occupier is a former tenant or guarantor under a lease and the current tenant has become insolvent.

(b) Effect on sub-tenancy — insolvency of immediate landlord

A corporate occupier may be served with a notice of disclaimer by its immediate landlord in relation to premises that it occupies as a sub-tenant. The service of a valid notice to disclaim on a sub-tenant has a very peculiar legal effect on the sub-tenancy. Although the headlease is immediately determined, the sub-tenant does not continue to occupy the premises on the basis of the sub-tenancy (as it would if the headlease was surrendered). Nor does it occupy on the basis of the defunct headlease. The sub-tenant is entitled to remain in the premises for the remainder of the term granted by the sub-tenancy provided that it complies with the terms of the headlease.

This can cause problems for the sub-tenant if the rent payable under the defunct headlease is higher than the rent that was payable under the sub-tenancy. If the sub-tenant falls into arrears, the superior landlord is entitled to forfeit and/or distrain for the rent. This is, however, subject to the sub-tenant's right to apply for relief from forfeiture under section 146(4) of the Law of Property Act 1925.

In view of the legal complexities, if served with a notice of disclaimer, it would be wise to consider the legal position carefully, and if the premises are important to the corporate occupier, it may wish to apply for a vesting order under section 181 of the Insolvency Act 1986. This application must be made within three months of service or of becoming aware of the notice of disclaimer. In view of the deadline, prompt action is called for. The court does have a discretion to extend the three month period but good reasons would have to be provided before such an application would be allowed.

If the sub-lease is protected under the Landlord and Tenant Act 1954, it is currently thought that an application for a vesting order is unnecessary because the sub-tenant's statutory protection will continue and its renewal rights remain. That being said the norm is now for subleases to be excluded from the protection of the Landlord and Tenant Act 1954.

(c) Legacy portfolio — sub-tenant's insolvency

There is very little a corporate occupier can do if it is served with a valid notice of disclaimer by the liquidator of its sub-tenant. If there has been no assignment or sub-underletting the sub-lease will simply come to an end on the date of the disclaimer. While the corporate occupier can prove the debt due in the liquidation or bankruptcy of the sub-tenant, it is unlikely to make any recovery (unless it is a solvent liquidation). The best advice in these circumstances is to re-market the premises as soon as possible.

If the sub-tenant has been wound up, but no notice of disclaimer has been served, a corporate occupier can serve a notice to elect on a liquidator requiring him to decide whether he will disclaim the sub-tenancy or not. A liquidator has 28 days in which to disclaim. If he fails to do so, he will be prevented from disclaiming at a later date. It is usually open to a landlord to forfeit (for rent arrears or insolvency) in these circumstances but if the tenant is still in occupation and/or leave of the court is required, the service of a notice to elect may be the quickest and cheapest route for the corporate occupier to obtain vacant possession.

The effect of section 178(4) of the Insolvency Act is to preserve claims against former tenants and guarantors and a corporate occupier should check to see if it can pursue a third party for past and future rent, and other liabilities, under the sub-lease.

(d) Original tenant and guarantor liabilities

The final situation where a corporate occupier may be affected by the service of a notice of disclaimer is where it is a former tenant or guarantor under a lease and the current tenant has become insolvent. In these circumstances it may be liable for past and future liabilities under the lease and may find itself being pursued by the current landlord. This is dealt with in Chapter 10.

11.3.5 Frustration/repudiation

The frustration of a contract occurs when a subsequent event occurs which is not envisaged and provided for by the contract and so significantly changes the nature of the contractual rights and obligations of the parties from what was reasonably contemplated that it is unjust to hold the parties to their obligations. If the principle applies to a contract both parties are released from further performance of the contract.

In *National Carriers* v *Panalpina (Northern)* [1981] AC 675 it was held that the doctrine of frustration could apply to a lease. However in practice, it will very rarely apply to a lease. The circumstances in which the principle might apply include where:

- the land has been destroyed (eg coastal erosion)
- the user clause requires the premises to be used for a single purpose which subsequently becomes illegal (eg licence premises where the licence is revoked).

If a lease is frustrated then it will terminate from the date of the frustrating event.

A repudiatory breach of a lease will entitle the party who is not in breach a choice whether to accept the repudiation and bring the lease to an end, or affirm the lease and sue for damages and/or an injunction. The use by tenants of the principle of repudiatory breach to try and get out of a lease has become fashionable in recent years. However, not all breaches of covenant by the landlord will be sufficiently serious to constitute repudiatory breach. To succeed a tenant must demonstrate that the landlord's unlawful action or inaction has deprived it of substantially the whole benefit that it was intended the tenant should obtain from the lease.

A case in which such a claim succeeded in the commercial arena was *Chartered Trust plc* v *Davies* [1997] 2 EGLR 83. Chartered Trust involved a shopping centre where the landlord maintained control over the common parts subject to the payment by the tenants of a service charge. The landlord was criticised by one tenant for not taking action to prevent the conduct of another tenant, who operated a pawnbrokers business from an adjoining unit. Entry to the pawnbroker's shop was restricted to one customer at a time, so that potential customers congregated outside, deterring visitors from the neighbouring shop. The Court of Appeal held that the conduct amounted to nuisance and the landlord's failure to enforce the covenant not to cause nuisance or keep the passageways clear, amounted to a derogation from grant. It was also held that the failure to act was so serious as to amount to a repudiation of the lease.

However in contrast, the attempt by a tenant to argue that a change in the original tenant mix letting policy resulting in an adverse trading figures amounted to derogation from grant and a repudiatory breach failed in *Petra Investment* v *Jeffrey Rogers* [2000] EGCS 66. It seems that only in exceptional cases will a court accept that the landlord's conduct is sufficiently serious to amount to a repudiatory breach and let the tenant off the hook.

11.3.6 Surrender

(a) General points

A surrender involves the tenant yielding up or relinquishing its estate in land (ie the lease) to the immediate landlord and is a further way in which a lease can be ended before the contractual termination date. However this differs from the other forms of early termination in that it does not involve the unilateral act of one party; it depends upon the consensual act of both.

A surrender can in some circumstances provide a corporate occupier with a useful way in which to get rid of premises which are surplus to requirements and a drain on resources. This may be a viable option where:

- the landlord wants or is happy to accept possession of the premises (eg where it wants to sell with vacant possession; would like to redevelop or can re-let at a higher rent and/or to a better covenant)
- the landlord is prepared to accept a reverse premium (ie a payment to take account of the lost rent which would have been paid if the lease had run its full course) to take back the premises early
- the corporate occupier has a break right (conditional or otherwise) approaching. As discussed above this may provide a corporate occupier with a clean and certain solution where there is a difficult conditional break right approaching.

In circumstances where the landlord wants possession for redevelopment it may be possible to negotiate a surrender at no cost or possibly for a small premium from the landlord. However, in most cases the payment of a premium will be required which is sufficiently attractive to make it worth the landlord's while. This will be purely a matter of negotiation and agreement. If the parties reach an agreement then it should be documented in a formal deed of surrender. If the parties are unable to reach an agreement the lease will continue because of the consensual nature of a surrender.

There are two types of surrender:

- express surrender — by deed
- implied surrender — by operation of law.

(b) Express surrender — by deed

Although it is possible for a surrender to take place by operation of law without the parties executing a formal deed of surrender, those principles can result in uncertainty and will not apply where one party simply changes its mind before acting on the agreement. It is therefore important in most circumstances to document the agreement by deed.

If there is likely to be a significant delay between the negotiated agreement and the actual date of surrender, the parties may wish to enter into an agreement for surrender to ensure the arrangement is binding pending the actual surrender. There are two important rules that may be of relevance.

(1) To be enforceable an agreement must be made in writing and incorporate all of the agreed terms and be signed by both parties (section 2 of the Law of Property (Miscellaneous Provisions) Act 1989). A purely oral agreement will be unenforceable (unless acted upon and the principles governing implied surrender apply).

(2) If the lease to be surrendered is protected by the Landlord and Tenant Act 1954, any agreement to surrender will be void unless a prescribed procedure is followed (section 38A). This mirrors the procedure that is required to exclude a tenancy for the protection of the 1954 Act:

   - *Ordinary procedure* — the landlord must serve a prescribed notice on the tenant (see schedule 3 of the Regulatory Reform (Business Tenancies Order) 2003) not less than 14 days before the tenant enters into an agreement for surrender *and* the tenant must make a declaration in a prescribed form (see schedule 4 of the 2003 Order) before entering into the agreement or
   - *Fast track procedure* — the prescribed notice must be served on the tenant before it enters into the agreement to surrender and the must make a statutory declaration in a prescribed form (see paragraph 7 to schedule 4 to the 2003 Order).

     The fast track procedure is appropriate in cases of urgency. It will, however, involve the tenant (or one of its officers) going to a solicitor's office to swear a statutory declaration.

It is important when negotiating or drafting a deed of surrender to ensure that both parties release each other from their obligations under the lease and any breaches of covenant whether past, present or future.

(c) Implied surrender by operation of law

In certain circumstances a lease can be surrendered without a formal deed. A corporate occupier may be able to take advantage of this principle (where it is the tenant and the premises are surplus) or may need to guard against it (in a legacy situation where it has sublet). It is important to be aware that a surrender does not depend on an intention of the parties.

An implied surrender takes place when one party does an act inconsistent with the continuation of the lease, and the other party concurs. There are two types of cases. The first is where the landlord is arguing that the lease has come to an end (where it wants vacant possession) and the tenant objects; the second (and more usual) is where the tenant argues the lease has been surrendered and the liability to pay rent brought to an end and the landlord disagrees.

The key to an implied surrender is that the act or acts, which are relied upon, must be unequivocal and the conduct of the parties must unambiguously amount to an acceptance that the lease has come to an end. There must be delivery by the tenant of possession coupled with the acceptance by the landlord of possession.

The following are acts which are usually relied on.

- *Acceptance of keys*: The offer and acceptance of the keys can be a strong indicator because they are a symbol of possession. If the landlord accepts the keys without reservation, then an implied surrender is likely. However, the determining factor is the basis on which the keys are accepted by the landlord. If the landlord has accepted them to inspect or carry out repairs; to show a prospective tenant around or by mistake then there will be no surrender.
- *Possession*: If the landlord occupies the premises for its own use or accepts the hand back of possession unconditionally, then a surrender will take place. If, however, the landlord changes the locks merely to secure the premises and/or takes other steps to protect the premises (eg issuing proceedings to evict squatters or carrying out works to make the premises wind and watertight) then it is unlikely the court will accept that there has been a surrender.
- *Re-letting*: If the landlord re-lets to a third party where the tenant consents then the earlier lease will be surrendered. Equally, if a new lease is granted to the tenant, the old lease will be surrendered by operation of law. The acceptance of rent from a sub-tenant may constitute a surrender of the headlease if the sub-tenant has been directed to do so by the tenant.

In most cases the party seeking to say the lease has been surrendered will point to a number of acts which it states show a surrender has occurred. A collection of acts, which, though equivocal if looked at in isolation, may collectively point to an unequivocal acceptance of possession by the landlord.

Ultimately, whether there has been a surrender or not in a particular case, is a question of fact. The outcome of such cases can be uncertain and usually will only be determined by a court after a full trial. Summary judgment applications are unusual in such cases.

(d) Legacy portfolio — protection of tenant v sub-tenant

If the sub-tenant vacates and retains the keys to the demised premises, to avoid allowing a sub-tenant off the (rental) hook by giving it grounds to argue there has been an implied surrender, the tenant should act promptly and take certain precautions.

- The tenant should immediately write an open letter to the sub-tenant stating that the tenant does not accept the lease has been surrendered; that the keys are held to the order of the sub-tenant, and that they are available for collection at any time.
- The tenant should continue to send out invoices for rent.
- If any access is required to the premises, the tenant should observe the requirements of the sub-lease; and in any event the sub-tenant should be notified of the need for access and the purpose of that access.
- The tenant should only carry out works to the premises if needed to keep them wind and watertight or secure the premises and the sub-tenant should be notified in writing. Any works

carried out by the tenant, which go beyond this, could result in an argument that the landlord has retaken possession.

(e) Reverse premium — approach

The amount paid to the landlord by way of compensation for an agreed surrender may well be very contentious. In a case where the landlord does not intend to redevelop and there is a long unexpired term, it is likely to require a substantial payment. There is no exact formula or rule of thumb, which can be applied. However, the approach of the House of Lords in *Re Park Air Services* [1999] 2 WLR 396 might be relevant. This case involved a tenant of a 25 year lease that was vastly overrented and that put itself into solvent liquidation causing the liquidators to immediately disclaim the lease. The case concerned the amount of compensation that was due to the landlord. The court suggested the following approach.

(1) The first step was to establish what the landlord would have received if the lease had not been disclaimed (or in this case surrendered). This figure is not simply a multiple of the unexpired term and the rent; it must be reduced to take account of costs that would not be recoverable from the tenant (eg works not falling within the tenants repairing covenant; management costs; surveyors and legal costs). It must also allow for the possibility of tenant default.
(2) The second stage was to establish the value of what is left to the landlord after the disclaimer (ie is the landlord able to re-let for the remainder of the period, and if so, at what rent?).
(3) Finally, a discount rate must be applied to the shortfall which makes an allowance for accelerated receipt by the landlord. The capital sum was calculated in *Park Air* by choosing a risk-free investment rate (gilt-edged securities) and then calculating the amount that would have been required by the landlord at the date of the disclaimer to allow it to receive the shortfall on the lease expiry date.

In many ways the calculation under *Park Air* mirrors the calculation a corporate occupier will carry out under an IAS37 assessment.

11.3.7 Effluxion of time

(a) General points

At common law, a lease granted for a fixed term will automatically expire by effluxion of time at midnight on the contractual expiry date (or term date) and without the need for the service of a formal notice. The tenant must give up possession of the premises on the expiry date unless it has statutory protection. The protection most commonly encountered by a corporate occupier is under Part II of the Landlord and Tenant Act 1954.

A sub-lease will also terminate on the expiry date unless the sub-tenant has statutory protection.

(b) Statutory protection under the Landlord and Tenant Act 1954

The effect of the 1954 Act and the detailed procedures necessary for termination are dealt with Chapter 13. In short if the security of tenure provision in the 1954 Act apply (ie sections 24–28), a tenancy can only be brought to an end by the termination procedures laid down in, or expressly approved by, that Act.

That procedure involves the service of a notice:

- under section 25 — by the landlord terminating the existing tenancy and indicating whether a new tenancy would be opposed and if so on what ground or
- under section 26 — by the tenant requesting a new tenancy or
- under section 27 — by the tenant stating that it does not want the existing tenancy to continue under the 1954 Act after the expiry of the contractual term.

In each case the tenant is free to, and if a section 27 notice is served must, vacate the premises on or before the contractual expiry date.

A landlord will, therefore, be entitled to possession of premises which had protection under the 1954 Act when:

- the tenant fails to issue a court application within the statutory time-limits
- the tenant fails to serve a court application (issued in time) within the statutory time-limits
- the tenant serves a valid section 27(1) or (2) notice
- the landlord establishes a ground of opposition to renewal under section 30(1)
- the tenant decides not to accept the terms for renewal ordered by the court.

(c) Excluded lease

Not all tenancies, which corporate occupiers are likely to encounter, will be protected by the security of tenure provisions contained in the 1954 Act. The most common exceptions are where:

- the parties have agreed that these provisions should be excluded and should not apply to a particular tenancy
- the premises are no longer occupied by the tenant for business purposes (in which case sections 24–28 will no longer apply).

If there is no protection, the position is as stated in (a) above. The tenant has no right to remain in the premises after the contractual expiry date.

11.4 Court proceedings

If an occupier's right to possession has been lawfully terminated, the landlord/licensor is entitled to re take possession. If the premises are empty, this may be simply by peaceably re-entering them and re-securing them (eg by changing the locks).

If the premises are not empty, because the previous occupier remains there unlawfully, the owner is nevertheless entitled to physically re-take possession. However, this right is curtailed by section 6(1) of the Criminal Law Act 1977 which makes it a criminal offence for a landowner to use or threaten violence to secure entry to property if there is someone present on the property who is opposed to entry. This restriction extends to the use of violence to property as well to the person. In practice this will prevent the exercise of the landowner's right while there is a person physically on the premises who opposes re-entry. If a landowner wants to exercise its self help remedy, the safest course is to instruct certificated bailiffs and wait until the occupier has left the premises for the day (eg at night).

There will be many situations where it is safer to issue court proceedings for possession (eg where there may be some doubt about the validity of the termination or where the occupier is still trading from the premises and the landowner does not wish to face a claim for unlawful eviction or a substantial claim for loss of profits).

There are essentially four different types of proceedings which a corporate occupier (as tenant or landlord) is likely to encounter in respect of termination.

11.4.1 Summary possession proceedings

Straightforward cases, which allege trespass but which do not involve a tenant who has remained in occupation following the expiry of its tenancy, can be dealt with by the issue of summary possession proceedings. Such cases include the termination of contractual licences. If successful (and no triable defence is raised), from the issue of proceedings to the enforcement of the possession order by the county court bailiffs can take as little as two to three weeks.

11.4.2 Ordinary possession proceedings

All other possession claims are likely to be dealt with under the ordinary possession proceedings. The landlord must issue a claim form containing (or attaching) particulars of claim which will be served on the defendant. A hearing date will be fixed when the claim is issued (usually two to three months later) and the defendant will be given the opportunity of serving a defence.

If the claimant considers that there is no sustainable defence, he can expedite the process and avoid the additional costs and delay involved with going to a full trial by making an application for summary judgment. This process, if successful, is likely to take between three to five months depending upon the individual court.

11.4.3 Declaratory proceedings

If the landlord and tenant are in dispute as to whether a lease has been terminated or not, the other way in which the issue may come before a court is by the landlord suing for rent for a period after the tenant alleges the lease came to an end; or by either or both parties applying for a declaration as to the status of the lease.

This usually arises in cases involving allegations of implied surrender or where the validity of a break right is challenged.

11.4.4 Forfeiture proceedings

These are discussed in Chapter 8.

11.5 Without prejudice marketing

Often, disputes concerning whether a lease has come to an end or not can lead to premises being sterilised for protracted periods while the parties litigate. This not in the best interests of either party. From the landlord's standpoint, if it loses the case, and the court decides the sub-lease was surrendered or a break clause properly exercised, it will have lost the opportunity to collect rent for the period of the dispute. Conversely, if the tenant loses, it will have to pay the rent (plus interest) for the period of the dispute and will have lost the opportunity to assign or sub-let the property (assuming the lease permits this).

The most sensible solution in these circumstances is for both parties to agree that the marketing of the premises will be without prejudice to each others rival contentions. If this can be agreed, both have the opportunity of finding a new tenant and/or assignee and/or sub-tenant while not prejudicing their right to prosecute their claim. If a new tenant who is acceptable to the landlord can be found quickly, the amount in dispute can be kept to a minimum thereby increasing the likelihood of an amicable settlement.

11.6 Practical checklists on termination

11.6.1 Prior to termination

There are number of points which a landlord should check prior to termination:

- What are the tenant's repairing and redecoration obligations?
- What are the tenant's re-instatement obligations? Is the landlord required to notify the tenant in writing before the lease expiry which works of re-instatement are required?
- If appropriate, a terminal schedule of dilapidations should be prepared and served on the tenant (see Chapter 15).
- If the tenant is to carry out works, ensure that the landlord's specifications are passed on to the tenant (eg paint colour; floor coverings).
- If there is to be a monetary payment in lieu of carrying out the works, negotiations should start as soon as possible to minimise delay in re-marketing premises.

Different principles may able where a tenant is seeking to exercise a break right (see Chapter 12).

The points, which a tenant will wish to check prior to termination, are likely to include:

- the removal of all tenant's effects and fixtures
- the extent of any likely terminal dilapidations claim, and the strategy to be used (ie do the works or seek to negotiate a cash settlement — see Chapter 15)
- seek to agree works of dilapidation necessary so as to minimise any delay in re-letting, which may result in a significant loss of rent claim.

11.6.2 Practical matters post termination

A landlord or its agent will need to take some or all of the following steps.

- Ensure that the premises are empty. If the tenant has left items behind a comprehensive list should be made and the items photographed and/or videoed for evidential purposes. The tenant should be given an opportunity to collect the belongings, and if necessary, a notice served under the Torts (Interference with Goods) Act 1977.
- The premises should be made secure. If there is a doubt as to whether all the sets of keys to the premises have been returned, the locks should be changed. The letterbox should be sealed. The

build up of post can pose a potential fire hazard. Furthermore, depending on the type of premises and its location, it may be necessary to board up some or all of the windows or fix security grilles. Consideration may also need to be given to installing a CCTV alarm or retaining security guards.

- The heating and water pipes should be drained down to prevent flood damage. This is especially important for office buildings and when a unit is repossessed from a tenant this should be done straight away.
- The insurers must be informed (which may involve notifying a superior landlord).
- The meters must be read and the relevant bodies will need to be informed that the premises are empty and a new billing address provided.
- The premises should be inspected on a regular basis. The frequency may be dictated by the insurer and/or the level of risk.

11.7 Precedents

11.7.1 Termination letter on behalf of the licensee

Dear Sirs

Re: Termination of Licence Agreement

We act for and on behalf of Fly By Night Ltd your current licensee of Unit 1, the Precinct, Womblington, under a licence agreement dated 1 June 2007.

It was an implied term of the agreement that the licensee could terminate the licence on reasonable notice. Our client no longer requires the premises and in the circumstances hereby notifies you of its intention to terminate the licence on 1 September 2007.

[***Alternatively***

Our client is entitled to terminate the licence on 4 weeks written notice pursuant to clause [ ] and we hereby notify you that our client intends to terminate the agreement and vacate the premises on 1 September 2007.]

Please could you acknowledge safe receipt of this letter by signing, dating and returning the copy notice enclosed.

Yours faithfully

11.7.2 Termination letter on behalf of licensor

Dear Sirs

Re: Termination of Licence Agreement

We act for and on behalf of Ta Ta Ltd your licensor of premises at Unit 1, the Precinct, Womblington, under a licence agreement dated 1 June 2007.

It was an implied term of the agreement that the licensor could terminate the licence on reasonable notice. Our client now requires the premises and in the circumstances hereby notifies you of its intention to terminate the licence on 1 September 2007.

[***Alternatively***

Our client is entitled to terminate the licence on 4 weeks written notice pursuant to clause [ ] and we hereby notify you that our client intends to terminate the agreement on 1 September 2007.]

Please note that our client will not accept licence fees after the above date. Any payments paid after that date will be accepted as compensation for trespass and will not give rise to the creation of a new periodic licence or tenancy.

Please could you acknowledge safe receipt of this letter by signing, dating and returning the copy notice enclosed. Please could you also contact the writer, Dodge E Geezer, to make the appropriate arrangements to ensure an orderly handing back of the premises.

Yours faithfully

11.7.3 Termination letter on behalf of the tenant at will

Dear Sirs

Re: Termination of Tenancy at Will

We act for and on behalf of Fly By Night Ltd your current occupier of Unit 1, the Precinct, Womblington.

On 1 June 2007 our client was allowed into occupation of the premises on a tenancy at will pending the negotiation of a lease. The parties have not been able to agree terms for the grant of a lease and our client now intends to vacate the premises and return the keys on 1 September 2007.

[***Alternatively***

Following the expiry of our client's previous 1954 Act excluded tenancy, our client was allowed to remain in occupation of the premises as a tenant at will pending the negotiation of a new lease. Unfortunately, the parties have not been able to agree terms for the grant of a new lease and our client now intends to vacate the premises on 1 September 2007.]

Please could you acknowledge safe receipt of this letter by signing, dating and returning the copy notice enclosed.

Yours faithfully

11.7.4 Termination letter on behalf of the landlord

Dear Sirs

Re: Termination of Tenancy at Will

We act for and on behalf of Ta Ta Ltd freehold owner of premises at Unit 1, the Precinct, Womblington.

Our client allowed you into occupation of the premises on a tenancy at will pending the negotiation of a lease. The parties have not been able to agree terms for the grant of a lease and our client now requires the return of the premises. In the circumstances please make arrangement to vacate the premises and return the keys to us by no later than 4pm on 1 September 2007.

[***Alternatively***

Following the expiry of your previous 1954 Act excluded tenancy, our client allowed you to remain in occupation of the premises as a tenant at will pending the negotiation of a new lease. Unfortunately, the parties have not been able to agree terms for the grant of a new lease and our client now requires the return of the premises. In the circumstances please make arrangement to vacate the premises and return the keys to us by no later than 4pm on 1 September 2007.]

Please note that our client will not accept rent after the above date. Any payments paid after that date will be accepted as compensation for trespass and will not give rise to the creation of a new tenancy.

Please could you acknowledge safe receipt of this letter by signing, dating and returning the copy notice enclosed. Please could you also contact the writer, Dodge E Geezer, to make the appropriate arrangements to ensure an orderly handing back of the premises.

Yours faithfully

11.7.5 Precedent notice to quit by tenant

To: Dreadful Investments Limited (Company Number 0012345) ("The Landlord")
Of: The Grange, Little Puddle on the Hill, Great Pile

From: Splurge Pump Limited (Company Number 0054321) ("The Tenant")
Of: Unit 1, Filthy Industrial Estate, Little Snoring,

We, as agents, acting for and on behalf of the Tenant hereby give you notice that the Tenant intends to quit and deliver up possession of premises know as Unit 1, Filthy Industrial Estate, Little Snoring ("the Lease") on 1 September 2007 or other the day on which the tenancy expires next after the end of of [one week][one month][one quarter][half year] — delete as appropriate] from the date of service of this notice"

Dated: [ ]

Signed . . . .
Duckin And Divin Managing Agents
1 Plushsville Tower
London
For and on behalf of the Tenant

I acknowledge receipt of this notice of which this is a true copy

Signed . . . . on 2007

For and on behalf of the Landlord

Print name . . . .

11.7.6 Notice to quit by landlord (periodic tenancy not protected by the Landlord and Tenant Act 1954)

To: Splurge Pump Limited (Company Number 0054321) ("The Tenant")
Of: Unit 1, Filthy Industrial Estate, Little Snoring,

From: Dreadful Investments Limited (Company Number 0012345) ("The Landlord")
Of: The Grange, Little Puddle on the Hill, Great Pile

We, as agents, acting for and on behalf of your Landlord hereby give you notice to quit and deliver up possession of premises know as Unit 1, Filthy Industrial Estate, Little Snoring ("the Lease") on 1 September 2007 or other the day on which the tenancy expires next after the end of of [week] [one month][one quarter] [half year] — delete as appropriate] from the date of service of this notice

Dated: [ ]

Signed . . . .
Duckin And Divin Managing Agents
1 Plushsville Tower
London
For and on behalf of the Landlord

I acknowledge receipt of this notice of which this is a true copy

Signed . . . . on 2007

For and on behalf of the Tenant

Print name . . . .

12 Operating Break Clauses

12.1 Introduction

A break clause or break right is an option granted to a landlord, or a tenant, or both, to unilaterally terminate a lease before its contractual expiry date. Break clauses can be very important to a corporate occupier by giving flexibility and providing an opportunity to end the lease, and more particularly, the liability to pay rent, before the contractual expiry date. However, the importance of their careful negotiation and drafting, and the potential difficulty of their operation are largely overlooked, sometimes at great potential cost.

This chapter starts by considering the pitfalls of a typical break clause, and then goes on to deal with the different types of conditional break clause and the problems which a tenant may face in trying to exercise one. The chapter then explores the technical difficulties involved with the drafting and service of break notices before concluding with a practical approach to break clauses from the corporate occupiers perspective of a tenant wishing to end a lease, a landlord wishing to frustrate their operation by a sub-tenant in a legacy portfolio situation, and finally from the perspective of a landlord seeking to exercise a break clause.

12.2.1 A typical break clause

There are almost as many types of break clauses as there are lawyers who draft them. However to provide an indication of a fairly typical break clause, the following clause has been extracted from an actual lease.

> Either party may determine this Lease on the Third Anniversary of the Term Commencement Date by giving the other party not less than six months and not more than 12 months prior written notice ... and provided that the said six months prior written notice has been given and (in the case of the Tenant exercising the break) the Tenant pays the rent reserved by the Lease and delivers vacant possession of the Premises in the state and condition required by this Lease on the determination date then this Lease shall determine ...

At first glance it may be thought that this clause is relatively straightforward. It is however a legal minefield for the tenant. There are five critical elements to the successful operation of this break clause by the tenant.

1. The tenant must correctly calculate the break date (ie the date when the lease can be terminated). The break clause could have simply specified a date, but instead a formula was used. This provides a potential trap for the unsuspecting tenant because if it calculates this date incorrectly, it may result in an invalid notice being served. In recent years, such a clause has resulted in a number of reported court decisions: see 12.5 below for a precedent notice and covering letter which addresses this difficulty.
2. A written notice must be served on the other party not less than six months and not more than 12 months before the break date.
3. The tenant must ensure there are no rent arrears on the break date.
4. The tenant must ensure that by the break date it has complied fully with its repairing, re-instatement and redecoration obligations under the lease.
5. The tenant must deliver up vacant possession on or before the break date.

If the tenant fails to satisfy any one of the five elements, he will fail to satisfy the conditions precedent and the lease will continue. If the landlord insists on its strict legal rights, at best the tenant may become embroiled in a lengthy (and uncertain) dispute that results in irrecoverable costs (a successful party usually only recovers between 50% and 60% of its actual legal costs), and at worst the court may decide that it has lost its right to break, leaving it with premises which are surplus to requirements, paying rent on a lease that may be difficult (or impossible) to sublet or assign, and a hefty bill for legal costs.

12.2 Compliance with conditions precedent

The difficulty of exercising a break clause ranges from being straightforward to being downright impossible. Who can exercise the break, the time and circumstances in which it can be exercised and the formalities involved are all governed by the terms of the lease, usually the break clause itself. In recent years a number of cases have been before the courts, some have been purely case specific and of little precedent value, but others have been of great significance to how a court will determine whether the essential conditions have been satisfied, and the break right properly exercised.

A break right is what lawyers, rather grandly, describe as a unilateral option. What this means in practical terms is that if a break right is made dependant upon the compliance with certain conditions precedents, they must be strictly complied with or the break will fail (see *Bass Holdings* v *Morton Music* [1987] 1 EGLR 214). If the tenant only partially complies with a condition, the break will fail.

12.2.1 Absolute compliance with covenants

The difficulties of satisfying a condition precedent that requires absolute compliance with tenant covenants are well documented (see *West Country Cleaners* v *Saly* [1966] 1 WLR 1485 and *Bairstow Eaves (Securities) Ltd* v *Ripley* [1992] 2 EGLR 47 where tenant options were lost as a result of minor breaches of redecoration covenants).

A striking example of this principle in operation should strike fear into the most steely of corporate occupiers, and raise doubts as to the level of flexibility they really have in their portfolio based on the existence of break clauses in their leases.

By way of example *Osborne Assets Ltd* v *Britannia Life Ltd* (1997) Liverpool CC (unreported)

Facts

- T occupied an office block under a 25 year lease at substantial rent with a break in the fifth year.
- The break clause was conditional on T complying absolutely with all of the tenant covenants in the lease.
- Following the break date L argue that T had failed to comply fully with its repairing and redecoration obligations.
- The judge at trial noticed that an invoice from T's painter and decorator showed that the premises had been painted with two coats of paint whilst the redecoration covenant required it to be painted with three coats of paint.

Held

- As the lease requires three coats of paint and the tenant had painted with two, it had not strictly complied with the condition precedent to the operation of the break clause and the lease therefore had not come to an end.
- The tenant was therefore on the hook to pay rent for a further 20 years!

In some cases therefore, while a corporate occupier may make assumptions and plan its business on the basis that it has a break right in its lease, the conditions which have to be fulfilled before the break can be operated may not, in practice, be capable of being fulfilled. This is sometimes referred to as the "illusory break right". From a portfolio management context, it is important to study a particular break clause in detail and make a judgment as to how difficult it may be for the corporate occupier to exercise it in practice. A corporate occupier should not simply assume that it can exercise a break clause because one is contained in the lease (which is often the case).

12.2.2 Qualified compliance with covenants

The harsh rule referred to above, together with the difficulty of achieving strict compliance where a landlord is determined to find fault, has led to attempts by tenants to avoid conditions altogether. Corporate occupiers should try and negotiate short-term leases or break clauses with no compliance problems. However, where tenant's have insufficient bargaining power, the alternative is to try and to water down the degree of compliance required (eg by requiring reasonable compliance with, or an absence of material breaches of, tenant covenants).

Achieving reasonable compliance or avoiding material breaches of tenant covenants can also be challenging. The difficulty for a tenant is knowing, in practical terms, what degree of compliance is required to satisfy a qualified condition. In *Reed Personnel Services* v *American Express* [1997] 1 EGLR 229 the tenant argued that outstanding repairs to the value of £21,000 in the context of an annual rent of £120,000 were not significant and did not constitute a material breach of its repairing obligations. The judge disagreed. An influential factor in the judge's view was that the tenant had made no attempt to comply with its repairing obligations; assuming erroneously that it would be able to negotiate a cash settlement in lieu of carrying out the repairs.

In contrast, *Commercial Union Life Assurance* v *Label Ink* L&TR 380 provided the corporate occupier with a small measure of comfort. In this case the tenant instructed a surveyor to advise it on what works would be required, and the tenant carried out most of those works before the break date. The judge held that the tenant had materially complied with its repairing obligations even though there were works outstanding (which would cost approximately £12,000).

In *Label Ink* the judge was influenced by the following factors:

- the tenant had made genuine efforts to comply, in particular by diligently carrying out works identified by his building surveyor
- the premises were left in a lettable condition i.e. the outstanding works were not likely to put off a prospective tenant (provided there was a rent free period or rent reduction to compensate)
- the landlord was not concerned with compliance but was looking to take advantage of the tenant's non-compliance to keep it on the hook.

Judge Rich QC in *Label Ink* when deciding whether the tenant had materially complied with its covenant was clearly influenced by the landlord's obstructive approach and desire to take advantage of any slip up by the tenant. He held that the test as to whether any breach was material, was whether in all the circumstances it would be fair and reasonable to deprive the tenant of the break right.

However, the position was reconsidered by the Court of Appeal in the leading case on material breach. *Fitzroy House Epworth Street (No 1) Ltd* v *Financial Times Ltd* [2006] 2 EGLR 13.

Facts

- This case involved a lease of a three-storey office block in central London with rents of £600,000 pa with a further six years to run on the lease (ie with a further rental liability of at least £3.6m).
- To break on 1 April 2004, the tenant was required to serve a break notice (which it did) and materially comply with its obligations under the lease.
- T's surveyor prepared a specification of works that was sent to the L's surveyor and a joint inspection sought, which was declined.
- L's surveyor also declined to inspect the progress of the works, despite the works carried out by the tenant costing in excess of £900,000.

HHJ Thornton QC (applying *Label Ink*) Held:

- That a breach of covenant is only material if it is fair and reasonable to refuse the tenant's right to break.
- The key points, which influenced the judge in concluding that the tenant had materially complied with its covenants, were:

  (1) the outstanding repairs were "insubstantial" at £20,000 including supervision fees and would not affect the L's ability to find a new T
  (2) T had taken all reasonable steps to comply with its repairing obligations and had followed a professional surveyor's advice as to what was required spending nearly £1m
  (3) T would have incorporated L's suggestions if it had asked
  (4) L had unreasonably declined to become involved in the T's attempts to agree a schedule of works and was trying to catch T out on a technicality
  (5) there was no diminution in value of the L's reversion as a result of the outstanding works
  (6) it would be unreasonable to T if unable to break, and L could prevent the break by its behaviour

Court of Appeal — Held

- Trial judge, and Judge Rich in *Label Ink*, had applied the wrong test. The word "material" could mean many things but "what is fair and reasonable between L and T" was not one of them.
- Items (2), (4) (6) were not relevant to whether T had materially complied with its covenants.
- Materiality had to be assessed by reference to the ability of L to relet or sell without delay or additional expenditure. "material" and "substantial" are interchangeable; "reasonable" connotes a different test.
- Given that the damage to L's reversion was "negligible or nil", and the outstanding defects had no effect on L's ability to relet, T *had* materially complied with its covenants.

It follows that the landlords conduct and motives leading up to a break will be irrelevant to the central issue whether any breaches of covenant are "material". Following *Fitzroy*, when considering whether a breach is or is not material a court will take into account:

- the extent of the breach
- the adverse effect (if any) on the landlord of any failure by the tenant to comply
- the landlord's ability to relet or sell the premises without delay or additional expenditure (or accepting a reduced rent or rent free as a result of breaches of covenant by the tenant)
- any diminution in value of the landlord's reversion.

Importantly, while the Court of Appeal considered that material and substantial breaches were governed by the same test, it was suggested that "reasonable compliance" involves a different test. It remains to be seen what approach the courts will now take on this issue. It is possible that Judge Rich's approach in *Label Ink* could be resurrected in a case of reasonable compliance.

12.3 Break notices

The service of a written notice to trigger the break is necessary in most cases. Time-limits for the service of break notices are of the essence and notices must be served within the specified time-limits. If the tenant fails to serve a notice in time or to ensure that a notice expires on the date provided for in the break clause, the notice will be ineffective and the break right will be lost.

The service of a valid break notices involves a very technical area of law and in most cases it would be wise to instruct a solicitor who has experience of serving break notices to draft and serve a notice. It is worth bearing in mind that if a solicitor makes a mistake, it may be possible to make a claim against his/her professional indemnity insurance policy. That being said it is far better to ensure a suitably qualified professional is dealing with the break, rather than trying to mitigate the liability if the break fails. The costs of going to court are high, as are the costs of trying to recoup monies from professional indemnity policies.

12.3.1 Common mistakes

It is perhaps surprising how often mistakes are made by tenants, or their advisors, when drafting and serving break notices. The checklist below may assist a tenant or its advisors to avoid the most common errors.

(a) Correctly identify the tenant

- The notice must be served on behalf of the legal owner of the lease. If the lease is registered, it must be served by the registered proprietor and not the beneficial owner (see *Brown & Root Technology Ltd* v *Sun Alliance* [1997] 1 EGLR 39).
- If the lease is registered, obtaining up to date office copies from the Land Registry will identify the legal owner of the tenancy, and will also identify any other discrepancies eg whether the lease is vested in one company but occupied by another group company.
- If the lease is unregistered, all deeds of assignment will have to be checked to identify the correct tenant.
- It is also prudent to carry out a company search (which can be done online at no cost) to check the company is still trading, has not changed it name and to establish its current registered office (which in most cases will be where the notice will have to be served — although it is important to consider any service of notice provisions in the lease).

(b) Correctly identify the landlord

- The notice must be served on the registered landlord.
- If the land is registered, office copies from the land registry will correctly identify the landlord.
- If the land is unregistered then the landlord's identity and address may appear on the last rent demand. Rent demands should, however, be treated with some caution as it is not uncommon for rent to be collected by a third party on behalf of the landlord.
- If there is any doubt about the identity of the landlord, it would be prudent to write to the recipient of the rent in good time asking it to clarify the position.
- Again, if the landlord is a company, a company search should be carried out to check that it is still trading and that there has been no change to its name and registered office.
- The practice of including a company number in a lease is a relatively recent one, so careful tracking of a company may be needed from the date of the lease to the date of service of the break notice. Changes of names or even the swapping of names within a group is not uncommon.

(c) Validly serve the notice

- It is important to check the lease for a clause dealing with service of notices under the lease, which may identify where and how a notice must (or can) be served.
- Most notices will be governed by section 196 of the Law of Property Act 1925 which provides that a notice will be validly served if left at the last known place of abode or business of the landlord in the UK.
- Although a notice which is sent by recorded delivery can be deemed to have been served even if a landlord did not in fact receive it (where section 196(4) of the Law of Property Act 1925 applies), corporate occupiers should be very wary about relying upon recorded delivery. The Royal Mail does not deal with the post in the way it used to and no longer tracks recorded delivery letters. They are sent out with the ordinary post and if a recipient is not available to sign for a letter, it will be returned to the depot with a note inviting the recipient to collect the letter at a later date. If that person fails to collect the letter, it will eventually be returned to the sender. A notice, which is returned undelivered, will *not* be deemed to be served (see section 196(4) of the LPA 1925). This should be contrasted with the position where notices are served under the Landlord and Tenant

Act 1954 (see 13.3.4). The obvious concern is that the notice may be returned undelivered *after* the time-limit has expired giving the corporate occupier no opportunity to serve another notice in time.

- Consequently, if the deadline for service of a valid break notice is approaching and it has not been possible to obtain a copy of the recipient's signature evidencing safe receipt, the only safe approach is to personally serve the notice or instruct a process server to do so.

(d) Comply with other requirements

- If a break premium is required when the notice is served, this should be paid at the same time. Failure to do so is likely to be fatal to the operation of the break.
- It is important to make sure that the landlord has cleared funds. This is best achieved by asking the landlord well before the notice is served for its bank details so that the premium can be transferred electronically to ensure safe receipt into its account.

12.3.2 Are mistakes fatal?

A great deal of court time has been spent considering break notices that are ambiguous or contain mistakes. Traditionally, the courts took an inflexible and strict approach to construing notices. However, a more relaxed approach now prevails following the House of Lords decision in *Mannai Investment Company Ltd* v *Eagle Star Life Assurance* [1997] 1 EGLR 57.

It was held in *Mannai* that a notice will be valid provided that it fulfils the requirements of the break clause and it would not have misled a "reasonable recipient". In *Mannai* the tenant was required to give not less than six months written notice of its intention to exercise the break notice expiring on "third anniversary of the term commencement date". The tenant's notice purported to expire on 12 January, when in fact, it should have stipulated 13 January. The House of Lords held that for the purposes of the test, the reasonable recipient is taken to be familiar with the terms of the lease, and as such, a reasonable recipient would not have been misled by such an obvious error.

Other examples of this principle have resulted in the following decisions.

- A notice sent in September 1974 purporting to terminate a lease in March 1973 was read as if it had intended to terminate the lease in March 1975 because that was obviously what was intended (*Carodine Properties* v *Aslan* [1976] 1 WLR 442).
- A notice, which specified the correct date, was valid where a covering letter contained references to a date that was obviously wrong (*Micorgrafix* v *The Working 8 Ltd* (1996) 71 P&CR 43).

However, prevention is better than a cure, and it goes without saying that careful and thorough drafting and service of a break notice can avoid serious problems later. While the application of the reasonable recipient test to the factual situation in *Mannai* was relatively easy, it will not always be the case and the application of the principle to less clear cases will by the very nature of the test, result in uncertainty until the matter is decided by the court.

12.3.3 Precedent break notice and covering letter

A common mistake is to wrongly identify the termination date where the lease defines this by reference to an anniversary of the term commencement date (eg see *Mannai*). This mistake can easily be avoided by:

- drafting a break notice that follows the wording of the break clause and
- writing a covering letter which follows the wording of the break clause, but also identifies what the tenant considers to be the actual date.

If the letter incorrectly identifies the date, the break will nevertheless be saved by a break notice which follows the exact wording of the break clause (see *Micorgrafix* v *The Working 8 Ltd* above).

A precedent break notice together with a precedent covering letter based on the typical break clause set out in section 12.2.1 above can be found at the end of this chapter at 12.5.

12.4 A practical approach to operating break clauses

Often corporate occupiers fail to deal with break rights early enough or take them sufficiently serious. Break opportunities are lost by a failure to carefully consider the break clause at an early stage and put in place an effective strategy. There is real opportunity for professional advisors to influence the outcome and, for example, to provide their client with a strategy to dispose of surplus property.

The importance of operating a break clause will be depend upon the length of the unexpired term and the annual cost of the property balanced against the financial strength of the tenant. However, the difficulty of exercising a break clause and the true financial cost is generally underestimated, particularly in a poor letting market.

In the 1980s and early 1990s it was not unusual for corporate occupiers to enter into 25 year leases with upward only rent reviews. However, research has showed that the average length of actual occupation by tenants is only six to seven years. That has left many tenants with premises that are over-rented, surplus to requirements and causing unwanted financial consequences. It is true to say that lease lengths have shortened in the last 15 years, but occupiers tend to have short memories once economic growth arrives, and we are once again seeing tenants taking leases of 15 years or more, when their business strategy is only four to five years. Since 1999 companies have had to make a provision for onerous leases under FRS12. Surplus leasehold properties are onerous leases and tenants must provide a reasonable estimate of that liability, so if a break clause is not operated successfully a provision for the liability is going to have to be made.

Where corporate occupiers do not have break rights, and/or are not able to negotiate a surrender, one solution has been for corporate occupiers to sublet on short term leases which often include break clauses. Thus, in the context of surplus properties, the exercise of a break clauses can arise where a corporate occupier has sublet and the subtenant wants to break, or where the corporate occupier itself wishes to terminate. Whilst the legal principles that govern these two situations are the same, the corporate occupier's strategy is likely to be different.

12.4.1 Break by tenant

How should a corporate occupier approach the exercise of a break clause dependant upon absolute, material/substantial or reasonable compliance with its covenants? While, the operation of such clauses will always be difficult, and to an extent uncertain, there are a number of steps that a corporate occupier can take which improve its prospects of compliance. The key to a successful outcome is devising a clear strategy at an early stage. Indeed, even before considering the tactics of a particular break, it is prudent for a corporate occupier to ensure that any property database includes the basic details of any break clauses and whether there are any compliance issues that are associated with their exercise. This exercise will identify the timescale that is needed for the following steps to implement a successful strategy.

(a) Step 1 — Compliance review

The corporate occupier should seek advice on what needs to be done, and by when. This will involve an audit of each and every covenant contained in the lease (and superior lease if appropriate). It quite literally means working through the lease clause by clause and drawing up a list of actions and dates when each action should be started and completed. It may also require a careful examination of any deeds of variation, licence for alterations and schedule of condition.

This step should be carried out in good time. A good starting point would be at least six months before the break notice can be served. Some corporate occupiers seek to penny pinch and fail to carry out such an essential preliminary step. However this overlooks the dramatic impact of failing to exercise the break properly.

(b) Step 2 — Break notice

As can be seen from 12.3 above, great care should be exercised to ensure that the notice is drafted correctly and is then validly served. It is important that the necessary investigations are carried out and the notice is drafted in good time.

(c) Step 3 — Break payment

The tenant, or its advisors, should serve the break notice in good time taking care to ensure that any conditions, such as the payment of a break premium, are complied with (see 12.3.1 (d) above).

(d) Step 4 — Schedule of works

In many cases the corporate occupier may be required to comply with repairing, redecoration and re-instatement covenants at or before the break date as a condition of the right to break, and for the remaining cases there will probably be a need to settle a dilapidations claim by the landlord after the lease has been broken. Again it is false economy not to deal with this properly. An independent building surveyor should be instructed to prepare a schedule of necessary works, or an assessment of any dilapidations claim that is likely to brought by the landlord, well ahead of critical dates. It may also be necessary at this stage to instruct a mechanical and/or electrical engineer depending on the premises and the corporate occupier's repairing and maintenance obligations under the lease.

Where the break is dependant on compliance with repairing obligations, in view of the context in which the exercise is being carried out, the schedule should endeavour to be accurate, with any doubt being resolved in the favour of the landlord (in contrast to the common practice of preparing a counter schedule for the purposes of negotiating a terminal dilapidations claim). This is particularly important if absolute compliance is required. Any savings by scrimping on repairs could be rapidly eaten up by legal fees if the landlord challenges compliance after the break date. Still worse, it could jeopardise the break altogether.

(e) Step 5 — Landlord's agreement

The landlord should be sent a copy of the schedule of works and be invited to agree or identify any other works that it believes are necessary. It should at the same time request any necessary consents (eg reinstatement, paint colour or carpet specification).

Every effort should be made to try and agree a schedule of works. This process can take some time, particularly with substantial premises, and this is yet another reason why a strategy should be initiated as early as possible.

(f) Step 6 — Paper trail

If, as is likely, the landlord fails to respond, it should be chased in writing on a regular basis. It is important to generate a paper trail that shows a genuine attempt by the tenant to comply with its covenants in the face of a lack of co-operation by the landlord. This correspondence can then be put in evidence in subsequent court proceedings provided that it is open correspondence ie not marked without prejudice and is not generated in the context of a negotiated settlement.

A court is unlikely to be impressed with a landlord who fails to act reasonably and seeks to take advantage of the situation. In *Label Ink* a tenant asked for the landlord's view on one aspect of proposed repairs. The landlord's response was that:

> it is for the tenant to determine what they must do to comply with the repairing obligations.

The judge said of this episode:

> ... the landlord was not so much concerned with compliance as with taking advantage of non-compliance ... That is an attitude which I think needs to be taken into account in assessing the materiality of the breach ...

Although the Court of Appeal in *Fitzroy* expressly rejected this point, the authors take the view that demonstrating that a landlord has been obstructive may assist a tenant seeking to break for the following reasons.

- *Fitzroy* only provided a test for identifying a "material breach" or a "substantial breach"; the Court of Appeal left open the test to be applied where "reasonable compliance" is required. Although unlikely, it possible that the *Label Ink* test could be applied in a case of "reasonable compliance".
- Even if a court does not expressly take the landlord's conduct into account, it could indirectly influence the judge and result in a merits based decision. At the very least, it could cause the judge to be sympatric to the tenants case and take a tough line on the landlord's case.

It may still assist, therefore, to ensure that the tenant's attempts to comply with its obligations are evidenced in open correspondence which can then be use in evidence to defend the tenant's position should the landlord choose to issue court proceedings.

(g) Step 7 — Undertaking the works

The corporate occupier must ensure that remedial works are started and completed in good time (with a margin of error built in). Advice should be obtained from a building surveyor on the length of time the works are likely to take. This period should allow for the vacation of the premises (if they are occupied) and for putting the works out to tender, carrying out the works and dealing with any snagging list.

If the corporate occupier is still in occupation there will be financial pressure for it to remain in the premises as long as possible to minimise the period for which it is still paying rent but the premises are empty and unproductive. This pressure should be carefully managed; from a commercial perspective it is better that the premises are empty for a few months before the break date to ensure compliance rather than being empty for many years afterwards if the break right is lost and the premises cannot be sub-let (through lack of demand or inflexible alienation provisions in the lease). This is especially the case when a tenant is taking new premises elsewhere and does not want double overheads. In addition the time taken to find, fit-out and move in to premises is often under estimated. This leads to time pressures and can lead to a failure to carry out properly all the works that are required to satisfy the break conditions.

Great care should be taken if the parties enter into negotiations which envisage the corporate occupier paying compensation in lieu of carrying out necessary works to the premises. A common landlord tactic in these circumstances is to string the tenant along in negotiations until it is too late for the works to be carried out. The landlord then withdraws from negotiations and the tenant is unable to comply with its obligations in time.

A counter tactic, which can be effective in an appropriate case, is for the corporate occupier to write to the landlord at the beginning of the process (eg when the break notice is served) setting a deadline for negotiations to be concluded and stressing that if a binding agreement is not in place by that date, the negotiations will be terminated and the works carried out. It goes without saying, that this tactic should only be used if it is thought that the landlord is aware of the conditional nature of the break clause. It would be folly to alert a landlord to the potential difficulties the tenant may face if the landlord may be blissfully unaware of these problems. The use of this tactic will depend upon a careful judgement as to whether the landlord is a well-informed corporate investment landlord, or a less well-informed private individual.

Once the works have been completed, the landlord should be invited to inspect the works and agree all necessary works have been carried out to the appropriate standard. If there is no agreement, the tenant's surveyor should be asked to prepare a completion report (including detailed photographs and video evidence) to show compliance. This should be prepared as close to the break date as possible. This will then form the basis of the evidence to be placed before the court should a landlord seek to challenge the validity of the break.

(g) Step 8 — Financial statement

The corporate occupier should request a statement of outstanding monies due and owing under the lease shortly before the break date. A careful check of the financial position should be undertaken and legal advice should be obtained on the amount of rent that should be paid on the last rent payment date.

A common error is for the tenant to pay an apportioned rent where the break date falls between rent payment dates. The Apportionment Act 1870 does not apply to rent payable in advance once it falls due (*Ellis* v *Rowbotham* [1900] 1 QB 740), and depending upon the precise wording of the lease, the failure to pay a full quarters rent could invalidate the break.

This potential mistake can be best illustrated by a simple example of the terms in a particular lease.

Example

- Rent is payable on the usual quarter days.
- The break date is midnight on 29 September.
- The end of the June rent period is midnight on 28 September and therefore a full quarters rent becomes due on 29 September even though the lease would come to an end at midnight on 29 September if the break is validly operated.
- Failure by the tenant to pay a full quarters rent on 29 September will invalidate the break.

In the authors experience there are many break clauses that have been drafted by property lawyers who have made this mistake. Corporate occupiers should beware because landlords are becoming increasingly aware of this tenant snare. While it is possible that the Court of Appeal could revisit the principles of *Ellis* v *Rowbotham*, unless and until this decision is reversed, it remains good law.

Tenants should also take care when calculating how much should be paid to a landlord to ensure that the rental obligation is discharged and/or how much should be paid by way of a break premium, when there is an outstanding rent review. The issue is should the payment be based on the old rent or the new rent that will be fixed when the review is complete? This will turn on the wording of the lease and in particular the break clause and the rent review. Usually, the rent payable and the basis for calculating the break premium will be the existing unreviewed rent. However, if the outstanding rent review is determined before the break date, the tenant should pay the shortfall rent plus interest as soon as possible (and in any event before the break date). Failure to do so could result in the tenant being in breach of covenant and the loss of the break right.

Label Ink graphically illustrates the importance of complying with financial obligations. Although the tenant satisfied the difficult task of persuading the court that it had materially complied with its repairing obligations, it slipped up by failing to pay an additional amount of rent required under the lease by the break date (representing six days rent from 25 December (rent quarter day) to 1 January (break date)) totalling £1800. A cheque dated 31 December was posted to the landlord on 30 December but was not received until 4 January. The judge held that this amounted to a material breach of covenant and therefore the break failed.

(i) Step 9 — Vacant possession

The tenant should ensure that vacant possession is handed back to the landlord or its agents on or before the break date. There can be practical problems, particularly where the tenant is responsible for security arrangements and where the landlord is trying to trip the tenant up.

There is no prescribed procedure for yielding up vacant possession of the premises: if there is a dispute the court has to look at the facts and consider whether the tenants actions, viewed objectively, demonstrated a clear intention to end the lease, and whether the landlord could go back into occupation without difficulty or objection (see *John Laing Construction Ltd* v *Amber Pass Ltd* [2004] 2 EGLR 128). However, it will always be a wise precaution for the tenant to tender the keys and require the landlord to decide whether it wants to make its own security arrangements. Again this should be documented in open correspondence.

(j) Achieving certainty

The removal of risk in break cases is very valuable, and the best way of achieving that objective is for the tenant to negotiate a surrender of the lease with the landlord. Therefore, in tandem with its compliance strategy, the corporate occupier should explore the possibility of a surrender of the lease on terms that it pays the landlord a premium in lieu of carrying out the necessary works. As referred to earlier, the tenant should set a deadline for reaching an agreement that allows it sufficient time to carry out the works should negotiations fail. If these negotiations are successful, the agreement needs to be formally documented. This will either take effect as a deed of surrender, or alternatively as a variation to the lease to remove the problematic conditions for the operation of the break clause.

In cases where compliance is likely to be particularly difficult, the corporate occupier should recognise there is value in achieving certainty over and above the cost of the works. Where the cost of compliance is likely to be very substantial, the tenant should consider offering a premium to the landlord to vary the break clause to remove the conditions precedent and buy out the risk of non-compliance. The costs of litigation to prove compliance are likely to be high and not all costs will be recoverable even if the tenant wins. However, this strategy is only likely to bear fruit if the negotiations take place at an early stage. Generally speaking, the closer to the break date that negotiations take place the less likely it is that they will be successful. Future planning is key.

12.4.2 Break by subtenant

A corporate occupier who has sublet will in one sense be in the same position as a landlord (freeholder). However, the significance of a subtenant having the right to break will usually be much greater for a Corporate Occupier because market conditions may make it difficult to find a new subtenant, or it may be left with a relatively short term that makes the property un-lettable. A corporate occupier in its capacity of landlord of surplus property can devise a strategy to try and foil a sub-tenant's attempt to exercise a break clause. This is illustrated by the real life example set out below.

- In 1992 L granted a lease of office premises to T for a term expiring in 2006.
- The space became surplus to T's requirements and in 2000 T granted ST a sublease for a term which expired a few days before the headlease.
- The sublease contained a break clause exercisable on 29 September 2004 and as the space became surplus to ST's needs, it served a break notice.
- ST's break right was conditional upon payment of all rent up and until the break date and delivery up on that date of "vacant possession of the Premises in the state and condition required by [the] Lease".

Absolute compliance with covenants was therefore required to ensure effective operation of the break clause. Those covenants included an obligation to repair and redecorate (subject to a schedule of condition) and to re-instate (requiring the removal of ST's fixtures unless T notified it to the contrary). ST therefore faced an uphill task. To successfully break, ST would require a careful and thorough strategy or a landlord who did not insist on strict compliance.

From T's perspective, market conditions and the shortness of its own lease made it unlikely that it would secure a new subtenant. It would face two years of substantial costs until the expiry of its lease (rent, rates, service charge and insurance premiums) on premises it would not use and it would have to make a substantial FRS12 provision.

The strategy a corporate occupier can use in these circumstances is influenced by the following factors.

- T and his advisors are not required to tip off ST as to the scale of the task facing it; but they must not act unprofessionally or obstruct ST in its attempts to comply. To do so may backfire and result in the disapproval of the court.
- If ST makes a request for a consent or information, then T should co-operate and respond promptly. All responses should be in writing and copies kept as evidence.
- T should obtain legal advice on the validity of the break notice.
- It may be unwise for T to carry out an inspection before the break date as this may alert ST to the need to consider its repairing/redecoration obligations.
- On the break date, T's advisors should carry out a full compliance review of the covenants in the sublease. This will involve checking whether all sums due under the lease have been paid in full (rent; service charge and insurance premiums) and carry out an inspection of the premises. If there is evidence that the premises have been handed back in a condition that does not comply with ST's covenants, an independent building surveyor should be instructed without delay to prepare a full report to include photographic (and possibly video) evidence.
- T should not accept the keys and vacant possession back unless and until it is satisfied that ST has complied with its covenants.

In the example discussed above, this approach was successful for T. ST did not implement a careful strategy and failed to satisfy the conditions precedent. It failed to carefully cross-reference the schedule of condition with the sublease resulting in partitions being removed that should have remained. The works that were carried out were not all completed by the break date and some works were not completed at all (eg ST failed to redecorate the toilets because it was thought (incorrectly) they fell outside the demise — a cursory glance at the lease plan would have prevented this error).

ST's residual liability led it to seek to surrender the premises. In the negotiations that followed the superior landlord was brought into the discussions and a surrender of both the sublease and headlease was negotiated with ST providing the full premium. T was therefore able to get out of an onerous lease at no cost and release the FR12 provision that had been made at the year-end.

Corporate occupiers should, therefore, be vigilant to ensure that a sub-tenant complies with its obligations to the letter. Often there will be a failure by the sub-tenant to comply, which may provide the corporate occupier with an opportunity to keep the sub-tenant on the hook or as with the example above, provide an opportunity to negotiate a surrender with the head landlord.

12.4.3 Break by the landlord

While most break clauses tend to be in favour of the tenant, it is not uncommon for there to be mutual break, or a break in favour of a landlord. There are a number of issues that a corporate occupier should be aware of:

- If the break right is made subject to a condition precedent (eg the landlord having an intention to redevelop the demised premises or occupy them for his own use) the landlord must have that intention bone fide to validly exercise the break right.
- A desire to redevelop may be less than an intention for the purpose of section 30(1) of the 1954 Act (see *Aberdeen Stay Houses* v *Crown Estate Commissioners* [1997] 2 EGLR 107) and therefore easier for the landlord to satisfy.

- If the tenant has the protection of the Landlord and Tenant Act 1954, in addition to serving a break notice, the landlord will have to serve a section 25 notice and have a valid ground to oppose the grant of a new lease under section 30(1). If the landlord does not establish one of the grounds of opposition, the court will grant a new lease to the tenant. It is important to remember that in such cases the landlord will have to litigate, unless the tenant co-operates, and that will take time and delay the landlord getting possession. For that reason on a potential development scheme it is rare for a landlord to grant a lease with a break clause that is protected by the 1954 Act. However a landlord may be constrained if the tenant was already in occupation and the break was inserted as part of a lease renewal.
- In a rising market therefore, a landlord could use a break clause as a means of increasing the rent (contrast the position with a tenant who cannot serve a section 26 request to double as a break clause to bring about a reduction in rent — see *Garston* v *Scottish Widows* [1998] 2 EGLR 73); although this would be a risky strategy for a landlord unless it is reasonably sure that the tenant will apply for a new tenancy or it believes that it can readily re-let at a higher rent to another tenant.
- If the corporate occupier is in the position of landlord in a legacy portfolio situation, and it wants to ensure a 'clean break', it should ensure that the break right is unconditional and that the lease is excluded from the Landlord and Tenant Act 1954.
- Where a tenant exercises a break right in respect of its own headlease, this will automatically determine the contractual term of a sub-lease. This is to be contrasted with the position where a head lease is determined by surrender (where the superior landlord is bound by pre-existing subtenancies — see section 139 of the Law of Property Act 1925). It is important for the corporate occupier to bear in mind the headlease break right when granting sub-tenancies. In particular:

  (1) The sub-lease should contain a corresponding break clause in favour of the landlord. Failure to do so may leave the corporate occupier exposed to liability to the sub-tenant for derogation from grant and/or breach of the covenant for quiet enjoyment.
  (2) The break right in the sub-lease should allow sufficient time for the corporate occupier to obtain vacant possession and to comply with any conditions precedent to the exercise of the break right in the headlease.
  (3) The sub-lease should be excluded from the Landlord and Tenant Act 1954.

12.5 Precedents

12.5.1 Covering letter

Dear Sirs

Re Operation of Tenant's Break Clause

We act for and on behalf of Splurge Pumps Ltd your current tenant of Unit 1, Filthy Industrial Estate, Little Snoring, under a lease dated 28 September 2004 between Dreadful Investments Limited and Original Tenant Limited.

Please note that our client intends to terminate the lease on the third anniversary of the term commencement date pursuant to clause 4(10) of the Lease. For the avoidance of doubt we enclose a formal break notice.

We calculate the termination date to be 28 September 2007 and we would be grateful if you could confirm that you agree.

We would be grateful if you could acknowledge safe receipt of the break notice by signing, dating and returning the copy notice enclosed.

Yours faithfully

12.5.2 Break notice

Notice to Terminate

To: Dreadful Investments Limited (Company Number 0012345) ("The Landlord")
Of: The Grange, Little Puddle on the Hill, Great Pile (Registered Office)

From: Splurge Pump Limited (Company Number 0054321) ("The Tenant")
Of: Unit 1, Filthy Industrial Estate, Little Snoring,

We, as agents, acting for and on behalf of the Tenant hereby give you notice that the Tenant wishes to terminate the lease dated 28 September 2004 between Dreadful Investments Limited and Original Tenant Limited of premises know as Unit 1, Filthy Industrial Estate, Little Snoring ("the Lease") on the third anniversary of the term commencement date pursuant to clause 4(10) of the Lease.

Dated: 1 January 2007

Signed ....
Duckin And Divin Managing Agents
1 Plushsville Tower
London
For and on behalf of the Tenant

I acknowledge receipt of this notice of which this is a true copy

Signed .... on 2007
For and on behalf of the Landlord

Print name ....

Lease Renewals 13

13.1 Overview

The Landlord and Tenant Act 1954 provides occupiers of business premises with security of tenure when their leases come to an end. There were strong policy reasons for providing occupiers with statutory protection in the 1950s. However, things have changed considerably in the last 50 years and there have been calls in some quarters for the 1954 Act to be repealed. This debate will run for some time to come, but at the moment the future of the 1954 Act seems assured, not least in view of amendments which were made to the 1954 Act in June 2004.

Part II of the 1954 Act provides security of tenure for occupiers of business tenancies unless the parties entered into a valid exclusion agreement before the tenancy was granted, or the tenancy is of a kind which is exempt or not protected by the 1954 Act. Where the 1954 Act applies, and the tenant remains in business occupation of the premises on the contractual expiry date, the tenancy will continue under section 24 unless and until it is terminated in accordance with the procedures contained in the 1954 Act. Quite simply, where the contractual rights to occupy the premises under the lease come to an end, the 1954 Act steps in and a statutory continuation tenancy provides a legal basis for the tenant's continued occupation until the procedures for renewal or termination have been completed.

However, the tenant's statutory rights are not absolute. It would be wrong in principle to keep a landlord out of its property indefinitely. When the statutory termination procedure is followed a tenant is entitled to apply to the county court for a new lease. The court is bound to grant a new lease unless the landlord is able to prove one or more grounds of opposition which are contained in section 30(1) of the 1954 Act.

Consequently, while an occupier can take some comfort from the fact that a lease it is about to take (by grant or assignment) is within the 1954 Act, it must always be conscious that a subsequent application for a new lease at the end of the existing contractual term could be defeated by a landlord. Typically such a challenge is on the grounds that the landlord intends to redevelop the premises or use them for its own business purposes. Clearly this is of paramount importance if a corporate occupier is planning to take an assignment of a "fag end" of a lease (ie with only a short period of the contractual term unexpired). It would be unwise to invest heavily in such premises unless and until a new lease is negotiated with the landlord. In such circumstances, the safest way of structuring a deal would be to enter into a tripartite deal between the landlord, tenant and the corporate occupier, whereby the original lease is surrendered and a new lease granted to the corporate occupier.

There are three battlegrounds which corporate occupiers are most likely to encounter on lease renewal:

- does the 1954 Act apply?
- does the landlord have a valid ground for opposition?
- what should the new terms of the tenancy be and in particular what should the new rent be?

As with most issues, the corporate occupier will have two possible perspectives. It will have the tenant perspective where it is occupying and trading from an operational property, and it may have the landlord perspective where the property is surplus to its needs and has been sub-let to generate sub-rent to be offset against the head-rent. The latter perspective will arise less with 1954 Act renewals than other topics discussed in this book for the simple reason that most head lease alienation clauses will stipulate that any sub-lease must be excluded from protection, and therefore the issue of lease renewal under the 1954 Act will not arise.

As with most issues a corporate occupier is likely to encounter, the key to achieving a successful outcome is planning a strategy early and being prepared for the process.

13.2 Does the 1954 Act apply?

An occupier will only have security of tenure if the occupation agreement satisfies the requirements of section 23 which provides that part II of the 1954 Act:

> ... applies to any tenancy where the property comprised in the tenancy is or includes premises which are occupied by the tenant and are so occupied for the purposes of a business carried on by him or for those and other purposes.

The key requirements are that on the relevant date:

- there must be a lease for a definite term or period
- the tenant (or a group company) must be in occupation of the leased premises or part of them and
- the tenant (or a group company) must be using the premises or part of them for business purposes.

In the vast majority of instances there will be no doubt whether the tenant has protection or not. The usual question will be whether the parties intended the tenancy to be within the 1954 Act or whether the parties agreed that it was to be excluded (see 3.2.3 (b)). In most cases this should be apparent from looking at a copy of the lease and any accompanying documents.

However, occasionally other issues can arise. In particular:

- Was the occupier granted a tenancy? A licence or a tenancy at will does not qualify for protection: see 3.2 and 3.3.2.
- Are the premises (or part of the premises) occupied by the tenant or another group company (within the meaning of section 42(2) of the 1954 Act)? If the tenant (or a group company) is not in business occupation of any part of the premises on the relevant date (usually the expiry date), then no protection can apply. Equally, if the tenant has allowed a third party into occupation, it will lose its protection unless it moves back into occupation and is using the premises for business purposes on the contractual expiry date.

- Is the tenancy of a nature specifically excluded by section 43 of the 1954 Act? The most commonly encountered exclusion is a tenancy for a term not exceeding six months: see 3.3.3.

The issues referred to above are the subject of a detailed body of law. In most circumstances a corporate occupier is unlikely to need advice on these issues and therefore they are not dealt with in detail here. If it is necessary to explore such issues, reference should be had to the leading book on this topic: *Renewal of Business Tenancies* by Reynolds & Clark (2nd ed).

13.3 Starting the renewal process

13.3.1 Continuation tenancy

Section 24(1) of the 1954 Act provides that a business tenancy (which throughout this chapter means a tenancy which satisfies the requirements of section 23 and is not excluded or exempt by sections 38 or 43) cannot come to an end unless terminated in accordance with the 1954 Act. Consequently, if the tenant remains in occupation of the premises for business purposes after the contractual expiry date there will be a statutory continuation of the tenancy on the same terms. Provided the tenant continues to pay the rent and comply with tenant covenants (so as to avoid forfeiture) the continuation tenant can, in theory, continue indefinitely. In practice, the continuation tenancy will usually be terminated by either the landlord if market rent significantly increases or by the tenant if it falls.

The various ways in which a business tenancy can come to an end on expiry of the contractual fixed term are considered at 11.3.7 (b) and below at 13.7.1.

The process of terminating a continuation tenancy and initiating a renewal is brought about by the:

- landlord serving a section 25 notice or
- tenant serving a section 26 request for a new tenancy.

13.3.2 Landlord's section 25 notice

Section 25(1) of the 1954 Act provides that the landlord may terminate a business tenancy by serving a prescribed notice on the tenant (unless a valid section 26 request has already been served on the landlord by the tenant). To terminate a tenancy the notice must comply with all of the requirements of section 25 and be validly served. A notice must fulfil the following requirements:

(a) Correctly identify the landlord

A section 25 notice will only be valid if it is served by the registered landlord, or if the interest is unregistered, by the owner of that interest. This may sound trite but mistakes can occur particularly where there are a number of companies in the same group and where transfers and assignments of reversions have taken place. If the landlord's interest is unregistered, a careful examination of the relevant deeds will be necessary. If the interest is registered, obtaining office copies will confirm the position. If the landlord is a company it is prudent to carry out a company search to ensure that it is still trading and that there has been no change to its name or registered address.

One problem that can arise in the context of registered land is where an assignee of the reversionary interest wishes to serve a section 25 notice immediately after transfer. The difficulty is that the assignee will not become the new landlord until its interest is registered. This can take a little time. It is doubtful whether a notice served by the assignee prior to the registration of its interest will be valid. The prudent course is to await completion of the registration before serving a section 25 notice to avoid any unnecessary validity arguments and the need for further without prejudice notices (and possibly additional proceedings).

Where a sub-lease has been granted a further issue to consider is which landlord (ie the superior or the immediate landlord) is entitled to serve a section 25 notice. The landlord who is entitled to serve and be served with notices under the 1954 Act is often referred to as the competent landlord. A competent landlord is identified by section 44 of the 1954 Act as:

(1) possessing an interest which is:

- the freehold or
- a business tenancy with an unexpired term of more than 14 months or
- a business tenancy with less than 14 months unexpired term and where no section 25 or 26 request has been served or
- a periodic tenancy where no notice has been served (ie common law notice to quit or under the 1954 Act) or
- a continuation tenancy under section 24 where no section 25 or section 26 request has been made.

(2) in the chain of tenancies next up from (or closest to) the business tenant.

Consequently, if L grants a lease to T, which in turn grants a sub-lease of the whole of the premises to ST for a term which terminates three days before the headlease, L will be the competent landlord. At the time when L considers serving a section 25 notice on ST (ie not more than 12 months before T's lease and ST's sub-lease expire), T is no longer in business occupation and does not have a business tenancy. Further, the next landlord up the chain will be L.

However, if ST's sub-lease expired three years before T's headlease, and T continues to occupy part of the premises, T would be the competent landlord.

(b) Correctly identify the tenant

If a section 25 notice is served on the wrong party, it will be invalid and of no effect. A careful examination of the lease and all assignments ought to correctly identify the tenant. If after carrying out this exercise it appears that a different party is in actual occupation then this should put the landlord on enquiry. If there has been an unlawful alienation the landlord, depending upon its objective, may wish to take steps to forfeit the lease instead. Alternatively, there may be a sub-tenant and the landlord may wish to review its tactics. It may also be necessary to serve a notice under section 40 of the 1954 Act on the tenant to ascertain who is actually in occupation and on what legal basis. A copy of the prescribed section 40 notice can be found at 13.8.1 below.

If there are joint legal tenants then a section 25 notice should be addressed to both tenants. If the tenant has been declared bankrupt then as the lease vests in the trustee in bankruptcy a section 25 notice should be addressed to him. Conversely, if the tenant is a company in liquidation, as the lease does not automatically vest in the liquidator, any section 25 notice should be addressed to and served on the tenant company.

(c) Prescribed form

The landlord must use a prescribed form. There are two forms of notice which are prescribed by the Landlord and Tenant Act 1954 Part II (Notices)(Regulations) 2004:

- form 1 — which should be used where a landlord does not oppose the grant of a new tenancy (see prescribed form at 13.8.2) and
- form 2 — which should be used where a landlord opposes renewal (see the prescribed form at 13.8.3).

If the landlord fails to use the correct prescribed form the notice will be invalid.

(d) Termination date

Unsurprisingly, to be valid the section 25 notice must specify a date when the current tenancy must come to an end. Care must be taken when completing this date. There are two factors which will influence the date which is included in the section 25 notice:

- it must be a date that complies with the various legal rules and
- tactical issues.

(i) *Legal rules*
The legal rules are contained in section 25:

- section 25(2) — the termination date must not be less than six months and not more than 12 months from the date of service of the notice
- section 25(3)(a) — in the case of a periodic tenancy, the termination date shall not be earlier than the earliest date the tenancy could have been terminated by the service of a notice to quit
- section 25(3)(b) — in the case of a periodic tenancy that requires more than six months notice to quit, the period prescribed by section 25(2) is extended to six months longer than the notice to quit
- section 25(4) — in the case of a fixed term tenancy, the termination date cannot be earlier than the date when the fixed term lease expires by effluxion of time or the operation of a break clause.

In most cases a corporate occupier will encounter it will simply be a question of identifying the contractual expiry date and working backwards (or in some cases forwards) from that date. For example, if the contractual term of a lease expires on 29 September 2009, the earliest that a landlord could serve a notice is 29 September 2008. The last date which the landlord could serve a section 25 notice to ensure it expired on the term date would be 29 March 2009. After that date, the term date ceases to be relevant and the landlord can serve a section 25 notice provided the termination date is not less than six nor more than 12 months from the date of service.

More difficult issues can arise in respect of periodic tenancies, particularly if they have come into existence by operation of law. It can be very difficult to identify the nature of the periodic tenancy and when it commenced. Both issues are critical to identifying the common law notice period for the purpose of section 25(3). If there is a possibility that a tenant has an annual tenancy, although only six months notice to quit is required, the notice must expire on the last day of a period of the tenancy or the first day of the new period. Consequently, if an annual periodic tenancy commenced on 1 January 2007, and a landlord decided to terminate it on 2 July 2007, the earliest termination date that could be stipulated is 31 December 2008, and the latest that a section 25 notice could be served is 31 June 2008. In short, the landlord would have to wait almost 18 months to terminate the tenancy. This is yet another reason why it is important for a landlord not to allow a tenant into occupation of premises on an informal basis.

(ii) *Tactical issues*

Prior to the 2003 reforms to the 1954 Act which were introduced in June 2004, there was considerable scope for the parties to gain tactical advantage by serving their notice first, and by stipulating a long (ie 12 month) notice or a short (ie six month) notice. This arose because of the way in which interim rent was calculated (typically 90% of the new rent). A tenant could benefit from a delay in the grant of a new tenancy.

The new rules regarding interim rent now mean that in a rising market (ie where rents are rising) a landlord may well benefit from delay, and is unlikely to be penalised in the interim rent which is awarded by the court (or agreed by the parties). Under section 24(B) of the 1954 Act interim rent is payable from the "earliest date that could have been specified" in either the section 25 notice or section 26 request. Section 24(C) has changed the valuation basis in most commonly encountered cases so that in most situations the interim rent will now be the new rent under the new tenancy.

Consequently, if the market is rising quickly, all things being equal a tenant is likely to want to complete the renewal quickly and serve a short notice (ie section 26 request). However, if a market is falling then a twelve month notice may be more appropriate.

(e) Landlord's attitude to new tenancy

A section 25 notice will be invalid unless it states whether or not the landlord is opposed to the grant of a new tenancy (section 25(6)). If the landlord opposes the grant of a new tenancy it must identify one or more of the grounds for opposition which the landlord proposes to rely upon (section 25(7)).

Section 25(8) provides that if the landlord does not oppose the grant of a new tenancy, the notice must contain the landlord's proposals as to:

(a) the property to be included in the new tenancy (ie is it the whole or part of the premises covered by the current tenancy
(b) the rent
(c) the other terms.

The requirement for the landlord to set out its proposals for the new tenancy in the section 25 notice was introduced in June 2004. While landlords' solicitors and agents do insert the landlord's proposals, the potential for challenge to the validity of a notice in an appropriate case (ie where there is a tactical reason for doing so) is significant. A common error is to overlook the need to specify whether the new tenancy being offered relates to the whole or part of the current premises. This is often overlooked because it is assumed that the tenant will take a lease of an identical demise.

A further potential line of attack could be the rent put forward by the landlord. Does this have to be a realistic assessment of the market rent at the time the notice is served? The answer to this question is yet to be decided. However, the prudent approach must be to put forward a realistic figure (albeit at the top end of the assessment) and to insert a caveat that the landlord reserves the right to claim a higher rent at trial should market rent increase after service of the notice.

13.3.3 Tenant's section 26 request

In certain circumstances a tenant can initiate the renewal process by serving a request for a new tenancy on the landlord under section 26 of the 1954 Act. However, the type of business tenancy to which section 26 applies is more restricted than section 25. Section 26 provides:

> A tenant's request for a new tenancy may be made where the current tenancy is a tenancy granted for a term of years exceeding one year, whether or not continued by section 24 of this Act, or granted for a term of years certain and thereafter from year to year.

It follows that a periodic tenant or a tenant which has been granted a lease for 12 months or less is not entitled to serve a section 26 request. Whilst such tenants can terminate their tenancies by the service of a common law notice to quit (or a section 27 notice in the case of a tenant with a term of 12 months or less) and vacate the premises, they cannot initiate a renewal and must wait until a landlord serves a section 25 notice.

A tenant is not entitled to serve a section 26 request where:

- the landlord has already served a valid section 25 notice
- the tenant has already served a notice to quit (periodic tenancy)
- the tenant has already served a section 27 notice (fixed term tenancy for twelve months or less) or
- the tenant has already served a valid section 26 request.

The legal requirements for a valid section 26 request are similar to a section 25 notice:

- the request must be served on the correct landlord ie the competent landlord see 13.3.2 (a) above
- it must correctly identify and be served by the tenant
- it must be in a prescribed form which is form 3 brought in by the 2004 Regulations (a precedent notice can be found at 13.8.4 below)
- it must specify a date on which the new tenancy is to begin which is not more than 12 nor less than six months after the service of the request and that date must not be earlier than the common law expiry date of the current tenancy
- it must identify the property which the tenant states will be comprised in the new tenancy (ie whole or part of the current premises) and must specify the tenant's proposals as to rent and other terms of the new tenancy (section 26(3)).

13.3.4 Service

The service of notices, requests and counter-notices under the 1954 Act is clearly very important. A notice will be valid if it is actually received by the recipient or its duly authorised recipient. It can therefore be sent by ordinary first class post. However, this can give rise to problems of proof and evidence if the intended recipient untruthfully denies receipt for his own purposes. Furthermore, what of the situation where the notice has been sent, but through failure of the postal system in was not received by the intended recipient?

With important notices, on which significant amounts of money may turn, there can sometimes be no substitute to instructing a process server to deliver the notice by hand to the intended recipient or its registered office. This is unlikely to cost more than £75–£150 in most cases, an amount which will be rapidly eaten up in legal fees arguing about service in correspondence afterwards and which will pale into insignificance if there is a subsequent court case about whether or not a notice was validly served. The benefit of service by a process server is that the sender can easily prove not only that the notice was served, but the time and date when it was actually received by the intended recipient. There has been a number of court cases in recent years focused solely on this peripheral but none the less critical issue.

The 1954 Act does provide a statutory mode of service for notices required to be served under the Act. Section 66(2) provides that section 23 of the Landlord and Tenant Act 1927 shall apply.

Section 23 provides that any notice may be served on the intended recipient:

> ... personally, or by leaving it for him at his last known place of abode in England and Wales, or by sending it through the post in a registered letter addressed to him there ... and in the case of a notice to a landlord, the person on whom it is to be served shall include any agent of the landlord duly authorised in that behalf.

The concept of registered post now in practice means recorded delivery (see: section 1 of the Recorded Delivery Service Act 1962).

The recorded delivery system has become increasingly unreliable over the last few years. Often it is not possible to obtain a copy of a signature to confirm safe receipt, and the track and trace system frequently indicates that the status of a particular letter is either pending or unknown. In relation to the service of break clauses and section 196 of the Law of Property Act 1925, this is unsatisfactory and cannot be relied upon (see 12.3.1 (c) above). However, where a notice can be and is served in accordance with section 23 of the 1927 Act, the courts have held that the risk of non receipt is on the intended recipient and the effective date of service is the date when the notice is actually sent, and not the date when it is in fact received (see: *CA Webber (Transport) Ltd* v *Railtrack* [2004] 14 EG 142 and *Beanby Estates* v *Egg Stores (Stamford Hill) Ltd* [2003] 3 EGLR 85). In short, provided that the sender of the notice sent by recorded delivery can prove the date when the notice was handed over to the post office, that is the end of the matter and the fact that the notice was not delivered to the intended recipient but was returned undelivered to the sender is irrelevant.

13.4 Protection of rights

Once the termination/renewal process has commenced there are certain actions which both the landlord and the tenant must take if they wish to protect their legal rights.

13.4.1 Landlord's counter-notice

If a landlord is served with a valid section 26 request for a new tenancy but intends to oppose the grant of a new tenancy, it *must* serve a counter-notice on the tenant within *two months* of being served with the tenant's request stipulating the ground or grounds in section 30 upon which the landlord will rely (section 26(6)). While the 2004 reforms did away with the need for a tenant to serve a counter-notice in response to a section 25 notice, the need for a landlord to serve a counter-notice to a section 26 request was retained.

Failure by a landlord to serve a valid counter-notice within the two month period will result in the loss of its right to oppose the grant of a new tenancy (although the landlord will be permitted to argue about the terms of the new tenancy).

The form of a counter-notice is not prescribed and a precedent notice can be found at 13.8.5.

13.4.2 Court proceedings or extension of time

Once a section 25 notice or section 26 request has been served the parties will usually either commence negotiations for the grant of a new lease if the renewal is unopposed or prepare for court proceedings if the lease renewal is opposed. The conduct of opposed and unopposed renewals is considered separately below. However, from the tenant's perspective, in both cases it *must* ensure that court proceedings are issued or an extension of time obtained in accordance with section 29B of the 1954 Act before the termination date or new tenancy commencement date provided in a section 25 or section a 26 notice (depending upon which has been served). Failure to do so will result in the current tenancy terminating on that date and the loss of the tenant's rights of renewal. This consequence arises by virtue of section 29A which provides that no court application can be made after the date provided in a section 25 notice or the section 26 request (or strictly speaking, the date immediately before the date stated in the request).

If a court application for renewal is validly issued, the tenancy will not terminate on the date stated in the section 25 notice or request, but will be continued by section 64 of the 1954 Act until three months after the court application is "finally disposed of".

If no court application is made, as stated above, the tenancy can be extended in accordance with section 29B which allows the parties to enter into an agreement to extend the statutory period for the issue of proceedings. To be valid, such an agreement must:

- be in writing (arguably)
- be agreed before the termination date or the expiry of a previous extension.

In theory such agreed extensions of time could continue indefinitely. In practice either the landlord or the tenant will at some stage reach the conclusion that further negotiation is unlikely to bear fruit and require the court to determine the issues which the parties are unable to agree.

13.5 Unopposed renewals

If the landlord does not oppose the grant of a new tenancy, this will either become apparent at the outset of the process and will be obvious from the section 25 notice, or will become clear within a short space of time from the service of a section 26 request (ie if no counter-notice is served within two months).

It customary for both parties (or at any rate their agents) to engage in negotiations concerning the proposed terms of the new lease. Those negotiations may be very straight forward and involve a simple discussion concerning the level of the new rent, or they may involve more complex and far reaching changes which take some time to progress. In more complex renewals the parties may elect to agree heads of terms to assist in clearly identifying all of the issues which exist between the parties.

Court proceedings are expensive and in the majority of cases not helpful to parties who have no real dispute and who are genuinely trying to negotiate new lease terms. It is for this reason, and the fact that courts under the Civil Procedure Rules will push cases forward promptly, that the 2004 reforms added section 29B to the 1954 Act giving the parties the ability to agree an extension of time for court proceedings to be issued. In many cases, therefore, it ought to be possible for the parties to agree and execute a new lease without court proceedings being issued. Both parties benefit from this by not wasting time, effort and money on unnecessary court proceedings.

However, there will always be some cases where it is not possible for the parties to agree new terms without external assistance. This is usually because:

- one or both parties are taking an unrealistic negotiating position on an issue or number of issues
- one of the parties is seeking to gain some perceived tactical advantage by delaying (eg where they think rent is likely to dramatically increase or decrease or in the case of a tenant, where it is looking to move into alternative accommodation which is not yet ready)
- there is a genuine dispute between the parties on a particular issue or issues
- one or both parties are not prioritising the process and therefore taking too long to respond.

If delay occurs advice may be sought as to what pressure can be put on the other side to expedite matters. However, before answering that question, it is often worth taking time to consider whether it is in that party's interests to take any action at all. For example, if rents have increased significantly since the last rent review but it is thought there might be a correction in the near future, it is unlikely to be in the tenant's interests to push the matter forward. Equally, if rents have not increased significantly, but it is thought they are picking up, or that a fresh favourable comparable is likely to emerge shortly, it may not be in the landlord's interest to press the renewal forward.

If, however, it is in the commercial interests of a party to push the renewal process forward, or in the case of the tenant, the landlord is not prepared to agree to an extension of time, renewal proceedings will need to be issued.

13.5.1 Substantive issues

If the parties cannot agree the terms of a new lease, the matter will have to be determined by a third party. Usually that third party is a county court judge. However, in recent years an alternative dispute resolution procedure has emerged in the form of expert determination or arbitration under the PACT scheme (see 13.5.2 (e) below).

As Lord Wilberforce noted in the leading case of *O'May* v *City of London Real Property Co Ltd* [1982] 1 EGLR 76, the 1954 Act is for the most part a discretionary Act, giving wide powers to a judge to grant and settle the terms on which a business tenant is to have a new lease.

The provision of the 1954 Act does however contain some guidance as to how the court should set about its task.

(a) Premises

Section 32(1) provides that a new tenancy shall be granted of the holding. The holding is that part of the premises originally granted which continues to be occupied by the tenant for the purposes of its business. Consequently, if the original lease granted a corporate occupier a tenancy of two floors in an office building, and at the relevant date the corporate occupier is only in business occupation of one of those floors, the holding will be the floor which it continues to occupy. It will only therefore be entitled to the grant of a new lease in respect of the floor it continues to occupy.

However, in the example given, if the landlord does not want its property sub-divided, it can under section 32(2) require the tenant to take a new lease of the entire premises demised by the existing tenancy. The landlord should specify this in its proposals for the new tenancy attached to its section 25 notice, or if the renewal process is initiated by a tenant's section 26 notice, in the landlord's acknowledgement of service.

(b) Duration

Section 33 of the 1954 Act provides that the new tenancy shall be for a term either agreed by the parties or for a duration which is determined by the court to be reasonable in all the circumstances not exceeding 15 years.

There is a great deal of case law involving disputes between landlords and tenants concerning lease length. Generally speaking, in recent years there has been an attempt by corporate occupiers in some sectors to reduce lease length to ensure flexibility. This tends to be opposed by landlords usually on investment grounds and convenience (ie avoiding the hassle of frequent lease renewals). It is difficult to draw out any common themes. Whether a tenant succeeds will depend upon the reason put forward for the shorter term. Often, expert evidence as to what is happening in the market place is given little weight by the courts. Inevitably, a court is likely to start from the position that the new lease should follow the lease duration of the existing lease (assuming it is 15 years or less) unless there is a compelling reason for a different duration.

What is a compelling reason will vary from case to case. A corporate occupier may point to the change in accountancy rules which took place in 1999 which means that once a property becomes surplus to its needs, the lease becomes a liability. As such it will be an onerous contract and will feature in a company's accounts as a provision for the potential liability. A lease of any significant length over five years, without an unconditional break clause, is unattractive and has the potential to have an adverse impact on the company's accounts. In addition, a corporate occupier may point to the problems of surplus leases (legacy portfolio) exacerbated by former tenant liability. An investor landlord's riposte is likely to be that the Landlord and Tenant (Covenants) Act has improved the lot of a former tenant and that a short lease will adversely affect the value of its reversion. On the latter point, it seems that a court will give little weight to an argument based on diminution in value unless that landlord can show that it is likely to sell its reversion within the foreseeable future.

Ultimately, the court has to balance the fairness to both parties and the issue can be very finely balanced and difficult for advisors to predict the outcome. As with all litigation, the corporate occupier will need to carry out a very careful cost benefit analysis taking into account the irrecoverable costs of successful litigation or the more onerous costs of losing, and weigh these against the benefit, in financial terms, of succeeding (see for example 15.2.3). Often when this exercise is carried out, one or both parties reach the conclusion that this is not an issue which alone justifies going to a full trial. If a landlord is not happy about accepting a five-year term, a compromise may be to have a 10 year term with an unconditional tenant only break option in year five.

An issue which is clothed with more certainty is whether a short lease should be granted where the premises are ripe for redevelopment. In circumstances where the landlord intends to redevelop premises in the foreseeable future, but is not in a position to make out the grounds for redevelopment at the time of the court hearing, the court is likely to grant a short lease, or a longer lease with a redevelopment break clause. While the primary purpose of the 1954 Act is to protect a business tenant, it is not the intention to impede redevelopment of land at the expense of the landlord.

(c) Other terms

Section 35 of the 1954 Act provides that the other terms of the new tenancy shall be the terms that are agreed by the parties, or if they are not agreed:

> may be determined by the court ... and the court shall have regard to the terms of the current tenancy and to all relevant circumstances.

Lord Wilberforce in *O'May* noted that this compelled something between an obligation to reproduce the existing terms and an unfettered right to substitute others. He went on to say that section 35 imposed:

> an onus upon a party seeking to introduce new, or substituted, or modified terms, with reasons appearing sufficient to the court

The approach of the court will not be to freeze or petrify the terms of the old lease, and the courts are ready to accept that lease terms should be updated to reflect developments in the law and modern commercial drafting. However, the starting position will be that a term should be the same as the old lease unless there is strong and cogent evidence why a term should be changed. When a modification is proposed by one party, the objections of the other will be considered and if the conflict is insoluble, the court will then determine the issue "according to fairness and justice".

In *O'May* the landlord sought to introduce a service charge provision which was radically different from the existing clause (restricted to paying a proportion of the heating and lighting expenses for the common parts of the building). The new clause required the tenant to contribute towards services, maintenance, repairs and redecoration for the common parts and the structure and exterior of the building. In return the landlord offered a reduction in rent. The tenant argued that the service charge obligation should be the same as the existing lease. The House of Lords agreed. The recast obligation involved a serious departure from the terms of the existing lease and the detriment to the tenant would be both real and serious.

(d) Rent

Section 34 of the 1954 Act directs the court on how to assess the rent for the new tenancy. It provides that the rent should be such amount as is agreed between the parties or in default:

> ... may be determined by the court to be that which, having regard to the terms of the tenancy (other than those relating to rent), the holding might reasonably be expected to be let in the open market by a willing lessor ...

The process is one of valuation. The judge will require valuation evidence from suitably experienced and qualified valuers to arrive at a new rent. As a matter of simple logic, a consideration of the rent should not take place until all the other terms are either agreed or fixed by the court. In practice this is not how negotiations tend to develop. However, if the matter goes to court, this is the approach which should be adopted by the parties and their valuers. It may be that some of the disputed terms do not affect rent. However, if others do, the respective valuers should put forward alternative rents based on the alternative terms contended for by the parties.

In carrying out the valuation process the valuer, and ultimately the judge, must disregard:

- the physical condition of the premises, where as is common, the existing tenancy contains a full tenant repairing obligation
- the tenant is in occupation
- any goodwill attached to the premises
- any effect on rent of certain improvements (dealt with in more detail below).
- the additional value resulting from the existence of a licence in the case of licensed premises.

Section 34 of the 1954 Act provides that improvements must be disregarded if the improvements were:

- carried out by the person who at the time was the tenant and
- not carried out under an obligation to the landlord to carry out the works and
- were carried out during the current tenancy or
  - they were completed not more than 21 years before the court application was made and
  - the holding or that part of it affected by the improvement has at all times been subject to a business tenancy within the 1925 Act and
  - at the termination of each tenancy, the tenant did not quit.

In most cases a valuation expert will arrive at his/her rental figure by considering evidence of lettings of comparable properties (see 7.2.1 which considers the different valuation methods used to ascertain market rent of demised premises). In reality, there will rarely be a comparable letting that matches all the features of the premises and the terms of the new tenancy. The task of the valuation expert is to isolate the characteristics of a comparable that have influenced the rent agreed in the open market, and apply those characteristics to the new tenancy. It may be necessary to make allowances where the subject property differs from the comparable. It is by no means an exact science and the process of compiling an expert report will go through various stages before it is disclosed. In high value cases where the parties are a long way apart on rent, it is particularly important to have a conference with counsel to test the expert's evidence before disclosure of their report. This conference will also give the corporate occupier and its legal team an opportunity to understand the reasoning for the valuation expert's view, and also consider the possible range of rents which the court may arrive at, and the factors which will affect that decision. This may enable a sensible offer to be made to the other party.

13.5.2 Renewal proceedings

(a) Issue and service

The reforms made in June 2004 mean that it is now possible for both the landlord and the tenant to issue renewal proceedings. Prior to June 2004, only the tenant could make an application for a new tenancy.

The rules governing the issue of renewal proceedings are contained in section 24(1), 29 and CPR Part 56 and can be summarised as follows.

- The landlord or the tenant can issue a renewal application at any time after the service of a section 25 notice, and must issue an application before the expiry date, or any agreement to extend time under section 29B.
- The landlord or the tenant can issue a renewal application where the tenant has served a section 26 request, but not before the landlord has served a counter-notice or two months from the date of service of the request which ever is the earlier. Consequently, if the landlord fails to serve a

counter-notice opposing renewal, the earliest that a renewal application can be made is two months after service of the request.

- Neither the landlord nor the tenant can issue a renewal application if the other has already done so and served that application.
- A renewal application cannot be issued if the landlord has already issued and served an application for termination of the current tenancy.
- If the landlord issues and serves a renewal application and the tenant does not want a new tenancy it can ask the court to dismiss the renewal application. However, the landlord cannot withdraw the application without the tenant's consent.

Once the claim form has been issued the proceedings can be served immediately or service can be delayed for up to two months from the issue of the claim form. It is not uncommon for a tenant to delay in serving proceedings, usually in the hope that the terms of a new lease can be completed within the two month period thereby avoiding unnecessary further legal costs. However, if the new lease is not completed it is very important that the proceedings are served within the two month period (unless an application is made to the court to extend this period) or the proceedings will be struck out and the tenant will lose its renewal rights.

A practical point to watch out for is that the county court will automatically serve the proceedings unless it is told to return the defendant's copy proceedings for service by the claimant. Given the importance of proper service, it is perhaps good practice for the claimant to serve the proceedings in any event. It is a well known fact that the courts are under resourced and mistakes do occur (see *Cranfield* v *Bridgegrove Ltd* [2003] 3 All ER 129 where the court mistakenly put the copy of the proceedings on the court file rather than serving it on the defendant. Fortunately for the claimant the court granted a subsequent retrospective application for service of the proceedings out of time).

(b) Stay of proceedings

The Civil Procedure Rules provide the landlord with a right to request a three month stay when served with renewal proceedings. If the landlord makes such a request it will be granted automatically and the landlord will not be required to file its acknowledgement of service. The purpose of the automatic stay is to provide the landlord with the opportunity to negotiate the terms of the new lease, although in view of the ability to grant an extension of time in which to issue proceedings, it seems this will be used in relatively few cases. If the landlord has refused an extension of time it seems unlikely in most cases that it will be minded to request a stay.

While only the landlord is entitled to request a stay, either party has the right to apply to lift the stay at any time.

(c) Acknowledgment of service

A defendant must serve an acknowledgment of service not more than 14 days after being served with a claim form. It is possible for the parties to agree an extension of time for the service of an acknowledgement or for the landlord to seek an extension from the court. In practice, if the claimant will not agree an extension in most cases it will be simpler to undertake the task of completing and serving an acknowledgement of service as in most cases it is a straightforward and simple task.

The content of an acknowledgement of service is prescribed by the CPR PD 56 — paragraphs 3.10

and 3.11 depending upon whether the landlord or tenant is the defendant. Where the landlord is defendant, it must state whether the landlord opposes renewal and if so on what ground and also what terms it proposes for the new tenancy in the event that the opposition fails. The landlord can also include a claim for interim rent in its acknowledgment of service. If the tenant is defendant it must set out the extent to which its proposals for the new tenancy terms differ from the landlord.

If the landlord fails to serve an acknowledgment of service, while it will be permitted to be heard at trial on what should constitute the new terms, it will not be permitted to oppose the grant of a new tenancy.

(d) Court directions

If the parties cannot agree on the news terms for the lease, either party or the court of its own motion can list the case for a case management conference. At the case management conference the district judge will make directions for the disposal of the claim at trial. The Post Action Protocol for renewal claims drafted by the Property Litigation Association contains standard directions for the disposal of renewal proceedings which are commonly used by the courts.

The specific directions made by a court will be tailored to the needs of the specific case. However, some or all of the following directions may be relevant:

- the landlord will usually be ordered to draft and serve on the tenant a 'travelling' lease which contains its proposed terms for the new tenancy within a specified period
- the tenant will then be required to amend the travelling lease in red with its proposed amendments and return it to the landlord by a given date
- the landlord will then be required to indicate in green on the travelling lease which amendments are disputed
- usually at this stage the key outstanding issues are rent and interim rent. The court will therefore order that each party be allowed to call one expert witness. The directions concerning expert evidence usually involve three stages:
    - the exchange of expert reports (including comparable evidence, plans and photographic evidence) by a given date
    - the experts to meet and speak on a without prejudice basis to try and agree their evidence or at least narrow the issues
    - the experts to agree by a given date a joint statement for the court stating the issues upon which they are agreed, and identifying the areas that are not agreed and the reason why.

While the travelling lease stage is likely to be necessary irrespective of the court proceedings, when the parties arrive at the stage of instructing a valuation expert the litigation process can become very much more expensive. Unfortunately, once court directions are made, the clock will be ticking and the cost of instructing a valuation expert to prepare a detailed report not that far away.

(e) Professional Arbitration on Court Terms ("PACT")

In the vast majority of renewal proceedings the key issue between the parties is what the rent should be under the new tenancy. Occasionally there are disputes relating to specific terms which one of the parties wishes to insert, or in respect of the precise wording of a particular clause.

In relation to rent, if the matter goes to full trial it will be the judge's task, after hearing both expert valuers give evidence, to determine what the new rent and interim rent should be. Without meaning to be discourteous to county court judges, their expertise and experience is unlikely to mean that they are particularly qualified to determine this issue. They deal with a broad diet of cases across the spectrum and will only rarely be asked to adjudicate on 1954 Act renewals.

There is an alternative to the very costly and expensive court process. If both parties agree, the court proceedings can be stayed and the issues referred to an expert or arbitrator under the PACT scheme which is available as an alternative dispute resolution procedure for unopposed renewals. The principal advantages of PACT are:

- all PACT experts and arbitrators are specialist surveyors and lawyers who practice in the area of commercial landlord and tenant
- it is much quicker than the court process and can be completed in a matter of weeks as opposed to many months in the case of proceedings
- it is vastly cheaper. Indeed, often the costs are limited to the costs of a letter of instruction; the PACT expert/arbitrator and any written submissions. The whole process can be less than a few thousand pounds. In contrast, the costs of a fully contested court hearing are likely to be measured in tens of thousands of pounds.

Take up has not been great under the PACT scheme. It is hard to understand why. At its best, the process can be over in a matter of weeks, at a fraction of the cost of a full trial, and the parties have the benefit of a decision which is of a much higher quality than that which is likely in the county court. It is often said that the loss of either party's ability to appeal is a drawback. However, reported appeals on unopposed renewals are almost non-existent so this concern is more theoretical than real.

(f) Protection on costs

If a renewal is not resolved by a PACT arbitration, a corporate occupier whether it is a tenant in an operational context, or a reluctant landlord in a surplus property context, should seek to protect its position on costs as early in the court process as possible.

The court has discretion at the end of a case whether or not to award costs. The starting position is that an award of costs should follow the event. In other words, the loser should pay the winner's costs. In unopposed renewals where there are a number of issues it can sometimes be difficult to determine who has won and who has lost. If a particular issue has taken up significant court time then the judge may award costs in favour of the victor of that issue.

In relation to rent, often the judge will fix a rent at a level which is between the two extremes put forward by the parties. In such a case there is a strong possibility the judge will order that each side should bear its own costs. If the figure she/he fixes is closer to one party's figure than the other then it is possible that the judge may order that the loser pay a proportion of the winner's costs.

A corporate occupier can attempt to protect its position on costs by making a written offer expressed to be "without prejudice save as to costs" (otherwise known as a Calderbank offer). If the court fixes a rent in the region of the offer, that offer can then be shown to the court when deciding the issue of costs. Importantly, the letter cannot be shown to the judge unless and until he/she has given judgment on the rental issue. It will not therefore play a role in fixing the rent.

CPR Part 36 contains a regime for making formal offers to settle in court proceedings. This is considered in detail at 15.2.2. However, from a defendant's point of view, an accepted Part 36 offer would make the defendant liable to pay the claimant's costs and therefore a Calderbank offer is more appropriate in the context of a lease renewal.

(g) Order to revoke/termination of continuation tenancy

At trial, a county court judge will determine the outstanding disputes concerning terms of the new lease, rent and interim rent and order that a new tenancy be granted.

By virtue of section 36(2) of the 1954 Act, the tenant has 14 days to decide whether or not to accept a new lease on the terms decided by the court. If it decides, for what ever reason, that it does not wish to take a new lease, it must make an application to the court within 14 days seeking an order revoking the order for the grant of a new tenancy. Provided the application is made within the time limit, the court must make a revocation order. However, the court will be given an opportunity to revoke or vary the original costs order and without a very good explanation for its conduct, the tenant may well be ordered to pay the whole of the landlord's costs of the proceedings.

If a revocation order is made, the parties can agree (or in default the court will determine) a new date for termination which will afford the landlord a reasonable opportunity to re-let or dispose of the premises. During this period the tenancy will not have protection under the 1954 Act.

13.6 Opposed renewals

The purpose of the 1954 Act is primarily to provide protection to an occupier of a business tenancy. However, the 1954 Act recognises that there are a number of circumstances where a court should refuse to grant a new tenancy and the landlord should be entitled to have its premises back.

The grounds upon which a landlord can oppose the grant of a new tenancy are contained in section 30 (1) of the 1954 Act. Those grounds are:

(a) the tenant ought not to be granted a new tenancy because of the state of the premises caused by breaches of its repairing obligations
(b) the tenant ought not to be granted a new tenancy because of its persistent delay in paying its rent
(c) the tenant ought not to be granted a new tenancy because of other substantial breaches of covenant, or any other reason connected to its management of the premises
(d) the landlord has offered to provide the tenant with suitable alternative accommodation on reasonable terms
(e) the existing tenancy was created by the sub-letting of part of the property comprised in the superior tenancy and the aggregate of the rents reasonably obtainable on separate lettings would be substantially less than the rents obtainable on letting the whole of the property
(f) redevelopment — on termination of the existing tenancy, the landlord intends to demolish or reconstruct the premises or carry out substantial works of construction, and could not reasonably do so without obtaining possession
(g) own use — on termination of the existing tenancy, the landlord intends to occupy the premises for the purposes of a business to be carried on by it.

The grounds of opposition fall into two broad categories: grounds which are fault based (a)–(c), and those which are not (d)–(g). The fault based grounds are rarely used, and even more rarely successful. The most common grounds of opposition, and the ones which most often succeed, are grounds (f) and (g).

A landlord who intends to oppose a renewal must not only indicate its intention to oppose a renewal in its section 25 notice, but it must also specify the ground or grounds upon which it relies. By the same token, if the renewal is initiated by the tenant serving a section 26 request, the landlord must serve a counter-notice on the tenant within two months of receiving the request (see 13.4.1 above). A precedent notice can be found at 13.8.5.

13.6.1 Fault based grounds of opposition — (a)–(c)

The landlord can oppose the grant of a new tenancy on the basis of the tenant's past behaviour in respect of the lease. It can rely upon the tenant's failure to comply with its repairing covenants and the state of the premises at the date of service of the section 25 notice or counter-notice and at the date of the hearing. It can also rely upon the tenant's persistent failure to pay rent or substantial breaches of other lease terms.

The fault based grounds of opposition are, however, subject to one important caveat. The court has discretion as to whether or not it should refuse to grant a new tenancy. This is evident from the use of the words "ought not". In practice, in particular in relation to persistent failure to pay rent, if the tenant can offer a reasonable explanation for the late payment of rent and reassure the court that it would not happen in the future, the court is likely to take a lenient view.

In reality these grounds only occasionally arise, and if they do, they tend not to be pursued by landlords at trial.

13.6.2 Ground (d) — suitable alternative accommodation

A landlord can oppose the grant of a new tenancy on the grounds that it has offered alternative accommodation to the tenant on reasonable lease terms.

To establish a ground (d) objection the landlord must establish that:

- it has offered alternative accommodation to the tenant
- the accommodation offered is suitable to the needs of the tenant. This will require a comparison between the existing premises and the premises offered. The court when considering suitability must have regard to "the nature and class" of the tenant's business, and the locality and extent of, and facilities afforded by, the existing premises
- the proposed terms of the new tenancy are "reasonable having regard to the terms of the current tenancy". Again, this will require a comparison between the terms of the existing tenancy and those of the new tenancy.

Importantly the ground is not discretionary. Consequently, once established the court must refuse to grant a new tenancy of the existing premises. However, the difficulties and uncertainties of offering suitable accommodation on reasonable terms often dissuade landlords from relying upon this ground. There will in most cases be substantial scope for argument and reasonable differing opinions (however for a case where the landlord did succeed see: *Knollys House Ltd* v *Sayer* [2006] PLSCS 55).

13.6.3 Ground (e) — sub-letting of part

The operation of this ground can be best illustrated by an example. L lets the whole of an office building to T. During the course of T's current tenancy it sub-lets the second floor to ST. If on renewal, the aggregate of the rent from the separate letting of the second floor and the remainder of the office building would be substantially less than the rent reasonably obtainable from letting the whole of the office building, L can oppose the grant of a new tenancy on ground (e) if it can also show that it requires the premises "for the purposes of letting or otherwise disposing" of the premises. Interestingly, the wording of ground (e) suggests that the court may have a residual discretion whether to refuse to grant a new tenancy even if the ground is made out. Given this is a non fault based ground it is unclear what factors would influence a court in the exercise of its discretion.

Importantly, if the landlord objects on this ground, it will, prima facie, be required to pay compensation to the tenant on quitting.

13.6.4 Ground (f) — redevelopment

This is the ground which a corporate occupier is most likely to encounter in an operational context (i.e. where it is the tenant). Since the 1954 Act was passed this has become a highly technical ground and generated a good deal of case law. From the corporate occupier's perspective, the service of a section 25 notice which indicates an intention by the landlord to oppose a renewal lease on ground (f) can be a nasty surprise giving rise to the possibility of re-locating and all the cost consequences and business upheaval that that entails. While a corporate occupier which quits the premises following a ground (f) opposition may be entitled to statutory compensation, that is unlikely to be adequate compensation to cover the expense of relocating and the disruption caused to its business. Consequently, it is important that the landlord's ground is examined carefully.

Ground (f) provides:

> that on the termination of the current tenancy the landlord intends to demolish or reconstruct the premises comprised in the holding or a substantial part of those premises or to carry out substantial work of construction on the holding or part thereof and that he could not reasonably do so without obtaining possession of the holding.

There are a number of essential issues which must be established and examined carefully.

(a) Necessary works

To rely upon ground (f) the landlord must intend to carry out one or more of the following types of works to the holding:

- demolition of the whole of the premises
- demolition of a substantial part
- reconstruction of the whole of the premises
- reconstruction of part
- substantial works of construction of the whole of the premises
- substantial works of construction on part.

There is a good deal of case law concerning whether the works which a landlord proposes to carry out qualify under one of the above work types. Ultimately it will be a question of fact and degree. It goes without saying that the types of works envisaged by ground (f) are reasonably substantial structural works.

(b) Who must carry out the works?

The landlord must intend to carry out the works itself (although it can of course engage building contractors to actually do the works on its behalf). If the true intention is for the landlord to sell the demised premises to a developer who in turn will develop, this is not sufficient and the landlord's ground (f) objection will fail.

A well advised landlord may seek to get around this problem in two ways:

- by transferring its reversion to the developer with the benefit of the section 25 notice leaving the developer as new landlord with the task of establishing the necessary intention at the hearing date or
- by structuring the deal so that the works are to be carried out by a tenant under a long building lease, provided that the landlord retains full control of the works (see *Gilmour Caterers Ltd* v *St Bartholomew's Hospital Governors* [1956] 1 QB 387)

A corporate occupier who is looking to defend a ground (f) objection should therefore look carefully at who it is proposed will carry out the works. If it is to be done under a building lease, the proposed obligations should be examined to ensure that it is not a sham. For the building lease option to satisfy the requirements of ground (f), it is essential that the landlord retains full control of the nature and design of the works which are to be carried out. This is usually achieved by a requirement that the landlord's surveyor must sign off all specifications of work before they are carried out.

(c) The need for possession

A landlord must not only establish that the works qualify, but it must also show that it could not reasonably carry out those works without obtaining the legal possession of the holding. The point is illustrated by the case of *Heath* v *Drown* [1973] AC 498.

Facts

- L intended to carry out extensive structural works to the front wall of the demised premises.
- The works were so extensive that T would have to move out of the premises for a number of months.
- The existing tenancy reserved a right for L to enter the premises for the purposes of carrying out any necessary repairs.

Held — House of Lords

- As the proposed works could be carried out by L under the reserved right of entry, it was not necessary for L to have possession of the holding to carry out the works.
- Possession means legal possession and not physical possession. It was not relevant T would be prevented from having access to the premises for the duration of the works.

The principle is not restricted to works of repair. In *Price* v *Esso Petroleum* [1980] 2 EGLR 58 the lease reserved a right for the landlord to enter to carry out improvements. It was held that this prevented the landlord from relying upon ground (f) in circumstances where the landlord intended to totally demolish an existing petrol station and replace it with a petrol filling station of a new design.

Even if the existing tenancy does not contain a reservation of the type referred to above, section 31A of the 1954 Act provides that a landlord's ground (f) opposition will fail if the tenant agrees to include a term in the new tenancy which would enable the landlord to reasonably carry out the works without obtaining possession. However, section 31A is restricted in its application by a proviso which requires that it must be possible for the works to be carried out "without interfering to a substantial extent or for a substantial time" with the tenant's business use. If the effect of carrying out the proposed works would be to exclude the tenant from the premises for a matter of months then a court is likely to hold that this would be substantial interference for a substantial time (see *Redfern* v *Reeves* [1978] 2 EGLR 52). In contrast, if the tenant will be kept out for only two weeks this is unlikely to be regarded as substantial (see *Cerex Jewels Ltd* v *Peachey Property Corporation plc* [1986] 2 EGLR 65).

Consequently, when faced with a ground (f) opposition, a corporate occupier should always consider carefully whether such a claim could be defeated by offering to accept a term in the new tenancy which would permit the landlord to carry out the proposed works.

(d) The landlord's intention

To succeed on ground (f) a landlord must show that it has the necessary intention to carry out the qualifying works. There are two separate aspects which must be demonstrated:

- the landlord must have made a firm and definite decision. A provisional desire is not enough and
- there must be a reasonable prospect of the landlord being able to implement that decision.

The first aspect was described by Asquith LJ in *Cunliffe* v *Goodman* [1950] 2 KB 237 as requiring the project to have:

> ... moved out of the zone of contemplation — out of the sphere of the tentative, the provisional and exploratory — into the valley of decision

The second aspect is concerned with the ability of the landlord to actually implement the decision and the likelihood that it will be able to overcome the necessary obstacles which stand between its decision and the works actually being carried out. The most common obstacles are planning permission and finance. However, less obvious hurdles may include access rights; restrictive covenants; rights to light impediments or the simple fact that the landlord does not own all of the land which is necessary to enable the project to go ahead. For a corporate occupier looking to defeat a ground (f) objection and secure a new lease, all of these possibilities should be investigated carefully.

Consequently, a judge must not only believe the landlord's evidence that it has a genuine and settled commitment to the project, but also be satisfied that it is a real possibility and not pie in the sky (see *Zarvos* v *Pradham* [2003] 2 EGLR 37). The motive of the landlord is irrelevant provided that it genuinely holds the intention to carry out the works.

The date when the intention must exist is the date of the court hearing. In *Betty's Cafes* v *Phillips Furnishings Stores Ltd* [1959] AC 20 it was held that it was not necessary for the landlord to have definitely decided to carry out works at the date that it served its section 25 notice provided that decision had been taken by the hearing date. Consequently, in that case the necessary intention was established by a company resolution to implement the works being passed during the course of the hearing. As it is the competent landlord at the date of the hearing whose intention is relevant, it is possible for the current landlord to rely upon a section 25 notice served by a previous landlord. The important issue is the intention of the competent landlord at the date of the court hearing.

(e) Evidence of the landlord's intention

If the landlord is a company, the usual way in which the subjective intention is established (ie the decision to carry out qualifying works) is by that company adducing the relevant resolution or board minutes which evidence a decision to proceed with the project and authorising the necessary expenditure. If the landlord is an individual then it will usually be necessary for the landlord to make a witness statement and if necessary, give evidence at trial.

The kinds of evidence which might demonstrate that there are reasonable prospects of the landlord's plans coming to fruition include:

- relevant correspondence/ internal memos relating to the project
- architect's plans
- planning permission or planning application
- application for building regulations consent/consent
- listed building consent (if necessary)
- mortgagee's consent if necessary
- feasibility study demonstrating viability of the scheme
- availability of finance for the project
- existence of draft building contract or building lease
- existence of working drawings and specifications
- existence of tender documents.

Whether or not a landlord has a realistic prospect of implementing a project is ultimately one of fact for the judge. Consequently, the above list is not exhaustive nor is it a requirement that the landlord be able to produce all, or even most, of the evidence referred to. However, the more evidence a landlord has, and the further down the road it has gone, the more difficult it will be for a tenant to defend a ground (f) opposition.

An issue which can arise is whether planning permission is fundamental to a landlord succeeding on ground (f). If the landlord does not have planning permission at the date of the hearing the appropriate test is whether a reasonable man, on the evidence, would believe that the landlord had a reasonable prospect of obtaining that permission (*Gregson* v *Cyril Lord* [1962] 3 All ER 907). The landlord will need to adduce some evidence demonstrating that there is a reasonable prospect of planning permission being granted. This may take the form of expert evidence from a planning consultant. If that evidence is doubted by the tenant, it may call its own planning expert to cast doubt on the prospect of planning permission being granted.

The burden on a landlord in relation to planning permission is not a heavy one. This is illustrated by the ground (g) opposition case of *Gatwick Parking Services Ltd* v *Sargent* [2000] 2 EGLR 45.

Facts

- T was a tenant of an off-airport car park.
- The planning permission imposed a condition limiting occupation and use of the demised premises to T personally.
- L opposed the grant of a new tenancy on ground (g) — on the basis that it intended to run its own car parking business from the site.
- Prior to the hearing the local authorities planning officers indicated that there was no longer a planning objection to removal of the condition.
- The county court judge rejected L's expert evidence that planning permission could in all likelihood be obtained.
- After the trial planning permission was granted. T commenced judicial review proceedings and L appealed against the court decision on ground (g).

Held — Court of Appeal (allowing L's appeal)

- L was not required to prove on balance of probabilities that permission would be granted.
- L merely had to prove that there was a real, as opposed to fanciful, prospect of permission being granted. The necessary intention had been established.

If a landlord obtains outline planning permission this is likely to be good evidence of the landlord's intention. However, there is one caveat to this. To satisfy ground (f) the landlord must demonstrate that it intends to carry out the qualifying works "on the termination of the current tenancy". While the courts have not taken an unduly strict approach to these words, a landlord is likely to be required to show that it will commence the works within a short period after obtaining possession. Usually a period of two to three months from the date of obtaining possession will suffice, but if the works are to be commenced outside this period then the landlord may run into difficulties and provide a tenant with a fruitful line of attack.

(f) Landlord's undertaking

One way in which a landlord may seek to beef up its position is to offer an undertaking to the court to carry out the proposed works within a certain period of time. This is likely to be treated as strong evidence that the landlord intends to carry out the intended works, although it will not be conclusive.

13.6.5 Ground (g) — own use

The final ground of opposition which a landlord may rely upon is ground (g) which provides:

> ... that on termination of the current tenancy the landlord intends to occupy the holding for the purposes, or partly for the purposes, of a business to be carried on by him ... or his residence.

Like ground (f), ground (g) has become a very technical ground with a formidable body of case law illustrating the court's approach and which it is not possible to cover in detail in this book. Ultimately, it will always be a question of fact and evidence. In relation to the necessary intention, the points made in respect of ground (f) also apply to ground (g).

There are a number of issues which a corporate occupier defending a ground (g) opposition should be particularly aware of.

(a) Who must occupy for business purposes?

Ground (g) clearly envisages that the proposed occupier should be the landlord and not a third party. However, the 2004 amendments to the 1954 Act inserted sections 30(1A) and 30(1B). These provisions provide that a landlord can also rely upon ground (g) where:

- a company in which the landlord is a controlling shareholder intends to occupy the premises for the purposes of a business or
- a controlling shareholder of the landlord company intends to occupy the premises for the purpose of its business.

(b) The five year rule

The landlord cannot rely upon ground (g) if the five year rule applies. The purpose of the five year rule is to prevent abuse of ground (g) by the landlord selling its reversion to a third party that wishes to occupy the premises shortly before the renewal process begins.

The five year rule is contained in section 30(2) and applies where:

> ... if the interest of the landlord, or an interest which has merged in that interest and but for that merger would be the interest of the landlord, was purchased or created after the beginning of the period of five years which ends with the termination of the current tenancy, and at all times since the purchase or creation thereof the holding has been [a business tenancy].

The wording of section 30 is fairly tortuous and not the easiest section to comprehend. However, in general terms, if the current landlord has become the landlord within five years of the termination date stipulated in the section 25 or 26 notice (ie working backwards), and the tenancy has at all times been within the 1954 Act, the rule will apply and the landlord will be barred from relying upon ground (g).

The 2004 amendments to the 1954 Act have extended the five year rule. Section 30(2A) provides that a landlord company in which there has been a change in the controlling interest in the company within five years of the date contained in the section 25 or 26 notice (again counting backwards), will be barred from relying upon ground (g) provided the tenancy has at all times remained a business tenancy. This provision prevents the owners of a landlord company selling the company as an alternative to selling the reversion so as to side-step the five year rule.

(c) Undertaking to the court

There is some doubt about whether an undertaking given by the landlord to the court that it will occupy the premises for the purpose of its own business is of any great weight. This arises from the unlikelihood of a subsequent court enforcing such an obligation (see 6.5.4 in the context of keep open clauses). The authors of *Renewal of Business Tenancies* suggest that the problem can be overcome by the landlord undertaking not to use the premises for an agreed period for any purposes other than a business carried on by the landlord. This negative form of undertaking avoids the problems of enforcement encountered with the positive obligation to carry out a business, whilst at the same time achieving the same purpose.

In some cases undertakings have been decisive. *Bentley & Skinner (Bond Street Jewellers) Ltd* v *Searchmap Ltd* [2003] EWHC 1621 was one such case. This case also illustrates what is likely to happen if the landlord subsequently changes its mind and decides, contrary to its undertaking, not to go ahead with its previously stated intention. The judge indicated that he would only be prepared to release the landlord from its intention if it consented to a new lease being granted to the tenant on new terms.

13.6.6 Procedure

(a) Application for termination

One of the changes introduced in 2004 was the right granted to a landlord to issue a court application for an order terminating the existing tenancy on one or more of the grounds of opposition contained in section 30(1) of the 1954 Act.

A landlord can issue an application as soon as it has served a section 25 notice, or has served a counter-notice to a section 26 request. It cannot, however, issue an application if the tenant (or indeed the landlord) has already issued and served a court application for renewal. Consequently, the landlord now has the ability to put the corporate occupier which wishes to remain, under immediate pressure to decide what approach to take.

(b) Request for details and disclosure

A corporate occupier served with a section 25 notice opposing renewal (possibly accompanied by an immediate application for termination) should write to the landlord in open correspondence requesting full details of the landlord's plans together with full disclosure of all documentary evidence. If the landlord fails to co-operate the corporate occupier may be able to obtain an adverse cost order against the landlord at a later date. The corporate occupier should then consider the issues identified above and scrutinise the landlord's evidence carefully to see if it has organised its affairs to satisfy the grounds relied upon. In most cases it will be possible to form a view on the landlord's prospects of success once full disclosure has taken place. If it appears that the landlord is likely to succeed, the corporate occupier may be better served by expending its valuable time and resources on re-locating rather than expensive and speculative litigation. However, if the landlord's evidence is thin on the ground or there are technical difficulties with its ground of opposition, it may be worth the corporate occupier fighting.

The other factor which may be of importance is timing. If the corporate occupier thinks re-locating may take significant time it may have little alternative but to issue renewal proceedings or defend termination proceedings to delay the landlord from obtaining possession for as long as possible. The downside to this approach will be that eventually when it loses or backs down, it may be required to pay a proportion of the landlord's costs in addition to its own costs.

(c) Renewal proceedings

In the event that a tenant issues and serves renewal proceedings before a landlord issues termination proceedings, the landlord will be required to file an acknowledgement of service not more than 14 days after being served with the claim form (see 13.5.2 (c) above).

To avoid wasting unnecessary time and costs, the court is likely to order that there be a trial of a preliminary issue, namely, whether the landlord is able to establish its ground or grounds of opposition (*Dutch Oven* v *Egham Estate and Investment Co* [1968] 1 WLR 1483). If the landlord succeeds then that will be the end of the case. If the landlord fails, the court will then give directions for a further hearing to determine the new terms of the renewal tenancy (see 13.5.2 (d) above).

13.6.7 Termination of continuation tenancy

If the landlord's ground of opposition is established, the existing tenancy and the tenant's right to occupy will terminate three months from the date on which the court application is finally disposed of (section 64 of the 1954 Act). In practice, this will be three months and 14 days after the court order, the 14 days representing the time in which a party can ordinarily seek permission and then lodge an appeal. If permission is granted, and an appeal issued, the continuation tenancy will not terminate until three months after the appeal is finally determined.

13.7 Termination, quitting and compensation

13.7.1 Termination by corporate occupier

The corporate occupier's objective may not be to renew a lease at the contractual expiry date, but rather to terminate the lease at the earliest possible date. This will certainly be the case if the premises have ceased to be operational and are surplus. The corporate occupier will want to ensure a continuation tenancy does not come into existence, and that the obligation to pay rent will come to an end on the contractual expiry date of the lease. The simplest and most effective way to do this is by serving a section 27(1) notice not later than three months before the contractual expiry date. A precedent notice can be found at 13.8.6 below. If this notice is served, no continuation tenancy will come into existence irrespective of whether the tenant is in occupation at the contractual expiry date or not (for example where the corporate occupier has underestimated the time it needs to vacate).

If, for whatever reason, the corporate occupier fails to serve a section 27(1) notice in time, provided that it has vacated and is not in business occupation of the premises on the contractual expiry date, a continuation tenancy will not come into existence (see *Esselte* v *Pearl Assurance plc* [1997] 1 EGLR 73). However, vacating premises in these circumstances can be a nerve-racking ordeal. If the corporate occupier fails to exit in time the landlord may well argue that a continuation tenancy has come into existence and that further rent is due. If the rental is significant, the prudent approach is to serve a without prejudice section 27(2) notice which enables a tenant to terminate a continuation tenancy on three month's notice. A precedent notice can be found at 13.8.7 below. If the corporate occupier vacates in time there will be no continuation tenancy and the notice will be unnecessary and of no effect. However, if it fails to vacate in time, the notice will at least have started the three month termination period and reduce the amount of rent which the tenant is required to pay under the continuation tenancy.

If a corporate occupier wishes to terminate a continuation tenancy after it has come into existence (ie after the contractual expiry period), it must also serve a section 27(2) which can be served both before and after a continuation tenancy has come into existence.

If a corporate occupier issues renewal proceedings but subsequently changes its mind, it must discontinue those proceedings. The continuation tenancy will then terminate three months after the notice of discontinuance by virtue of section 64 of the 1954 Act.

The other possible modes of termination arise where no section 27 notice is served and:

- the landlord has served a section 25 notice but both tenant and landlord have failed to issue and serve an application for a new tenancy. The tenancy will terminate on the date specified in the notice
- the tenant has served a section 26 request but both tenant and landlord have failed to issue and serve an application for a renewal tenancy. The tenancy will terminate immediately before the date specified in the section 26 request.

13.7.2 Compensation

If a tenant is prevented from obtaining a new tenancy as a result of a valid ground (e), (f) or (g) opposition then the tenant may be entitled to compensation from the landlord on quitting the premises. Section 37(1A)–(1C) of the 1954 Act provide three separate cases where a tenant is entitled to compensation:

(1A) where a tenant makes an application for a new tenancy and the court is prevented from granting a new tenancy because of grounds (e), (f) or (g) of section 30(1) or
(1B) where the landlord makes an application for termination and the court is prevented from granting a new tenancy because of grounds (e), (f) or (g) of section 30(1) or
(1C) where the landlord serves a section 25 notice or a counter-notice to a tenant's section 26 request relying upon grounds (e), (f) or (g) and
- no court application is made by either party or
- an application is made but is subsequently withdrawn.

Where a tenant is entitled to compensation, the amount of compensation payable will be determined by the length of time the tenant has been in occupation under a business tenancy, and the rateable value of the premises. The rules on compensation are complex and their detail beyond the scope of this book. However, in general terms, if a tenant has been in business occupation of the whole tenancy for at least 14 years calculated backwards from the date of termination, the tenant will be entitle to two times the rateable value of the premises. If the tenant has been in business occupation for less than 14 years, it will be entitled to one times the rateable value of the premises. In calculating the 14 year period a tenant can take into account the period of occupation of a former tenant provided it was a successor to the business.

There is one circumstance where the tenant may not be entitled to compensation. It is possible to exclude the obligation to pay compensation for disturbance by a contractual term in the lease. However, by virtue of section 38(2) of the 1954 Act, an exclusion will be void and of no effect if the tenant has been in business occupation for at least five years calculated backwards from the date that the tenant "is to quit the holding". Again, in calculating this period, any period of business occupation of a previous tenant can be taken into account provided that the current tenant is a successor to the previous tenant's business.

A corporate occupier hoping to claim compensation where the lease contains a compensation exclusion clause should take care to organise its departure to ensure that there is not a significant void period prior to the end of the tenancy. If there is, a landlord may argue that the tenant has not been in business occupation of the premises for the "whole of the five years immediately preceding" the date the tenant is to quit the premises. This argument failed in *Baccihocchi* v *Acadmenic Agency Ltd* [1998] 1 WLR 1313 where the tenant was wrongly advised about the termination date and vacated 12 days early. However, if the period was significantly longer, depending on the facts, it could succeed.

13.8 Prescribed forms and precedents

13.8.1 Section 40 notice

LANDLORD'S REQUEST FOR INFORMATION ABOUT OCCUPATION AND SUB-TENANCIES

Section 40(1) of the Landlord and Tenant Act 1954

To: (*insert name and address of tenant*)

From: (*insert name and address of landlord*)

1. This notice relates to the following premises: (*insert address or description of premises*)

2. I give you notice under section 40(1) of the Landlord and Tenant Act 1954 that I require you to provide information —

(a) by answering questions (1) to (3) in the Table below;
(b) if you answer "yes" to question (2), by giving me the name and address of the person or persons concerned;
(c) if you answer "yes" to question (3), by also answering questions (4) to (10) in the Table below;
(d) if you answer "no" to question (8), by giving me the name and address of the sub-tenant; and
(e) if you answer "yes" to question (10), by giving me details of the notice or request.

TABLE

(1) Do you occupy the premises or any part of them wholly or partly for the purposes of a business that is carried on by you?
(2) To the best of your knowledge and belief, does any other person own an interest in reversion in any part of the premises?
(3) Does your tenancy have effect subject to any sub-tenancy on which your tenancy is immediately expectant?
(4) What premises are comprised in the sub-tenancy?
(5) For what term does it have effect or, if it is terminable by notice, by what notice can it be terminated?
(6) What is the rent payable under it?
(7) Who is the sub-tenant?
(8) To the best of your knowledge and belief, is the sub-tenant in occupation of the premises or of part of the premises comprised in the sub-tenancy?
(9) Is an agreement in force excluding, in relation to the sub-tenancy, the provisions of sections 24 to 28 of the Landlord and Tenant Act 1954?
(10) Has a notice been given under section 25 or 26(6) of that Act, or has a request been made under section 26 of that Act, in relation to the sub-tenancy?

3. You must give the information concerned in writing and within the period of one month beginning with the date of service of this notice.

4. Please send all correspondence about this notice to:

Name:

Address:

Signed: Date:
*[Landlord] *[on behalf of the landlord] *delete whichever is inapplicable

IMPORTANT NOTE FOR THE TENANT

This notice contains some words and phrases that you may not understand. The Notes below should help you, but it would be wise to seek professional advice, for example, from a solicitor or surveyor, before responding to this notice.

Once you have provided the information required by this notice, you must correct it if you realise that it is not, or is no longer, correct. This obligation lasts for six months from the date of service of this notice, but an exception is explained in the next paragraph. If you need to correct information already given, you must do so within one month of becoming aware that the information is incorrect.

The obligation will cease if, after transferring your tenancy, you notify the landlord of the transfer and of the name and address of the person to whom your tenancy has been transferred.

If you fail to comply with the requirements of this notice, or the obligation mentioned above, you may face civil proceedings for breach of the statutory duty that arises under section 40 of the Landlord and Tenant Act 1954. In any such proceedings a court may order you to comply with that duty and may make an award of damages.

NOTES

The sections mentioned below are sections of the Landlord and Tenant Act 1954, as amended, (most recently by the Regulatory Reform (Business Tenancies) (England and Wales) Order 2003)

Purpose of this notice

Your landlord (or, if he or she is a tenant, possibly your landlord's landlord) has sent you this notice in order to obtain information about your occupation and that of any sub-tenants. This information may be relevant to the taking of steps to end or renew your business tenancy.

Time-limit for replying

You must provide the relevant information within one month of the date of service of this notice (section 40(1), (2) and (5)).

Information required

You do not have to give your answers on this form; you may use a separate sheet for this purpose. The notice requires you to provide, in writing, information in the form of answers to questions (1) to (3) in the Table above and, if you answer "yes" to question (3), also to provide information in the form of answers to questions (4) to (10) in that Table. Depending on your answer to question (2) and, if applicable in your case, questions (8) and (10), you must also provide the information referred to in paragraph 2(b), (d) and (e) of this notice. Question (2) refers to a person who owns an interest in reversion. You should answer "yes" to this question if you know or believe that there is a person who receives, or is entitled to receive, rent in respect of any part of the premises (other than the landlord who served this notice).

When you answer questions about sub-tenants, please bear in mind that, for these purposes, a sub-tenant includes a person retaining possession of premises by virtue of the Rent (Agriculture) Act 1976 or the Rent Act 1977 after the coming to an end of a sub-tenancy, and "sub-tenancy" includes a right so to retain possession (section 40(8)).

You should keep a copy of your answers and of any other information provided in response to questions (2), (8) or (10) above.

If, once you have given this information, you realise that it is not, or is no longer, correct, you must give the correct information within one month of becoming aware that the previous information is incorrect. Subject to the next paragraph, your duty to correct any information that you have already given continues for six months after you receive this notice (section 40(5)). You should give the correct information to the landlord who gave you this notice unless you receive notice of the transfer of his or her interest, and of the name and address of the person to whom that interest has been transferred. In that case, the correct information must be given to that person.

If you transfer your tenancy within the period of six months referred to above, your duty to correct information already given will cease if you notify the landlord of the transfer and of the name and address of the person to whom your tenancy has been transferred.

If you do not provide the information requested, or fail to correct information that you have provided earlier, after realising that it is not, or is no longer, correct, proceedings may be taken against you and you may have to pay damages (section 40B).

If you are in any doubt about the information that you should give, get immediate advice from a solicitor or a surveyor.

Validity of this notice
The landlord who has given you this notice may not be the landlord to whom you pay your rent (sections 44 and 67). This does not necessarily mean that the notice is invalid.

If you have any doubts about whether this notice is valid, get advice immediately from a solicitor or a surveyor.

Further information
An explanation of the main points to consider when renewing or ending a business tenancy, "Renewing and Ending Business Leases: a Guide for Tenants and Landlords", can be found at *www.odpm.gov.uk*. Printed copies of the explanation, but not of this form, are available from 1 June 2004 from Free Literature, PO Box 236, Wetherby, West Yorkshire, LS23 7NB (0870 1226 236).

13.8.2 Section 25 notice where landlord does not oppose renewal

LANDLORD'S NOTICE ENDING A BUSINESS TENANCY WITH PROPOSALS FOR A NEW ONE

Section 25 of the Landlord and Tenant Act 1954

IMPORTANT NOTE FOR THE LANDLORD: If you are willing to grant a new tenancy, complete this form and send it to the tenant. If you wish to oppose the grant of a new tenancy, use form 2 in Schedule 2 to the Landlord and Tenant Act 1954, Part 2 (Notices) Regulations 2004 or, where the tenant may be entitled to acquire the freehold or an extended lease, form 7 in that Schedule, instead of this form.

To: (*insert name and address of tenant*)

From: (*insert name and address of landlord*)

1. This notice applies to the following property: *(insert address or description of property*).

2. I am giving you notice under section 25 of the Landlord and Tenant Act 1954 to end your tenancy on *(insert date*).

3. I am not opposed to granting you a new tenancy. You will find my proposals for the new tenancy, which we can discuss, in the Schedule to this notice.

4. If we cannot agree on all the terms of a new tenancy, either you or I may ask the court to order the grant of a new tenancy and settle the terms on which we cannot agree.

5. If you wish to ask the court for a new tenancy you must do so by the date in paragraph 2, unless we agree in writing to a later date and do so before the date in paragraph 2.

6. Please send all correspondence about this notice to:

Name:

Address:

Signed: Date:
*[Landlord] *[On behalf of the landlord] *[Mortgagee] *[On behalf of the mortgagee]
*(delete if inapplicable)

SCHEDULE

LANDLORD'S PROPOSALS FOR A NEW TENANCY

(*attach or insert proposed terms of the new tenancy*)

IMPORTANT NOTE FOR THE TENANT
This Notice is intended to bring your tenancy to an end. If you want to continue to occupy your property after the date specified in paragraph 2 you must act quickly. If you are in any doubt about the action that you should take, get advice immediately from a solicitor or a surveyor.

The landlord is prepared to offer you a new tenancy and has set out proposed terms in the Schedule to this notice. You are not bound to accept these terms. They are merely suggestions as a basis for negotiation. In the event of disagreement, ultimately the court would settle the terms of the new tenancy.

It would be wise to seek professional advice before agreeing to accept the landlord's terms or putting forward your own proposals.

NOTES

The sections mentioned below are sections of the Landlord and Tenant Act 1954, as amended, (most recently by the Regulatory Reform (Business Tenancies) (England and Wales) Order 2003).

Ending of tenancy and grant of new tenancy
This notice is intended to bring your tenancy to an end on the date given in paragraph 2. Section 25 contains rules about the date that the landlord can put in that paragraph.

However, your landlord is prepared to offer you a new tenancy and has set out proposals for it in the Schedule to this notice (section 25(8)). You are not obliged to accept these proposals and may put forward your own.

If you and your landlord are unable to agree terms either one of you may apply to the court. You may not apply to the court if your landlord has already done so (section 24(2A)). If you wish to apply to the court you must do so by the date given in paragraph 2 of this notice, unless you and your landlord have agreed in writing to extend the deadline (sections 29A and 29B).

The court will settle the rent and other terms of the new tenancy or those on which you and your landlord cannot agree (sections 34 and 35). If you apply to the court your tenancy will continue after the date shown in paragraph 2 of this notice while your application is being considered (section 24).

If you are in any doubt about what action you should take, get advice immediately from a solicitor or a surveyor.

Negotiating a new tenancy
Most tenancies are renewed by negotiation. You and your landlord may agree in writing to extend the deadline for making an application to the court while negotiations continue. Either you or your landlord can ask the court to fix the rent that you will have to pay while the tenancy continues (sections 24A to 24D).

You may only stay in the property after the date in paragraph 2 (or if we have agreed in writing to a later date, that date), if by then you or the landlord has asked the court to order the grant of a new tenancy.

If you do try to agree a new tenancy with your landlord remember:

- that your present tenancy will not continue after the date in paragraph 2 of this notice without the agreement in writing mentioned above, unless you have applied to the court or your landlord has done so, and
- that you will lose your right to apply to the court once the deadline in paragraph 2 of this notice has passed, unless there is a written agreement extending the deadline.

Validity of this notice
The landlord who has given you this notice may not be the landlord to whom you pay your rent (sections 44 and 67). This does not necessarily mean that the notice is invalid.

If you have any doubts about whether this notice is valid, get advice immediately from a solicitor or a surveyor.

Further information
An explanation of the main points to consider when renewing or ending a business tenancy, "Renewing and Ending Business Leases: a Guide for Tenants and Landlords", can be found at *www.odpm.gov.uk*. Printed copies of the explanation, but not of this form, are available from 1st June 2004 from Free Literature, PO Box 236, Wetherby, West Yorkshire, LS23 7NB (0870 1226 236).

13.8.3 Section 25 notice where landlord opposes renewal

LANDLORD'S NOTICE ENDING A BUSINESS TENANCY AND REASONS FOR REFUSING A NEW ONE

Section 25 of the Landlord and Tenant Act 1954

IMPORTANT NOTE FOR THE LANDLORD: If you wish to oppose the grant of a new tenancy on any of the grounds in section 30(1) of the Landlord and Tenant Act 1954, complete this form and send it to the tenant. If the tenant may be entitled to acquire the freehold or an extended lease, use form 7 in Schedule 2 to the Landlord and Tenant Act 1954, Part 2 (Notices) Regulations 2004 instead of this form.

To: *(insert name and address of tenant)*

From: *(insert name and address of landlord)*

1. This notice relates to the following property: (insert address or description of property)

2. I am giving you notice under section 25 of the Landlord and Tenant Act 1954 to end your tenancy on (*insert date*).

3. I am opposed to the grant of a new tenancy.

4. You may ask the court to order the grant of a new tenancy. If you do, I will oppose your application on the ground(s) mentioned in paragraph(s)* of section 30(1) of that Act. I draw your attention to the Table in the Notes below, which sets out all the grounds of opposition.

*(*insert letter(s) of the paragraph(s) relied on*)

5. If you wish to ask the court for a new tenancy you must do so before the date in paragraph 2 unless, before that date, we agree in writing to a later date.

6. I can ask the court to order the ending of your tenancy without granting you a new tenancy. I may have to pay you compensation if I have relied only on one or more of the grounds mentioned in paragraphs (e), (f) and (g) of section 30(1). If I ask the court to end your tenancy, you can challenge my application.

7. Please send all correspondence about this notice to:

Name:

Address:

Signed: Date:

*[Landlord] *[On behalf of the landlord] *[Mortgagee] *[On behalf of the mortgagee]
(**delete if inapplicable*)

IMPORTANT NOTE FOR THE TENANT
This notice is intended to bring your tenancy to an end on the date specified in paragraph 2.

Your landlord is not prepared to offer you a new tenancy. You will not get a new tenancy unless you successfully challenge in court the grounds on which your landlord opposes the grant of a new tenancy.

If you want to continue to occupy your property you must act quickly. The notes below should help you to decide what action you now need to take. If you want to challenge your landlord's refusal to renew your tenancy, get advice immediately from a solicitor or a surveyor.

NOTES

The sections mentioned below are sections of the Landlord and Tenant Act 1954, as amended, (most recently by the Regulatory Reform (Business Tenancies) (England and Wales) Order 2003)

Ending of your tenancy
This notice is intended to bring your tenancy to an end on the date given in paragraph 2. Section 25 contains rules about the date that the landlord can put in that paragraph.

Your landlord is not prepared to offer you a new tenancy. If you want a new tenancy you will need to apply to the court for a new tenancy and successfully challenge the landlord's grounds for opposition (see the section below headed "Landlord's opposition to new tenancy"). If you wish to apply to the court you must do so before the date given in paragraph 2 of this notice, unless you and your landlord have agreed in writing, before that date, to extend the deadline (sections 29A and 29B).

If you apply to the court your tenancy will continue after the date given in paragraph 2 of this notice while your application is being considered (section 24). You may not apply to the court if your landlord has already done so (section 24(2A) and (2B)).

You may only stay in the property after the date given in paragraph 2 (or such later date as you and the landlord may have agreed in writing) if before that date you have asked the court to order the grant of a new tenancy or the landlord has asked the court to order the ending of your tenancy without granting you a new one.

If you are in any doubt about what action you should take, get advice immediately from a solicitor or a surveyor.

Landlord's opposition to new tenancy
If you apply to the court for a new tenancy, the landlord can only oppose your application on one or more of the grounds set out in section 30(1). If you match the letter(s) specified in paragraph 4 of this notice with those in the first column in the Table below, you can see from the second column the ground(s) on which the landlord relies.

Paragraph of section 30(1)	*Grounds*
(a)	Where under the current tenancy the tenant has any obligations as respects the repair and maintenance of the holding, that the tenant ought not to be granted a new tenancy in view of the state of repair of the holding, being a state resulting from the tenant's failure to comply with the said obligations.
(b)	That the tenant ought not to be granted a new tenancy in view of his persistent delay in paying rent which has become due.
(c)	That the tenant ought not to be granted a new tenancy in view of other substantial breaches by him of his obligations under the current tenancy, or for any other reason connected with the tenant's use or management of the holding.
(d)	That the landlord has offered and is willing to provide or secure the provision of alternative accommodation for the tenant, that the terms on which the alternative accommodation is available are reasonable having regard to the terms of the current tenancy and to all other relevant circumstances, and that the accommodation and the time at which it will be available are suitable for the tenant's requirements (including the requirement to preserve goodwill) having regard to the nature and class of his business and to the situation and extent of, and facilities afforded by, the holding.
(e)	Where the current tenancy was created by the sub-letting of part only of the property comprised in a superior tenancy and the landlord is the owner of an interest in reversion expectant on the termination of that superior tenancy, that the aggregate of the rents reasonably obtainable on separate lettings of the holding and the remainder of that property would be substantially less than the rent reasonably obtainable on a letting of that property as a whole, that on the termination of the current tenancy the landlord requires possession of the holding for the purposes of letting or otherwise disposing of the said property as a whole, and that in view thereof the tenant ought not to be granted a new tenancy.
(f)	That on the termination of the current tenancy the landlord intends to demolish or reconstruct the premises comprised in the holding or a substantial part of those premises or to carry out substantial work of construction on the holding or part thereof and that he could not reasonably do so without obtaining possession of the holding.
(g)	On the termination of the current tenancy the landlord intends to occupy the holding for the purposes, or partly for the purposes, of a business to be carried on by him therein, or as his residence.

In this Table "the holding" means the property that is the subject of the tenancy.

In ground (e), "the landlord is the owner an interest in reversion expectant on the termination of that superior tenancy" means that the landlord has an interest in the property that will entitle him or her, when your immediate landlord's tenancy comes to an end, to exercise certain rights and obligations in relation to the property that are currently exercisable by your immediate landlord.

If the landlord relies on ground (f), the court can sometimes still grant a new tenancy if certain conditions set out in section 31A are met.

If the landlord relies on ground (g), please note that "the landlord" may have an extended meaning. Where a landlord has a controlling interest in a company then either the landlord or the company can rely on ground (g). Where the landlord is a company and a person has a controlling interest in that company then either of them can rely on ground (g) (section 30(1A) and (1B)). A person has a "controlling interest" in a company if, had he been a company, the other company would have been its subsidiary (section 46(2)).

The landlord must normally have been the landlord for at least five years before he or she can rely on ground (g).

Compensation
If you cannot get a new tenancy solely because one or more of grounds (e), (f) and (g) applies, you may be entitled to compensation under section 37. If your landlord has opposed your application on any of the other grounds as well as (e), (f) or (g) you can only get compensation if the court's refusal to grant a new tenancy is based solely on one or more of grounds (e), (f) and (g). In other words, you cannot get compensation under section 37 if the court has refused your tenancy on other grounds, even if one or more of grounds (e), (f) and (g) also applies.

If your landlord is an authority possessing compulsory purchase powers (such as a local authority) you may be entitled to a disturbance payment under Part 3 of the Land Compensation Act 1973.

Validity of this notice
The landlord who has given you this notice may not be the landlord to whom you pay your rent (sections 44 and 67). This does not necessarily mean that the notice is invalid.

If you have any doubts about whether this notice is valid, get advice immediately from a solicitor or a surveyor.

Further information
An explanation of the main points to consider when renewing or ending a business tenancy, "Renewing and Ending Business Leases: a Guide for Tenants and Landlords", can be found at *www.odpm.gov.uk*. Printed copies of the explanation, but not of this form, are available from 1st June 2004 from Free Literature, PO Box 236, Wetherby, West Yorkshire, LS23 7NB (0870 1226 236).

13.8.4 Tenant's section 26 request

TENANT'S REQUEST FOR A NEW BUSINESS TENANCY

Section 26 of the Landlord and Tenant Act 1954

To *(insert name and address of landlord)*:

From *(insert name and address of tenant)*:

1. This notice relates to the following property: (insert address or description of property).

2. I am giving you notice under section 26 of the Landlord and Tenant Act 1954 that I request a new tenancy beginning on (insert date).

3. You will find my proposals for the new tenancy, which we can discuss, in the Schedule to this notice.

4. If we cannot agree on all the terms of a new tenancy, either you or I may ask the court to order the grant of a new tenancy and settle the terms on which we cannot agree.

5. If you wish to ask the court to order the grant of a new tenancy you must do so by the date in paragraph 2, unless we agree in writing to a later date and do so before the date in paragraph 2.

6. You may oppose my request for a new tenancy only on one or more of the grounds set out in section 30(1) of the Landlord and Tenant Act 1954. You must tell me what your grounds are within two months of receiving this notice. If you miss this deadline you will not be able to oppose renewal of my tenancy and you will have to grant me a new tenancy.

7. Please send all correspondence about this notice to:

Name:

Address:

Signed: Date:
*[Tenant] *[On behalf of the tenant] (*delete whichever is inapplicable)

SCHEDULE

TENANT'S PROPOSALS FOR A NEW TENANCY

(*attach or insert proposed terms of the new tenancy*)

IMPORTANT NOTE FOR THE LANDLORD
This notice requests a new tenancy of your property or part of it. If you want to oppose this request you must act quickly.

Read the notice and all the Notes carefully. It would be wise to seek professional advice.

NOTES

The sections mentioned below are sections of the Landlord and Tenant Act 1954, as amended, (most recently by the Regulatory Reform (Business Tenancies) (England and Wales) Order 2003)

Tenant's request for a new tenancy
This request by your tenant for a new tenancy brings his or her current tenancy to an end on the day before the date mentioned in paragraph 2 of this notice. Section 26 contains rules about the date that the tenant can put in paragraph 2 of this notice.

Your tenant can apply to the court under section 24 for a new tenancy. You may apply for a new tenancy yourself, under the same section, but not if your tenant has already served an application. Once an application has been made to the court, your tenant's cur-rent tenancy will continue after the date mentioned in paragraph 2 while the application is being considered by the court. Either you or your tenant can ask the court to fix the rent which your tenant will have to pay whilst the tenancy continues (sections 24A to 24D). The court will settle any terms of a new tenancy on which you and your tenant disagree (sections 34 and 35).

Time-limit for opposing your tenant's request
If you do not want to grant a new tenancy, you have two months from the making of your tenant's request in which to notify him or her that you will oppose any application made to the court for a new tenancy. You do not need a special form to do this, but the notice must be in writing and it must state on which of the grounds set out in section 30(1) you will oppose the application. If you do not use the same wording of the ground (or grounds), as set out below, your notice may be ineffective.

If there has been any delay in your seeing this notice, you may need to act very quickly. If you are in any doubt about what action you should take, get advice immediately from a solicitor or a surveyor.

Grounds for opposing tenant's application
If you wish to oppose the renewal of the tenancy, you can do so by opposing your tenant's application to the court, or by making your own application to the court for termination without renewal. However, you can only oppose your tenant's application, or apply for termination without renewal, on one or more of the grounds set out in section 30(1). These grounds are set out below. You will only be able to rely on the ground(s) of opposition that you have mentioned in your written notice to your tenant.

In this Table "the holding" means the property that is the subject of the tenancy.

Paragraph of section 30(1)	*Grounds*
(a)	Where under the current tenancy the tenant has any obligations as respects the repair and maintenance of the holding, that the tenant ought not to be granted a new tenancy in view of the state of repair of the holding, being a state resulting from the tenant's failure to comply with the said obligations.
(b)	That the tenant ought not to be granted a new tenancy in view of his persistent delay in paying rent which has become due.
(c)	That the tenant ought not to be granted a new tenancy in view of other substantial breaches by him of his obligations under the current tenancy, or for any other reason connected with the tenant's use or management of the holding.
(d)	That the landlord has offered and is willing to provide or secure the provision of alternative accommodation for the tenant, that the terms on which the alternative accommodation is available are reasonable having regard to the terms of the current tenancy and to all other relevant circumstances, and that the accommodation and the time at which it will be available are suitable for the tenant's requirements (including the requirement to preserve goodwill) having regard to the nature and class of his business and to the situation and extent of, and facilities afforded by, the holding.
(e)	Where the current tenancy was created by the sub-letting of part only of the property comprised in a superior tenancy and the landlord is the owner of an interest in reversion expectant on the termination of that superior tenancy, that the aggregate of the rents reasonably obtainable on separate lettings of the holding and the remainder of that property would be substantially less than the rent reasonably obtainable on a letting of that property as a whole, that on the termination of the current tenancy the landlord requires possession of the holding for the purposes of letting or otherwise disposing of the said property as a whole, and that in view thereof the tenant ought not to be granted a new tenancy.

(f) That on the termination of the current tenancy the landlord intends to demolish or reconstruct the premises comprised in the holding or a substantial part of those premises or to carry out substantial work of construction on the holding or part thereof and that he could not reasonably do so without obtaining possession of the holding.

(g) On the termination of the current tenancy the landlord intends to occupy the holding for the purposes, or partly for the purposes, of a business to be carried on by him therein, or as his residence.

Compensation

If your tenant cannot get a new tenancy solely because one or more of grounds (e), (f) and (g) applies, he or she is entitled to compensation under section 37. If you have opposed your tenant's application on any of the other grounds mentioned in section 30(1), as well as on one or more of grounds (e), (f) and (g), your tenant can only get compensation if the court's refusal to grant a new tenancy is based solely on ground (e), (f) or (g). In other words, your tenant cannot get compensation under section 37 if the court has refused the tenancy on other grounds, even if one or more of grounds (e), (f) and (g) also applies.

If you are an authority possessing compulsory purchase powers (such as a local authority), your tenant may be entitled to a disturbance payment under Part 3 of the Land Compensation Act 1973.

Negotiating a new tenancy

Most tenancies are renewed by negotiation and your tenant has set out proposals for the new tenancy in paragraph 3 of this notice. You are not obliged to accept these proposals and may put forward your own. You and your tenant may agree in writing to extend the deadline for making an application to the court while negotiations continue. Your tenant may not apply to the court for a new tenancy until two months have passed from the date of the making of the request contained in this notice, unless you have already given notice opposing your tenant's request as mentioned in paragraph 6 of this notice (section 29A(3)).

If you try to agree a new tenancy with your tenant, remember:

- that one of you will need to apply to the court before the date in paragraph 2 of this notice, unless you both agree to extend the period for making an application.
- that any such agreement must be in writing and must be made before the date in paragraph 2 (sections 29A and 29B).

Validity of this notice

The tenant who has given you this notice may not be the person from whom you receive rent (sections 44 and 67). This does not necessarily mean that the notice is invalid.

If you have any doubts about whether this notice is valid, get advice immediately from a solicitor or a surveyor.

Further information

An explanation of the main points to consider when renewing or ending a business tenancy, "Renewing and Ending Business Leases: a Guide for Tenants and Landlords", can be found at www.odpm.gov.uk. Printed copies of the explanation, but not of this form, are available from 1st June 2004 from Free Literature, PO Box 236, Wetherby, West Yorkshire, LS23 7NB (0870 1226 236).

13.8.5 Landlord's counter-notice opposing grant of new tenancy

Landlord's counter-notice

To: [*insert Tenant's name and address*]

From: [*insert Landlord's name and address*]

Take notice that the Landlord will oppose an application to the court for the grant of a new tenancy of [*insert address of demises premises*] on ground(s) [*insert ground or grounds*] of section 30(1) of the Landlord and Tenant Act 1954.

For the avoidance of doubt this counter-notice is in response to the Tenant's request for a new tenancy dated [insert date] and is served pursuant to section 26(6) of the Landlord and Tenant Act 1954.

Dated: [*insert date*]

Signed . . . .
For and on behalf of the Landlord

I acknowledge receipt of this notice of which this is a true copy

Signed . . . . on 2007
For and on behalf of the Tenant

Print name . . . .

13.8.6 Tenant's section 27(1) notice

Notice pursuant to section 27(1) of the Landlord and Tenant Act 1954

To: [*insert Landlord's name and address*] "the Landlord"

From: [insert Tenant's name and address] "the Tenant"

1. Take notice that the Tenant does not desire its tenancy under a lease dated [*Insert date*] between [*Insert original parties to lease*] to be continued by section 24 of the Landlord and Tenant Act 1954.

2. This notice is given pursuant to section 27(1) of the Landlord and Tenant Act 1954.

Dated: [*insert date*]

Signed . . . .
For and on behalf of the Tenant

I acknowledge receipt of this notice of which this is a true copy

Signed . . . . on 2007
For and on behalf of the Landlord

Print name . . . .

13.8.7 Tenant's section 27(2) notice

Notice pursuant to section 27(2) of the Landlord and Tenant Act 1954

To: [*insert Landlord's name and address*] "the Landlord"

From: [*insert Tenant's name and address*] "the Tenant"

1. Take notice that the Tenant under a lease dated [*Insert date*] between [*Insert original parties to lease*] and continued by section 24 of the Landlord and Tenant Act 1954 wishes to terminate the continuation tenancy on [*Insert date not less than three months from date of service*] or if later, other the day which is three months from the date of receipt of this notice by the landlord.

2. This notice is given pursuant to section 27(1) of the Landlord and Tenant Act 1954.

Dated: *[insert date*]

Signed ....
For and on behalf of the Tenant

I acknowledge receipt of this notice of which this is a true copy

Signed .... on 2007
For and on behalf of the Landlord

Print name ....

Dilapidations — The Principles 14

14.1 Introduction

This is the first of two chapters on the subject of dilapidations. Dilapidations can be a major cost issue to a corporate occupier and should be dealt with carefully. The impact of dilapidations on the total cost of accommodation over the life of the lease is often overlooked. As a consequence the corporate occupier rarely budgets for the costs in advance. Furthermore, dilapidations can often be regarded simply as a building surveyor issue, whereas there are usually other factors which can be equally important. As such a corporate occupier can lose the opportunity to reduce the overall cost to its business by failing to adopt the right approach and in particular by being proactive, rather than reactive.

The concept of dilapidations is a uniquely British topic and executives of overseas companies can struggle to understand the effect on their properties. It can be especially difficult for US owned companies to comprehend dilapidations. As a Director of an international US owned company recently opined:

> So I have to pay to fit the unit out, pay the landlord rent through the term and finally I have to pay the landlord for him to put it back in to the condition it was originally, before it could be used. What does the landlord do for his money?

Welcome to the FRI lease.

This chapter considers the legal obligations which corporate occupiers are likely to be under in relation to the upkeep, maintenance and reinstatement of leased premises during, and in particular at the end of, a lease. There are a number of obligations. They are predominantly to be found in the lease or any licences for alterations, but can also be found in statute and occasionally implied by law. By way of shorthand, the breaches of such obligations are often described as dilapidations.

There is a tendency by corporate occupiers to give little thought to dilapidations issues until faced with a claim after the premises have been vacated and the lease has come to an end. However, a dilapidations claim at the end of a 25 year lease is likely to represent a significant drain on the resources of most businesses. Careful planning, early evaluation and the implementation of a sensible strategy can mean that a corporate occupier is able to budget sensibly for a claim, and in some cases, make significant cost savings. For example, where there is a substantial terminal dilapidations claim and there is unlikely to be a defence (eg redevelopment), a corporate occupier could make a 35%–45% cost saving by carrying out the necessary works and handing the premises back to the landlord in compliance with its obligations. This may avoid the higher repair costs of the landlord carrying out the works and then seeking reimbursement from the former tenant, and also avoid consequential losses (eg loss rent) and hefty legal costs of defending the landlord's claim. Furthermore, in some case where a tenant has continued to use premises which have been allowed to fall into a poor state of repair, (resulting in all the operational problems which disrepair can entail) it may have made more commercial sense for the corporate occupier to have done the works earlier, perhaps as part of a cycle of planned maintenance, and then had some benefit from the works.

There are also steps that can be taken at the outset, when a lease is granted, that can result in significant cost savings in the event of a subsequent claim by the landlord. Often, the desire to keep professional costs to a minimum or the need to find and occupy premises quickly means that little thought is given to minimising any future dilapidations claims. This may result in such claims being significantly higher than might otherwise have been the case. It can also lead to an unrealistic estimate of the overall cost of the accommodation and mean that the corporate occupier has a nasty (and unbudgeted for) surprise at the end of its lease.

The chapter starts by identifying the most frequently encountered obligations and a summary of the general principles that apply in relation to liability. The chapter then addresses the most difficult aspect of dilapidations, the heads of damages which are recoverable in a terminal dilapidations claim. The chapter concludes by considering the availability of legal remedies during the term. Chapter 15 then goes on to give some practical tips on how a corporate occupier might approach a terminal dilapidations claim both as tenant and landlord (where it has a legacy portfolio).

14.2 Liability — types of obligation

Most legal obligations which a corporate occupier is likely to encounter are contained in, and imposed by, the lease (or in the case of reinstatement, a licence for alterations). However, there may also be statutory requirements imposed on a corporate occupier outside the general law of landlord and tenant and not necessarily regulated by the landlord.

A schedule of dilapidations served by a landlord is likely to allege breaches of some or all of the following covenants:

- to put and keep the premises in repair
- to decorate the premises periodically
- to maintain mechanical and electrical plant and installations
- to comply with statute
- to yield up vacant possession of the premises to the landlord at the end of the lease in compliance with the lease covenants
- to reinstate alterations.

14.2.1 Obligation to keep in repair

The most significant obligation in dilapidations claims is usually the repairing covenant. The wording of repairing covenants varies from lease to lease. It is always important, therefore, when serving or being served with a schedule of dilapidations to consider the wording of the repairing covenant very carefully.

Most commonly, a repairing covenant will require a tenant to "well and substantially repair" or to keep the premises "in good and substantial repair". There are many cases which have considered the precise meaning of these words against a specific factual situation. Whether a breach of a repairing covenant has occurred can be a question of fact, expert evidence, and possibly law. Invariably an experienced building surveyor will be required. A good working relationship between the building surveyor and the lawyer will be important, and it is important that the corporate occupier's building surveyor is familiar with the legal principles which govern such claims.

There are a number of key legal principles which apply to a covenant to repair.

(a) To put into repair

An important point, often overlooked, is that an obligation to keep in repair imposes an obligation to put the premises in repair. Consequently, a tenant who takes premises in poor repair cannot keep the premises in that poor state, but is required to put them into a proper state of repair (see *Proudfoot* v *Hart* (1890) LR 25 QBD 42). A corporate occupier should, therefore, obtain advice from a building surveyor on the existing condition of premises before entering into a full repairing lease to identify what works it can expect to carry out during or at the end of its lease, and importantly what those works are likely to cost. Failure to do so may mean that the tenant pays the price for the landlord's or a previous tenant's failure to repair.

(b) Is there a breach of covenant?

The first matter to consider when investigating a potential breach of covenant is whether the possible defect can fall within the physical extent of the repairing covenant. In most cases this will be obvious. However, where the corporate occupier occupies only part of the premises and there are common parts, the potential defect may fall outside the definition of the premises for the purpose of the lease. Generally draftsmen do not inspect the premises to be let and as a consequence mistakes can occur. It is always worth considering the definition of the premises to see if the defect falls outside of the repairing covenant.

If the potential defect is in a part of the building or on land covered by the repairing covenant, the next stage is to consider whether the defect can properly be described as a repair. In *Brew Brothers Ltd* v *Snax Ross Ltd* [1970] 1 QB 612 the court considered that repair was an ordinary English word that took its precise meaning in any case from the context in which it is used. Its meaning may change from case to case. In determining its meaning in a given case, the court will consider, among other things, the language of the particular covenant and the lease as a whole, the commercial relationship between the parties, the state of the premises at the time of the letting, and any other circumstances that may colour the way in which the word is used.

The issue is ultimately whether the ordinary English speaker would consider the word repair as used in the covenant would appropriately describe the work which has to be done. The test, however formulated, will always be one of degree, and therefore in some cases it will be a matter of impression on which different surveyors (and lawyers) could reasonably give different views. This inherent uncertainty can make evaluating dilapidations claims difficult. Nevertheless, there are a number of factors which can assist in determining whether a defect constitutes an actionable breach of the repairing covenant.

(i) *Deterioration of the state of the premises*

To succeed, the party seeking to enforce the repairing covenant must prove that the state of the premises has deteriorated from an early better state. There must be proof of damage or deterioration to the premises: *Quick* v *Taff Ely Borough Council* [1985] 2 EGLR 50.

Facts

- T occupied a house which suffered from severe condensation.
- L was under an obligation to "keep in repair the structure and exterior" of the house.
- T claimed breach of covenant.
- The condensation was caused by a combination of the windows lacking adequate insulation and the internal temperature being too low.
- The problem could be alleviated by replacing existing metal windows with new timber or UPVC frames.

Held — Court of Appeal

- Although there was evidence of extensive damage to internal decorations, furnishings, bedding and clothing, there was no evidence of damage to the subject matter of the repairing covenant ie the structure or exterior of the house.
- Liability could only arise where a physical condition existed which called for repair. There was no breach of repairing covenant.

Quick v *Taff Ely* was followed by the Court of Appeal in *Post Office* v *Aquarius Properties Ltd* [1987] 1 EGLR 40. In that case, the basement of premises became flooded when the water table rose. This was due to a defect in the construction of the building. The landlord argued that to keep the premises in good and substantial repair the tenant had to remedy the inherent defect by carrying out extensive works to the basement. The cost of those works was £175,000 (where the current rent was £11,250 pa and the capital value of the building was £687,000). The Court of Appeal held that there was no breach of the repairing covenant because there was no evidence that the flooding had caused damage to the building or that there had been any deterioration in its condition since it had been built.

In relation to plant, it is often argued that if the plant has reached the end of its recommended lifespan as suggested by industry guidelines, it is incumbent on the tenant to replace that plant under the repairing covenant. This is not necessarily the case however. The key issue is whether or not the plant is out of repair. If it is not, the tenant cannot be required to repair or replace the plant, however old it may be.

(ii) *Standard of repair required*

Once damage to the premises has been identified, the next question is whether that damage is sufficient to warrant a repair under the covenant. A covenant to keep in substantial repair does not require the tenant to put the premises in to perfect repair or pristine condition (see *Proudfoot* v *Hart* and *Riverside Property Investments Ltd* v *Blackhawk Automotive* [2005] 1 EGLR 114). This is a very common misconception and can lead to the inclusion in a schedule of dilapidations of very minor items which do not constitute actionable breaches of repairing covenant.

The correct test was formulated over a century ago in Proudfoot v Hart and has stood the test of time. Lopes LJ in that case said that "good tenantable repair", (which is no different to a covenant to keep premises in substantial repair) means:

> such repair, as having regard to the age, character, and locality of the [premises], would make it reasonably fit for the occupation of a reasonably-minded tenant of the class who would be likely to take it.

A number of important points arise from this test:

- The relevant date for considering the age, character and locality of the premises is the commencement of the lease (see *Lister* v *Lane* [1893] 2 QB 212).
- The standard of repair is likely to be less onerous with an old building than a relatively new building. An incoming tenant will not expect an industrial warehouse that is 60 years old to be in the same condition of repair as one that is two years old.
- A building that is likely to be occupied by an office tenant will need to be in a better state of repair than a building that is to be use as an industrial warehouse.
- The standard of repair expected of high class offices in the City of London is likely to be higher than office accommodation above a shop in Romford.

The test is very flexible and is applied in a modified form depending upon the character of the premises and the likely tenant. For example, in *Blackhawk Automotive* which involved an industrial unit in Kent, the appropriate standard of repair to be applied was that of:

> an intending occupier of an industrial warehouse building, with modern construction, who judges repair reasonably by reference to his intended use of the premises.

It is important to note that the parties can expressly agree to a higher standard of repair. In some cases repairing covenants expressly provide for the standard of repair required (eg to keep the premises in repair to a standard consistent with their use as high class offices).

(iii) *Works which go beyond repair*

One of the most difficult issues which may arise is whether the required works go beyond the scope of the repairing obligation and beyond that intended by the parties when entering into the lease and the repairing covenant. This issue has come before the courts on many occasions and the argument (usually put by the tenant required to carry out works) has taken a number of forms:

- the works go beyond repair and constitute an "improvement"
- the repairing obligation does not extend to remedying an inherent defect
- the works are so extensive that they go beyond the contemplation of the parties when entering into the repairing covenant
- carrying out the required works would result in the tenant giving back to the landlord a wholly different property from that which was let.

The case law is not always consistent which can make advising the parties difficult. However, it is now clear that any attempt to formulate a defence based solely on the proposition that required works would constitute an improvement, or alternatively, on the basis of a doctrine of inherent defect, will fail.

It has been recognised time and again by the courts that if the only practical way of carrying out a repair involves replacement, then the tenant is under an obligation to do so (see for example *Lurcott* v *Wakely* [1911] 1 KB 905). At one level, most repairs will involve an element of replacement.

For example, if a roof fails it may be necessary to replace rotten timber with sound wood, or substitute new tiles or slates for cracked, broken or missing ones. Identifying whether the necessary works involve replacement and therefore constitute an improvement is a red-herring and irrelevant to the task of determining liability. A tenant's plea of improvement is therefore, unlikely to cut any ice with the court.

Equally, merely identifying that works are necessary because of an inherent defect in the original design or construction of a building will not assist the tenant. It was thought at one time that a tenant could never be required to remedy an inherent or structural defect. This proposition was roundly rejected in *Ravenseft Properties Ltd* v *Davstone* [1979[1 EGLR 54.

The correct test to determine whether the required works would exceed the repairing obligation contained in a lease is a matter of degree: can that which the tenant is being asked to do be properly described as repair, or would it involve giving back to the landlord a wholly different property from that which was let? (see *Lister* v *Lane* [1893] 2 QB and *Ravenseft Properties*).

In most cases a tenant's plea that it is being required to carry out works which go beyond those contemplated as repairs by the parties is likely to fail.

The defence *succeeded* in the following cases:

Lister v *Lane* [1893] 2 QB 212

Facts

- T was required to "well and sufficiently and substantially, repair, uphold, sustain, maintain, amend and keep" a house.
- One of the walls (which was bulging) was pulled down following the service of dangerous structures notice.
- The foundations of the house were merely a timber platform resting on boggy soil. The bulging of the wall had been caused by the rotting of the timber platform.

Held — Court of Appeal

- A covenant to repair does not require the tenant to give a very different thing back to the landlord from that which the tenant took.
- The necessary underpinning would have required the tenant to dig down through the mud some 17 ft to the gravel and then building up to the brickwork of the house. Those works could not be described as 'repairs', but would amount to making a new and entirely different house.
- The change of circumstances which arose could not have been foreseen by the parties and it was not reasonable to construe the repairing covenant as extending to the change of circumstances.

Sotherby v *Grundy* [1947] 2 All ER 761

- This involved a condemned house built in 1861 without footings or on defective footings.
- The house could only be repaired by underpinning and substituting a new foundation.
- T not liable under repairing covenant.

Halliard Property Co Ltd v *Nicholas Clarke Investments* [1984] 1 EGLR 45

- T was not required to rebuild the back wall of a 'jerry-build' utility room constructed in the back yard of a terraced house.

- The cost of rebuilding the structure would be more than one-third of the total cost of rebuilding the entire premises.
- If T was liable to do the works, L would be given back a structure entirely different from the unstable structure which had existed at the start of the lease.

Examples of cases where tenants have *not* succeeded include:

Lurcott v *Wakely* [1911] 1 KB 905

- T was under obligation to "substantially repair and keep in thorough repair and good condition" a house.
- A dangerous structures notice was served requiring the occupiers to take down the front external wall.
- The condition of the wall was caused by old age and could not have been repaired without the wall being rebuilt.
- T argued that the repair did not mean renewal and the natural operation of time and its affect on the building fell on L.

Held — Court of Appeal

- Repair often involves replacing worn out parts of a building. The distinction between repair and renewal is not material.
- The portion of the wall which had to be replaced was 24 ft in front, which in the context of a building which extended back 100 ft would not mean that rebuilding the wall would change the character or nature of the building.

Ravenseft Properties v *Davstone* [1979] 1 EGLR 54

Facts

- T covenanted "well and sufficiently to repair renew rebuild amend" the premises.
- The premises had been constructed in concrete with an external stone cladding but without expansion joints. The omission occurred because the different coefficients in the expansion of stone and concrete (and therefore the need for expansion joints) had not been appreciated when the building had been constructed.
- The stones in the cladding on one elevation became loose and in danger of falling.
- T argued that where the need for remedial work arises out of an inherent defect, the damage caused by that defect can never fall within a repairing covenant. Alternatively, if T has to repair the damage, it does not have to remedy the inherent defect.

Held

- There is no doctrine of inherent defect. The test is one of degree. Can the works be described as a repair, or would it involve giving back to the landlord a wholly different thing from which was let?
- The proportion of the cost of work bears to the value of the premises (or cost of rebuilding) may be relevant. The cost of inserting expansion joints and repairing the inherent defect was £5,000. The cost of re-fixing the stones and repairing the damage to the building caused by the inherent defect was £50,000. The cost of rebuilding the whole building was £3 m.

- The works could not therefore be said to result in the premise becoming a wholly different character.
- Whether a tenant is required to remedy the inherent defect is a question of degree. If, as was the case here, no competent engineer would permit the replacement of the cladding without putting in expansion joints, it was the only way of repairing the building and therefore the tenant was liable. The cost was trivial and would not amount to a building of different character being handed back to the landlord.

Elite Investments v *Bainbridge Silencers* [1986] 2 EGLR 43

- T covenanted "well and substantially to repair, replace... mend and keep the demised premises and the ... roof" of two industrial units.
- On expiry of the lease, the galvanised steel sheeted roof of one of the units had deteriorated beyond repair.
- The cost of replacing the roof was £84,000 compared with an estimated cost of £1m to replace the whole building.

Held
- The replacement of the roof would not involve T giving back to L something different. T would merely be handing back an industrial unit with a new roof.
- The replacement of the roof constituted a repair.

In applying the fact and degree test it is difficult to be definitive about what factors are likely to influence a judge. It will always turn on the facts of a particular case. However, the following factors may be significant:

(1) the nature of the building
(2) the wording of the lease and the repairing covenant
(3) the state of the premises at the time the lease was granted
(4) the age of the building and its prospective life
(5) the nature and extent of the defect and the remedial works
(6) the probable cost of works compared with:
  - the commercial relationship between the parties such as the cost of the annual rent and the length of the term
  - the cost of rebuilding the premises as a whole.

In *Sotherby* v *Grundy*, in which the court rejected a suggestion that underpinning a house was a repair, the judge concluded that putting in a new foundation would alter the nature of the leased premises turning a building that would not last more than 80 years (by that stage the building was already over 80 years old) into a building that would last for a further 100 years.

Consequently, it is unlikely that a court would consider a corporate occupier liable to carry out very substantial structural works to premises held on a short lease where those premises are approaching the end of their prospective life and where the cost of that work would equate to the cost of rebuilding the premises. However, in most cases it will be difficult for a tenant to succeed with such a defence.

It may be possible for a tenant, in negotiating the terms of a lease, to ensure that the definition of repair in the lease expressly excludes any defect which constitutes, or arises as a result of, an inherent defect. A tenant will not, however, be able to ensure it avoids the possibility of carrying out substantial and costly works of renewal in discharging its repairing obligation. These potential costs need to be budgeted for, or at least taken into account, when deciding whether to take a lease or in negotiating rental levels or rent free periods. It emphasises the need for a comprehensive acquisition report by a competent building surveyor before making a decision and/or finalising negotiations. This is most important with older premises. However, even modern buildings can suffer from inherent defects.

(iv) *Patch repair* v *replacement*

An issue which arises often in dilapidations claims is whether a tenant can comply with its repairing obligations by carrying out running repairs or whether replacement or renewal is required. A variation on this theme is where there are a number of ways of repairing a defect and the landlord and tenant disagree which is appropriate. Naturally, as the tenant is paying, it is almost certainly going to choose the least expensive option. The landlord, on the other hand, may want the tenant to carry out (or pay for) a more expensive and longer lasting solution.

These issues are predominantly matters for expert evidence. However, there are essentially three legal points which may arise.

- If there is a dispute between replacement and repair, replacement will only be necessary if repair is not reasonably or sensibly possible (see *Ultraworth* v *General Accident Fire & Life Assurance Corporation* [2000] 2 EGLR 115 and *Riverside Property Investments Ltd* v *Blackhawk Automotive* [2005] 1EGLR 114).
- If the expert evidence suggests there are a number of ways in which the repairing covenant can properly be performed, the party who is required to carry out the repair can choose which method to utilise. If that party is the tenant, it is very likely to choose the cheapest option and it cannot be criticised for doing so (see *Ultraworth*).
- However, if the tenant decides not to carry out a necessary repair before the end of the lease term (in effect electing to pay damages instead), the landlord can chose the method of repair (even if this is not the cheapest option) provided it acts reasonably in doing so. The tenant cannot insist on the landlord adopting the cheapest method of repair (see *Flour Daniel Properties Ltd* v *Shortlands Investments Ltd* [2001] 2 EGLR 103).

(c) Wider words — condition; renewal; amend; rebuild

It is not uncommon for a repairing covenant to contain additional words which appear to go beyond repair. For example, in *Post Office* v *Aquarius* the covenant included:

> well and substantially to repair ... *amend* ... *renew* ... and keep in good and substantial repair *and condition*

Do these words impose a greater obligation on the party responsible for carrying out the works? In that case the landlord focussed on repair and did not argue that the wider words would impose a more onerous obligation. The Court of Appeal expressed no view.

However, this point was considered in *Credit Suisse* v *Begass Nominees Ltd* [1994] 1 EGLR 76 where the covenant provided:

> to maintain repair *amend renew* ... and otherwise *keep in good and tenantable condition*

The judge took the view that amend and renew were capable of requiring works which went beyond repair.

He also suggested that "otherwise keep in good and tenantable condition" had to be considered as a separate covenant, and could substantially extend beyond a covenant to repair. Points to bear in mind when considering a covenant to keep premises in a specified condition include that:

- it is not necessary to prove that the premises are damaged and have deteriorated from an earlier better state
- the premises must be put into tenantable condition (if they are not in that condition)
- the standard required is that which would make it reasonably fit for the occupation of a reasonably minded tenant of the class likely to take it (having regard to the age, character and locality of the premises)
- this test applies at the date that the covenant is being considered, and is not judged at the lease commencement date
- it is unclear whether this word can impose an obligation on a party to hand over premises which are of a wholly different character (it was not necessary to decide this point in *Credit Suisse*).

This is an area of law that is developing and therefore its principles are not clear or settled. There are many cases where the courts have not given a wider meaning to such words and it will always be a question of construction on the particular wording of a repairing covenant. However, a corporate occupier when entering into a lease should be alive to these words, and the extent to which they potentially may impose a greater burden than a simple covenant to repair.

Of course, in some cases a landlord may insist on a very onerous repairing covenant. If it does, and the corporate occupier decides to go ahead anyway, it should not overlook the onerous nature of the repairing covenant at rent review. In *Norwich Union Life Insurance Society* v *British Railways Board* [1987] 2 EGLR 137 the covenant not only required the tenant to keep the premises in good and substantial repair and condition, but it also required "where necessary to rebuild reconstruct or replace". It was held that this could extend to rebuilding the whole premises, and therefore was an especially onerous provision which warranted a discount on rent review.

14.2.2 Redecoration covenant

A corporate occupier is likely to encounter two types of obligation to decorate:

- an obligation to redecorate as part of its repairing obligation (eg if repair work is required to replace blown plaster following an ingress of water, the tenant will be required to redecorate as part of repairing the damage)
- an obligation to decorate that arises under an express covenant to paint and decorate the premises at regular periods and immediately before the termination of the lease.

If the landlord relies upon the repairing covenant, the rules and principles which apply to repairing covenants will govern the position.

An express covenant to periodically redecorate is not hindered by any such restraints and operates in a different way. A typical covenant is:

> ... in every third year and in the last year of the Term howsoever determined to paint or otherwise treat as the same may require all the outside wood metal and cement work of the buildings on the demised premises (including all new buildings which may at any time during the said term be erected on and all additions made to the demised premises) usually requiring to be painted or otherwise treated with three coats of good paint or other suitable materials of the best quality ...

There are two principal points to be aware of:

- it imposes an obligation to redecorate even if no redecoration is required (although if the premises are in a good decorative state a landlord's claim for the cost of redecoration may fail if the court considers the landlord's decision to redecorate was unreasonable)
- the standard of decoration required will be that expected of the type of tenant likely to occupy the premises having regard to the age, character and locality of the premises.

14.2.3 To comply with statute

Most leases will impose an obligation on the tenant to comply with statute, regulations or notices which apply to the premises during the currency of the lease. Some statutes which will apply may require the tenant to carry out works to the premises in certain circumstances. A good example of the obligations that might apply is those arising under Parts II and III of the Disability Discrimination Act 1995, dealing respectively with obligations imposed upon employers and providers of services to the public. The 1995 Act and the Disability Discrimination Act 1995 (Amendment) Regulations 2003 provide that if premises were occupied for either employment, or for the provision of services to members of the public, then the tenant must carry out an audit and consider what, if any, works are required to the premises. The obligation on a tenant is to take such steps as are reasonable in all the circumstances to overcome physical obstacles. This may require, for example, a tenant to construct toilets for disabled people, or provide a disabled access.

If the tenant fails to construct disabled toilets or provide a disabled access in circumstances where it was necessary under the 1995 Act, a landlord is likely to include the cost of constructing them as part of its terminal dilapidations claim. It will argue that the tenant's failure to do so means that the landlord will have to do so, or alternatively, it will have to give an incoming tenant a rental discount to offset the capital cost of constructing them.

14.2.4 Alterations and reinstatement

Building surveyors often refer to reinstatement to cover two separate and distinct breaches of covenant.

The first type of breach occurs where the tenant carries out an alteration which is prohibited by the alterations covenant in the lease, or alternatively, where there is a qualified restriction, it has failed to obtain the landlord's prior (written) consent. In such circumstances the landlord may be able, as part of its dilapidations claim, to seek damages for this breach of covenant. Those damages will usually be the cost of reversing the unlawful alterations. There are two issues that a tenant facing such a claim should be alive to:

- is there any evidence to suggest that the breach of covenant was waived? (which is unlikely but possible)

- is the claim for damages statute barred? If the unlawful alterations were carried out more than 12 years before the termination of the lease, the landlord's claim may be statute barred. However, if, as sometimes is the case, the lease entitles the landlord to require unlawful alterations to be reversed and the premises reinstated to their original unaltered state, the landlord will not face a limitation problem provided it serves the necessary notice on the tenant. The breach in such cases is a failure to reinstate unlawful alterations at the end of the term; not carrying out unlawful alterations which took place more than 12 years before.

The second type of breach arises where consent has been obtained for certain alterations but subject to the landlord having a right to require that the premises be reinstated to their original unaltered state at the end of the lease, and the tenant has failed to reinstate. The key issues to consider with this potential breach are set out below.

- The nature of obligation will depend on the wording of the express covenant. This may be contained in the lease, or more likely in the licence for alterations. Occasionally it is contained in both (and sometimes the details are contradictory where a provision in the lease has been overlooked by the draftsman of the licence).
- Depending on the terms of the lease/licence, the obligation to reinstate will either be activated by a notice from the landlord requiring reinstatement, or alternatively, it will be automatic unless the landlord serves a notice requiring the tenant *not* to reinstate.
- In the latter case, while on the face of it this would appear to be favourable to the tenant, difficulties may arise in a case where the tenant decides to do the necessary works (to avoid a potentially more costly terminal dilapidations claim) and the landlord serves its notice *after* reinstatement works have been completed. In those circumstances, and prior to carrying out the work, it would be prudent for the tenant to write to the landlord indicating that unless it serves a notice within a given (reasonable) period, it intends to carry out the reinstatement works. If a landlord fails to respond, stands by and watches the tenant carry out reinstatement works, and then serves a notice indicating that reinstatement is not required, the landlord is likely to be estopped from relying upon its strict legal rights.
- If the reinstatement clause requires activation by a landlord's notice, failure to serve a notice before the lease expiry date will usually mean that the landlord's right has been lost. It is important, therefore, that in a legacy situation, a landlord corporate occupier should instruct a building surveyor well before the lease expiry date and if appropriate, serve a notice on the sub-tenant to reinstate.
- Usually, the notice does not need to take any specific form. Consequently, if the landlord serves a schedule of dilapidations which includes works of reinstatement, this will be adequate notice, even if the landlord does not specifically refer to reinstatement in correspondence (see *Westminster City Council* v *HSBC Bank plc* [2003] 1 EGLR 62).
- If the tenant is notified of the requirement to reinstate too late to reasonably carry out the works, this is unlikely to provide a defence to a landlord's subsequent claim for reinstatement costs. In this situation, provided the tenant has put the works in motion, the legal position (perhaps surprisingly) appears to be that the tenant is entitled to remain in the premises for so long as is reasonably necessary to complete the works.

14.2.5 Schedule of condition

It is not uncommon for a tenant's repairing obligation to be qualified or watered down by reference to a photographic schedule of condition attached to the lease. This practice is most prominent in a legacy situation, where the premises have become surplus to a corporate occupier's requirements with a number of years to run to lease expiry, and the corporate occupier has no alternative but to sub-let the premises to offset its rental loss. In those circumstances the corporate occupier may not be in a strong negotiating position, and a potential sub-tenant may not be prepared to take on full repairing and redecoration obligations.

The principal reason for this is a general reluctance by corporate occupiers to spend money on surplus property to bring it up to a good condition, preferring to wait until a tenant has been found rather than undertaking it speculatively. There is logic in such an approach as it avoids spending money before a subtenant is found. However, it does not assist in the process of letting the space and usually when a subtenant is found it will want to go into occupation without delay. The simplest approach is to record the disrepair by means of a schedule of condition, but this can give rise to a future management problem. In other cases the corporate occupier creates a mismatch in repairing obligations (ie between the sublease and the headlease) by only carrying out part of the work which is required to bring the premises up to the standard of repair required by an incoming tenant. A classic example of this is in an office suite, where a corporate occupier has fitted the space out with a kitchen and the installation of partitions. When the premises become surplus to the corporate occupier's requirements, the decision is taken, often with the support of the letting agent, to redecorate it in its current configuration rather than strip out the partitions. The incoming tenant's demise includes the partitions so at expiry of the headlease and the sublease, the subtenant is merely required to redecorate while the corporate occupier may be required by the superior landlord to strip out the partitions.

Where the parties agree that the tenant should have limited obligations, the repairing and decoration obligations in the lease will be in a modified form, qualified by words such as:

> ... save that the Tenant is not required to repair or redecorate the premises to, and yield them up to the Landlord in, any better condition or state than is evidenced by the schedule of condition dated... and prepared by ... and attached to the Lease as schedule 1.

If the parties are considering restricting a tenant's obligation by the use of a schedule of condition it is very important that:

- if the letting is to be an underlease, the parties check to ensure that a diluted sub-tenant obligation will not breach the alienation clause in the headlease (see *Allied Dunbar plc* v *Homebase Ltd* [2000] L&TR 27 and 5.6.1)
- the landlord ensures that the photographic schedule is thorough and accompanied by a written report or a written narrative explaining what is shown on each of the photographs
- both parties ensure that the schedule is incorporated into the underlease (it is not uncommon for the schedule to be attached and lost).

While this practice may appear to be a neat pragmatic solution to ensure that surplus premises can be sub-let as quickly as possible, there is no doubt that if a dilapidations dispute arises at the end of the sub-lease, the schedule of condition will complicate the liability position and in most cases cause increased costs and delay in reaching a settlement. Schedules of condition tend to be rushed affairs, and often they provide insufficient detail to the landlord and tenant's surveyors who many years later have to determine the condition of the premises when the sub-lease was entered into, compared with their condition at the termination of the lease. It is unlikely that the surveyor who prepared the schedule of condition will be around at the end of the lease to give evidence as to the condition of the premises at lease commencement, and even if he is around, he is unlikely to remember any more detail than is contained in his original schedule which may have been prepared many years earlier.

A schedule of condition can provide a sub-tenant with an additional advantage that is often overlooked. This can be demonstrated by a simple example. A corporate occupier takes a 25 year lease of industrial premises in 1980, expiring in 2005. In 1995 (15 years into the term), the corporate occupier sublets the whole of the premises for the remainder of the term, less a few days. The sub-tenant, not surprisingly, only agrees to repairing and decoration obligations which are restricted to keeping the premises in their 1995 condition, as evidenced by a schedule of condition. In 2005 the head landlord serves a very extensive schedule of dilapidations on the corporate occupier, which in turn, serves a diluted schedule of dilapidations on the sub-tenant.

The problem which the corporate occupier may face is that many items which it seeks to pass on to the sub-tenant may be unsustainable because the works which the sub-tenant was required to carry out under the watered down covenant would be superseded by the more extensive works which the corporate occupier was required to carry out under its full repairing and decoration covenants (contained in the headlease). For example, if the sub-tenant was under an obligation to repair a concrete parking area(which in 1995 had already been patched up with tarmac repairs by the corporate occupier) it may only be required to carry out patch repairs (with tarmac) to satisfy its repairing obligation. However, the corporate occupier may be required to carry out the more onerous task of completely renewing the parking area to satisfy the headlease obligation. In those circumstances the sub-tenant works will be entirely superceded by the works which the corporate occupier must carry out and therefore, the damages recoverable from the sub-tenant will be nil (see 14.3.2 (a) (i) and (ii) below).

The effect of this principle can dramatically reduce a sub-tenant's liability where the sub-lease and headlease are co-terminus, and may beg the question whether a watered down obligation offers any real value to the corporate occupier in its role as reluctant landlord.

14.3 Assessment of damages

One of the most contentious areas of dispute in a terminal dilapidations claim can be the amount of damages payable. Quite often, building surveyors will agree that there have been breaches of covenant by the tenant, but are unable to agree the quantum of damages which the landlord is entitled to recover. This is a highly technical area of law, and this section is intended to merely identify the key areas which corporate occupiers may need to focus on, and offer practical tips. For a detailed legal analysis, regard should be had to more detailed books on dilapidations.

The correct approach to quantifying damages is to:

1. identify the damages which are *prima facie* recoverable at common law and
2. consider if those damages would be reduced by the statutory cap contained in section 18(1) of the Landlord and Tenant Act 1927 or
3. if the breach is one to which section 18(1) of the 1927 Act does not apply (eg reinstatement), consider whether the correct measure should be the cost of the cure or whether this would be unreasonable or disproportionate in the circumstances.

14.3.1 Common law damages

The normal rules of damages apply to breaches of repairing, decoration and reinstatement obligations. A court will endeavour to put a landlord in the position it would have been in but for the breach of covenant(s) by the tenant. Put another way, it should be compensated for any loss that was foreseeable and was caused by the breach of covenant(s). If the landlord would have suffered a loss in any event that particular loss will not be recoverable. The most common example of this is where a landlord would have suffered a loss of rent what ever condition the premises were handed back in because there was no market for the premises at the end of the lease.

At common law the measure of damages for dilapidations is:

- the cost of remedial works to comply with the tenant's covenants minus
- a credit for betterment (ie substitution of new for old) plus
- any foreseeable consequential losses (see *Joyner* v *Weeks* [1891] 2 QB 31).

There are a number of issues which can arise when considering common law damages.

(a) Cost of works

At common law the landlord is entitled to recover the proper and reasonable cost of the works which the tenant should have carried out to comply with its covenants.

Clearly, as a matter of liability, the appropriate remedial works need to be identified first. This has been dealt with in section 14.2.1 (b) above. The issue of competing types of repair arises often; the landlord arguing for renewal (eg a roof) and the tenant arguing that a patch repair will suffice. In the case of such a dispute, it is sensible for the purposes of the dispute, for the parties' experts to price both possibilities.

There are two principal ways that the cost of works can be determined. The first, and to some extent the most accurate approach, is for a landlord to put the works out to tender, select an appropriate contractor and then carry out the necessary works. The landlord is then in a position to sue on the contractor's invoices and seek to recover the actual cost of carrying out the works.

An alternative approach, which is often adopted in practice, is for the landlord's building surveyor to include an estimate of the likely cost of the necessary works in his schedule of dilapidations. The cost of works, in circumstances where those works have not yet been carried out, is a matter of expert evidence. The parties building surveyors are likely to have regard to recognised price books and also draw on their experience and possibly informal discussions with local contractors. Needless to say that providing an estimate of costs is not an exact science and there is likely to be significant scope for challenge by the tenant's building surveyor.

There has been some academic debate as to whether, as a matter of law, it is still appropriate to make a deduction for betterment. The debate in large part turns on a liability issue and in particular whether the only means of repair involves replacement or renewal. It is quite clear that if the only means of repair involves giving the landlord back an improved and more valuable building, no deduction should be made for betterment. Betterment rarely arises in practice and perhaps will only be an issue where a landlord, after lease expiry, carries out a more expensive method of repair (which results in betterment) than a tenant would have been required to carry out to comply with its repairing obligation. A possible example of this is where the tenant's repairing obligation is qualified by a schedule of condition. In those circumstances, the landlord will only be able to recover the cost of the repair which would not have involved betterment.

(b) Recovery of VAT

Where a landlord carries out the necessary works and pays VAT to the contractors it will be recoverable from the tenant where:

- the landlord is VAT registered and has elected not to charge VAT on the leased premises or
- the landlord is not VAT registered (see *Drummond* v *S&U Stores Ltd* [1981] 1 EGLR 42).

A landlord cannot recover VAT as part of its damages where the landlord is VAT registered and it has waived the exemption on the premises. In those circumstances it will be able to recover the VAT paid to its contractors and therefore will not suffer a loss in respect of VAT.

If the landlord has not carried out the necessary works, a court will only award a sum to compensate it for the likely outlay for irrecoverable VAT if it is satisfied that there is a real likelihood of the landlord actually carrying out the works (see *Elite Investments Ltd* v *Bainbridge (Silencers) Ltd* [1987] 2 EGLR 50).

(c) Consequential losses

In addition to the cost of works, a landlord is entitled to claim consequential losses provided those amounts are foreseeable and caused by the breach of covenant(s). The most commonly claimed consequential losses include the following:

- loss of rent
- loss of non domestic rates
- loss of service charges
- cost of borrowing money to pay for works (ie interest on finance)
- security costs
- surveying/engineering fees
- legal fees.

All of the above heads of claim are potentially recoverable. Whether they are legitimate in a particular claim will depend upon the particular circumstances of the case.

(i) *Loss of rent*
A landlord may be entitled to claim loss of rent for any period the premises are unavailable for reletting while the landlord carries out works which the tenant has failed to do (see *Joyner* v *Weekes* and *Drummond* v *S&U Stores Ltd* [1981] 1 EGLR 42). However, there are a number of issues to consider.

- A landlord will not succeed unless it can prove that its loss of rent was *caused by* the tenant's failure to carry out the works before the termination of the lease. If the landlord would have suffered a loss in any event (eg where the market is flat and where it would take many months to find a new tenant even if the premises were handed back in compliance with covenants) loss of rent will not be recoverable.
- The period of any claim for loss of rent must be examined closely. This will be a matter of expert evidence for the parties' building surveyors. The key question is how long the works are likely to take. This should include the whole process, which in a substantial claim, may

include preparing tender specifications; going out to tender; selecting a contractor and completing the works. However, in a case where the tenant has indicated that it does not intend to renew its lease, but vacate at the end of the term, it may be reasonable for the landlord to commence preparations *before* the end of the lease if it is obvious that the tenant does not intend to carry out the works itself (see *Drummond* v *S&U Stores Ltd*). In those circumstances the court may only award loss of rent for the period of the actual works (which could not be carried out before the end of the lease), but discounting any preparations which could have been carried out before the end of the lease.

- Quite often, from the landlord's perspective, two expert witnesses will be necessary to determine whether, and to what extent, loss of rent is recoverable. First, a building surveyor will need to estimate the period that the premises will be unoccupied whilst the necessary works are completed. Second, the landlord's valuer will have to consider how long it would have taken to find a new tenant in the prevailing market assuming that the premises had been handed back in repair. The landlord will only be able to recover loss of rent for any period of works that exceed any marketing void period.

(ii) *Rates and service charge*

If the landlord has a good claim for loss of rent, it may also be able to recover the non domestic rates which it is required to pay for the period that the premises are empty while works are carried out. Previously the issue has not arisen with empty industrial premises which were exempt for the period they are vacant, whereas other commercial premises had a six month grace period before non domestic rates were payable. That is changing with the proposed legislation making its way through Parliament that would remove empty rates relief altogether.

A further possible head of loss is the payment of service charges. This will only arise if the landlord is itself a tenant and looks to pass on any service charge requests from a superior landlord. For example, if the premises are situated on an industrial estate and all tenants contribute to the cost of the services (e.g. security barriers and maintenance of common roads), a landlord will have to pay those costs until the repair works are complete and it can re-let the premises to a sub-tenant. In those circumstances, and subject to the points made above in respect of loss of rent, the landlord's loss is potentially recoverable.

(iii) *Cost of finance*

If a landlord, acting reasonably, decides to carry out works which a tenant was required, but failed, to do, it may need to borrow funds to carry out those works. Alternatively, it might take money out of its business to finance the works. Either way the landlord will suffer additional consequential losses. In the former case it will be the interest that it is required to pay on the borrowed money. In the latter case the cost to the landlord will be a reduction in profit that would have been generated by that capital sum if it had been used in its business, and not to finance the works the tenant wrongfully failed to carry out.

There seems no reason in principle why either should not form a head of loss which is recoverable from the tenant. In most cases, however, it will be much easier to prove a claim based on interest payments on borrowed money, which in most cases will merely require the production of the appropriate bank statements. A court is likely to scrutinise very carefully a loss of profit claims that would significantly exceed usual interest rates on borrowing.

(iv) *Security costs*
If the landlord, acting reasonably, considers that additional security measures are required whilst the premises remain empty and the works are carried out, these ought to be recoverable from the tenant. Where substantial works are carried out it may be necessary to erect temporary fencing and overnight security cover to protect the premises and the materials on site.

(v) *Fees*
In a terminal dilapidations case the fees charged by the landlord's experts and lawyers can be a very significant proportion of the amount claimed. It is therefore important for corporate occupiers to be clear as to what fees they are entitled to recover as a matter of law, and perhaps more realistically, what fees they are likely to recover in the cut and thrust of negotiation. Most claims are settled out of court, and where they are, the first casualty of the usual horse trading tends to be professional fees. It follows that there can be a reality gap between what fees are legally recoverable and what fees a landlord is likely to recover following a negotiation. In view of the irrecoverable cost of litigation which a winning party is likely to suffer (see 15.2.3), holding out for fees is rarely a cost effective tactic except in the most substantial of cases.
There are two different routes to the recovery of professional fees.

Fees recoverable as part of the claim
Some fees which are incurred by a landlord are recoverable as a consequential loss flowing from the breach of covenant(s). These may include the reasonable costs of:

- instructing a surveyor to inspect the property and prepare a schedule of dilapidations, but only where the lease expressly provides for recovery of these fees, an example of such a covenant being:

  the tenant covenants .... to pay the landlord's reasonable and proper professional fees in connection with the preparation and service of a schedule of dilapidations

- investigatory work to identify the extent of the necessary works (eg CCTV of defective drains), but only where it has already been ascertained that there is a breach
- instructing a solicitor to serve the schedule of dilapidations (provided the wording of the express costs covenant extends to this)
- preparing a specification of works and putting those works out to tender (including the consideration of tenders and appointment of the contractor)
- supervising the works.

However, a landlord will not be able to recover the costs of:

- a surveyor's inspection and service of a schedule of dilapidations if there is no express cost recovery provision in the lease (see *Lloyds Bank Ltd* v *Lake* [1961] 1 WLR 884)
- a section 146 notice served towards the end of the lease (when there is no realistic prospect of the lease being forfeited). If the lease does not contain an express costs provision which provides for the recovery of fees for serving a terminal schedule of dilapidations (which is less prevalent than a clause entitling the landlord to recover the cost of serving a section 146 notice), the landlord may serve a section 146 with the sole purpose of seeking to recover the

costs of serving the schedule of dilapidations. Corporate occupiers should resist this tactic; it is very unlikely that such a practice will be upheld by the court

- testing (eg by mechanical or electrical consultant) in the absence of evidence of disrepair, unless the lease contains an express costs recovery provision that extends to such investigatory costs (see *Commercial Union Life Assurance Co Ltd* v *Label Ink Ltd* [2001] L&TR 29)
- negotiation of dilapidations claim — landlords often include additional surveying fees (usually expressed as a percentage of the claim value) to represent the cost of negotiating a settlement; it is strongly arguable that these costs are not a necessary consequence of a tenant's breach of covenant and therefore should not be recoverable as a separate head of loss.

Fees recovered as costs in the litigation

If the parties are not able to settle a terminal dilapidations claim, the landlord may choose to issue proceedings for damages and pursue its claim in the courts. If it is successful, the court may order that the tenant pay or contribute towards the landlord's costs.

The costs order is capable of extending to the reasonable and proper costs of:

- complying with the Property Litigation Pre-Action Protocol (see 15.2.2 (b) below)
- the preparation, issue and service of the court proceedings
- solicitor's and counsel's fees incurred during the conduct of the litigation
- surveyor's fees incurred in the preparation of a Scott Schedule (provided that the court has ordered that one be prepared)
- surveyor's fees for advising on the claim, preparing an expert report, attending any without prejudice meetings ordered by the court (for the purposes of narrowing down the issues), preparing joint reports and attending at court to give evidence.

The factors which the court can take into account when considering an application for costs and the orders which it can make are dealt with in more detail at 15.1 below.

14.3.2 Section 18(1) of the Landlord and Tenant Act 1927

Once a party has calculated the *prima facie* common law damages, it must then go on to consider the extent to which that those damages might be reduced (or in some cases even extinguished) by section 18(1) of the 1927 Act.

Section 18(1) of the 1927 Act was introduced to mitigate a perceived injustice which could arise in some cases by the common law rule that a landlord was entitled to recover the cost of works as opposed to any diminution in value to its reversion caused by the breach of covenant. Where a landlord intended to do the necessary works, the cost of doing those works represented its true loss. However, if a landlord did not intend to do remedial works, its actual loss could be negligible. In *Joyner* v *Weeks*, for example, the landlord granted a lease to a new tenant under which the new tenant covenanted to put the premises into repair. The new tenant did carry out the necessary remedial works and therefore the landlord's actual loss was nil. Nevertheless, the Court of Appeal held that the proper measure of damages payable by the former tenant was the cost of works plus consequential losses.

The landlord in *Joyner* v *Weeks* therefore received a windfall. He received damages equivalent to the cost of works from the outgoing tenant, and the works were carried out by the incoming tenant at no expense to the landlord. The landlord was therefore able to pocket the damages. Section 18(1) was passed to address this injustice.

Section 18(1) can be conveniently broken down into two separate limbs.

The first limb

> Damages for breach of covenant... to keep or put premises in repair during the currency of a lease, or to leave or put premises in repair at the termination of a lease... shall in no case exceed the amount (if any) by which the value of the reversion (whether immediate or not) in the premises is diminished owing to the breach of covenant ...

The second limb

> ... and in particular no damage shall be recovered for a breach of any such covenant... to leave or put premises in repair at the termination of a lease, if it is shown that the premises, in whatever state of repair they might be, would at or shortly after the termination of the tenancy have been or be pulled down, or such structural alterations made... as would render valueless the repairs covered by the covenant ...

It is necessary to look in detail at the two limbs.

(a) The first limb

The effect of the first limb is to impose a cap on the amount of damages which a landlord can recover. In claims to which section 18(1) applies, if the common law damages exceed the diminution in value of the landlord's reversion, it will only be able to recover the lower amount (ie the landlord's actual loss represented by the reduction in the value of its reversion). However, it is only a cap and therefore if there were to be a case (which would be very unusual) where the diminution in value of the landlord's reversion caused by the disrepair exceeded the cost of remedial works, the landlord would still only be entitled to recover the cost of works figure and not the higher diminution in value figure.

There are a number of important points to consider when applying the first limb:

- Section 18 specifically refers to a covenant to keep, put or leave premises in repair. It is a popular misconception that the statutory cap applies to all breaches of covenant usually relied upon in dilapidation claims (see 14.2 above). The prevailing view is that it does not. The cap probably does not apply to a covenant to redecorate except when ancillary to a repair (see *Latimer* v *Carney* [2006] 50 EG 86 in which the point was fully argued but the Court of Appeal expressed no opinion on whether section 18 applied to a periodic decoration covenant); a covenant to reinstate; a covenant to cleanse or a covenant to comply with statute. A landlord's advisors should be astute to a tenant who claims a blanket reliance on section 18.
- The section 18 cap will apply to any consequential losses which arise as a result of a tenant's failure to repair (eg VAT on works; loss of rent and fees).
- The diminution in value of the landlord's reversion must be assessed as at the termination date of the lease and is a matter of expert evidence for a qualified surveyor who specialises in valuation. If the tenant raises a section 18 defence, it is important for the landlord and tenant to find a valuer who has experience and a clear understanding of how the section 18 cap applies.

It is suggested that the correct approach when applying section 18 should be as follows:

Stage 1 calculate the common law heads of damages by reference to each category of covenant breaches
Stage 2 identify the heads of loss to which the section 18 cap potentially applies and total the cost of works
Stage 3 carry out a valuation of the landlord's reversion on the lease termination date assuming the tenant had complied with its repairing obligations
Stage 4 carry out a second valuation of the landlord's reversion on the basis of the actual state of repair on the termination date (being careful to exclude reference to breaches which do not fall within the repairing covenant)
Stage 5 if the difference between 3 and 4 is less than 2; total damages recoverable should be that difference (ie the diminution in value) *plus* the damages to which the section 18 cap does not apply (eg reinstatement/redecoration), provided it would be reasonable for the landlord to carry out such works
Stage 6 if the difference between 3 and 4 is the same (or more) than 2, the correct measure of damages should be 1 (ie the common law damages).

The section 18 test which involves carrying out two different valuations on the termination date (ie in repair and out of repair) is easily stated. However, in practice, the methodology and approach adopted by a valuer may be less clear. How the comparative valuations should be carried out is a matter of expert judgment for the valuer.

At first glance the valuer's task appears to be a relatively simple comparative exercise. However, in most cases it will be impossible to find comparable evidence of similar premises in a similar state of disrepair. Most leases of commercial premises assume that the premises being let are in good repair. Most lease renewal and rent review comparables will therefore be of little assistance to a valuer attempting to determine the out of repair valuation.

A valuer's approach to this task is likely to be influenced by factors such as:

- the type of purchaser most likely to buy the premises
- what the purchaser would be likely to do with the premises
- whether the purchaser is likely to do all or some of the works
- or whether the purchaser is likely to carry out structural alterations (eg subdivision of large 1970s warehouse into smaller units)
- what the total cost for carrying out works would be.

The legal and valuation complexities involved in the application of the section 18 cap are beyond the ambit of this book. However, there are a number of practical issues which corporate occupiers and their advisors should be aware of, and there are a number of common circumstances.

(i) *Landlord carries out, or intends to carry out, repairs*
One of the most effective ways in which a landlord can defeat a tenant's defence based on section 18 is to put the works of repair out to tender, and then carry out or indicate its intention to carry out those repairs. Where the landlord carries out (or intends to carry out) repairs which the tenant has failed to do, the cost of carrying out those repairs plus consequential losses will be the *prima facie* evidence of the damage to the landlord's reversion (see *Jones* v *Herxheimer* [1950] 2 KB 106). The court will use this figure as the starting point, and in the absence of strong evidence to the contrary from the tenant's valuer, this is likely to be accepted as a genuine measure of the landlord's loss.

In theory it is possible for a tenant to succeed with a section 18 defence in circumstances where the landlord carries out, or indicates an intention to carry out works of repair. In practice it is extremely difficult to succeed and would require very clear and strong valuation evidence from the tenant. The only way in which a tenant could succeed is if it were able to demonstrate that the landlord's decision to carry out some or all of the works was unreasonable in all the circumstances because the hypothetical and prudent purchaser would not have done so.

For example, in the case of a substantial 1980s industrial warehouse, a tenant may seek to argue that the age, nature and configuration of the premises are such that they would only attract a tenant in the secondary warehouse market. Such a tenant may only be concerned with the premises being functional and watertight, and not concerned with some of the more cosmetic works which the landlord may intend to carry out (eg replacing cladding to ensure uniformity, internal or external decorations or resurfacing access roads). If the tenant's valuer can satisfy the court that carrying out such works would not improve the bid rent, and therefore it is unlikely a hypothetical purchaser would pay for such works, the court may reach the conclusion that the landlord's decision to carry out those works was unreasonable and cap the damages accordingly.

The theory of such an argument is flawless. However, persuading a court to accept such an analysis in circumstances where the landlord produces invoices evidencing that it has actually carried out the works, is quite another matter. We are aware of no decided case where a tenant has successfully defended a claim relying on the section 18 cap where the landlord has carried out the works of repair. The reason may lie in the nature of the evidence before the court. On the one hand, it has the clear and incontrovertible evidence of expenditure by the landlord, on the other it has the opinion of a valuer about a hypothetical set of circumstances. Some judges regard valuation as something approaching a black art, and are very sceptical about its weight. This may explain a reluctance to accept such arguments advanced by tenants in the face of actual works carried out by a landlord.

In many cases a landlord may carry out repairs at the same time as it carries out other works of refurbishment. In such cases it is important that the landlord separates out those works of repair for which the tenant is liable from improvements for which the tenant is not liable (see *Latimer* v *Carney* [2006] 50 EG 86). The landlord is also not entitled to recover damages for any items that have been superceded by the refurbishments. Failure to do so can leave the landlord in difficulties. In *Latimer*, the landlord who failed to provide evidence of the actual costs of repairs, instead relying upon estimates prepared before the repairs/refurbishments were carried out, suffered a 60% discount in the costs of works claimed because of this failure.

(ii) *Landlord does not intend to carry out repairs*

A section 18 defence is likely to be most fruitful for a tenant where the landlord has no intention of carrying out the works which the tenant failed to do. The cost of works may bear little or no resemblance to the actual loss suffered by the landlord, being the damage to its reversion.

This is particularly the case where the premises are ripe for redevelopment and a hypothetical purchaser would be likely to demolish or radically redevelop the premises. Here the hypothetical purchaser is unlikely to alter his bid on the basis of the condition of the premises. A good example of this is *Ultraworth Ltd* v *General Accident Fire & Life* [2000] 2 EGLR 115

- L brought a terminal dilapidations claim in respect of a five-storey office.
- L argued that the air-conditioning and heating system required replacing at over £400,000.

- T argued that system could be reconditioned at a cost of £100,000 but that the diminution in value to the L's reversion was nil.
- L had no intention of carrying out the works.
- In March 1999 the premises were purchased by a developer for £1m.

Held
- L was not entitled to a new air-conditioning and heating system just because it was new at the beginning of lease; repairs/reconditioning would suffice provided it left the system in working order.
- As no one would pay more than a developer for the property (who would be unconcerned by the disrepair), the value of the reversion was £1m and the value had not been affected by the disrepair.

In *Ultraworth* the judge said that where a landlord did not intend to carry out the works the correct measure of damages was the diminution in value of the reversion. There were four alternative valuation scenarios before the court which gives an idea of just how complex, unpredictable and expensive such cases can be. The court valued the freehold reversion on the following scenarios:

1. that it would be purchased by an investor for multiple lettings in its unrefurbished state or
2. that it would be purchased by a purchaser who would refurbish the premises before re-letting or
3. that it would be purchased by an owner-occupier for office use or
4. that it would be purchased by a developer.

In relation to option 2, the landlord's out of repair valuation was £760,000. The judge noted that the premises had in fact been sold for £1m eight months after the termination date and concluded that the landlord could show no loss. This is an important point and was emphasised by Judge Bowsher QC in *Shortlands Investments Ltd* v *Cargill plc* [1995] 1 EGLR 51 when he said:

> I am very conscious that the function of damages in civil actions is to compensate the plaintiff and not punish the wrongdoer. That is a principle which has been forgotten all too often in modern cases. Plainly, if the plaintiffs have suffered no damage, or only minimal damage, one does not begin to assess damages

In practice this principle means that a landlord must show some actual loss before the court will go on to carry out the complex assessment exercise referred to above. In *Shortlands* an inducement had been paid to a new tenant so it was clear that there was some damage. In *Ultraworth* the landlord, which did not carry out the works or provide an inducement to a new tenant, was unable to demonstrate that it had suffered any loss on selling the premises to a developer.

It is perhaps of no great surprise that the court in *Ultraworth* reached the conclusion it did. Option 1–3 all involved hypothesis. Option 4, in contrast, involved what actually happened. The judge was of the view this was admissible, even though it occurred after the termination date. Indeed he treated it as the best evidence of the value of the freehold at the end of the term. Interestingly, the judge heard evidence from the purchaser who stated that the intended redevelopment involved taking the building back to a shell, emphasising that the state of repair of the building was irrelevant to the purchaser.

Consequently, where the tenant can show that the premises have a shell or site value, a court is more likely to conclude there is no diminution in value of the landlord's reversion.

If a hypothetical purchaser would carry out works of refurbishment, modernisation or subdivision, it is important to identify the extent to which the repairs which the outgoing tenant should have carried out will survive the improvements. If the repairs would be wholly superceded by the upgrading of the building, then the diminution in value will be nil. If some of the repairs will survive the upgrading works and will be carried out by the hypothetical purchaser, the landlord will have suffered a loss, and that loss may well be the cost of carrying out the survival items. In a refurbishment case a valuer should carry out a "residual valuation" to identify the works which will survive any upgrading a hypothetical purchaser would carry out.

A good example of a case where a residual valuation played a part is *Shortlands Investments Ltd* v *Cargill plc* [1995] 1 EGLR 51.

Facts

- L had a 99 year lease of 150,000 sq ft of office premises at a rent of £3.4m.
- L brought a claim against T on expiry of the underlease for terminal dilapidations.
- By the termination date L had not made a decision whether to carry out some or all of the works and by the hearing date had re-let part of the premises to a new subtenant.
- L had paid to the new sub-tenant an inducement of £690,000 being an estimate of the amount required to bring the premises up to "normally accepted letting condition".
- In the event, the new sub-tenant put the premises into a far higher standard of fit out than previously, but did carry out some of the remedial works which were the responsibility of T.
- L argued that it was entitled to damages which represented costs of works plus consequential losses in the sum of £300,000.
- T argued admitted works to the tune of £50,000 were required, but denied any sum was due because:
  - any incoming tenant would require a new fit-out which would supercede any disrepair alternatively
  - if there was disrepair, this did not result in a diminution in value of L's reversion which had a negative value because it was an onerous lease.

Held

- L had suffered loss as a result of the disrepair; this was evident from the fact that it had to pay an inducement to the new tenant of part based on the condition of the premises.
- Notwithstanding the negative value of L's reversion, L had suffered a loss. The incentive which would have to be paid to an incoming tenant (or by the same token to a hypothetical purchaser) was greater than it would be if the premises were in repair. It was foreseeable at the termination date that an incoming tenant would use the disrepair as a bargaining point.
- As T was only liable for some of the items of disrepair, the correct way of assessing diminution in value was to examine the cost of repairs and then apply the section 18 cap.
- On the evidence, T's case that it was inevitable that a new tenant would carry out a complete fit out which would supercede the repairs was rejected. Indeed, in the event, some parts of the premises which were re-let were not subject to such a fit out.
- The cost of works (plus consequential losses) were £355,000. Applying the residual approach, the diminution in value was £295,000. T was entitled to the lower figure.

Where the landlord does not intend to carry out the remedial works, the following principles apply.

- The landlord must demonstrate that it has suffered a more than nominal loss. This may be:
  - that the premises would have been worth more if they had been handed back in repair
  - an additional inducement would have to be paid to account for the disrepair
  - an inherited subtenant may have poor covenant strength and unlikely to be in a position to carry out the works at the end of the sublease.
- Assuming that it can be shown the landlord has suffered a loss, when considering the appropriate valuation scenario, the court is likely to be influenced significantly by what has actually happened after the termination date provided it was foreseeable.
- If the valuation scenario involves a hypothetical purchaser refurbishing, upgrading or subdividing the premises, a residual valuation will need to be carried out.
- The tenant's valuer may be able to challenge some or all of the items contained in the landlord's schedule on the basis that an ingoing tenant would not require, or a hypothetical purchaser carry out, those works. This approach requires a valuer to consider each item in turn and consider whether the works would be required by a hypothetical purchaser.

The fact that a landlord does not intend to carry out remedial works is likely to give a tenant a better opportunity to challenge a terminal dilapidations claim than where the landlord carries out works. However, the key to a successful challenge is for the tenant to engage a realistic and experienced valuer who has a detailed knowledge of the relevant market in the locality of the premises (unless the premises are of such a nature that there is a national market).

(iii) *New tenant to carry out repairs*

There is a potentially confusing rule which has developed in the case law to the effect that a landlord's damages cannot be directly affected (ie reduced or extinguished) by what actually happens after the termination date (see for example *Smiley* v *Townshed* [1950] 2 KB 311 and *Haviland* v *Long* [1952] 2 QB 80). The justification for this rule is that the termination date is the valuation date for the purposes of section 18 and therefore the critical issue is what the hypothetical purchaser would have done with the premises at that date; not what actually happened to the premises after that date. However, in some circumstances the courts have taken into account post termination events where those facts were operative or potential at the lease expiry date (see *Family Management* v *Grey* [1980] 1 EGLR 46 and *Crown Estate Commissioners* v *Town Investments Ltd* [1992] EGLR 61).

Shaw LJ in *Family Management* suggested that the rule at first glance appeared "not only cryptic but possibly self-contradictory". While the courts have tried to rationalise the distinction, it is very difficult to apply in practice and the casual observer could be forgiven for thinking that in some cases the application of the rule is heavily influenced by the merits of the case. Whatever the semantic niceties, it is quite clear that the courts do, in some cases, have regard to the events which occur after the termination date. When a court wants to have regard to a supervening event, it tends to be justified on the basis that what has subsequently happened is the best evidence of what the hypothetical purchaser would have done with the premises.

A good example where the court will have regard to post termination events is where the landlord, following the termination date, has granted a new lease to a new tenant. The landlord may have negotiated a full repairing lease with the new tenant, in return for either a rent free period, or a rent reduction for a specific period. It is clear that the amount of the rental inducement given to the new tenant (plus consequential losses) represents the landlord's actual loss caused by the tenant's failure to comply with it covenants. However, the application of the post termination rule, makes the approach less straight forward (although in most cases it will end with the same result).

The rental inducement plus consequential losses will be the correct measure of damages where:

- the hypothetical purchaser would have re-let the premises and
- a tenant would have insisted on a rent free period to represent the cost of repairs (as opposed to fitting out works).

The post termination evidence of the actual deal negotiated by the landlord with a new tenant is likely to be admissible as cogent evidence of how the market would have dealt with the premises at the term date.

In practice, unless the tenant is able to point to other tenants who would have taken a new lease for a lesser inducement (or no inducement at all), or alternatively, that because the market was flat, a rent inducement would have been required in any event, the correct measure of damages will be the amount of the rental inducement plus any consequential losses.

(iv) *Sublet premises*

If at the termination date the whole of the premises are sublet and the subtenant is seeking a renewal in accordance with the Landlord and Tenant Act 1954, this must be taken into account when considering diminution in value.

This principle is illustrated by *Family Management* v *Gray* [1980] 1 EGLR 46.

Facts

- T's lease of a building which comprised residential premises (upstairs) and commercial premises (ground floor) terminated in December 1974.
- T had sublet the commercial premises to two business subtenants.
- The subleases contained full repairing covenants.
- T and STs had failed to comply with repairing obligations.
- In June 1974 L had served unopposed section 25 notices on STs.
- New leases were granted to STs which included full repairing covenants.

Held — Court of Appeal

- The court must take into account the renewals; it cannot treat the valuation exercise as if the landlord was in possession.
- It was entitled to consider the post termination evidence that the renewals had gone ahead and that the sub-leases contained full repairing obligations.
- As the disrepair would be disregarded for the purposes of fixing a new rent, L had suffered no diminution to his reversion, and no actual monetary loss by virtue of the disrepair.

This decision in part explains why landlords tend not to serve a schedule of dilapidations where the tenant is seeking a renewal lease, or alternatively, where they inherit sub-tenants who are likely to renew. Consequently, if a corporate occupier is fortunate enough to find sub-tenants who are likely to stay after the expiry of the head lease (either as statutory tenants seeking renewal under the 1954 Act or on new leases agreed by the landlord) it may avoid a costly dilapidations claim. Conversely, if a landlord does serve a terminal schedule of dilapidations in such circumstances, the corporate occupier may well have a very good defence on the basis that the landlord has suffered no diminution in value of its reversion.

In *Family Management* the repairing obligations in the headlease matched those contained in the renewed sub-leases. However in *Crown Estate Commissioners* v *Town Investments Ltd* [1992] EGLR 61 this was not the case and the court accepted that the landlord might be able to show that its reversionary interest had diminished in value. Other possible answers to the *Family Management* defence are:

- the premises are only partially sub-let; the unoccupied parts could give rise to a diminution in the landlord's reversion by reason of the disrepair
- the hypothetical purchaser at the term date would consider the sub-tenants were unlikely to renew (and they did, in fact, not renew)
- the disrepair is serious and the sub-tenant's financial position is weak such that a hypothetical purchaser may discount its bid to take account of the likelihood that it will ultimately have to bear the cost of repair.

(v) *Consequential losses*

Consequential losses at common law are dealt with at section 14.3.1 (c) above. The extent to which consequential losses are recoverable after applying the section 18 cap will depend upon whether these expenses would be incurred by the hypothetical purchaser, and the purchase price reduced to reflect those losses.

Where the diminution in value is assessed by reference to the cost of works, and the landlord would have been in a position to relet immediately if the tenant had carried out remedial works, it is likely that the landlord will be able to recover loss of rent for the period of the works (including in some cases the delay caused by the need to prepare a specification, putting the works out to tender, and selecting a contractor — see 14.3.1 (c)(i) above). It is justified on the basis that a hypothetical purchaser would reduce the purchase price it was prepared to pay by an amount corresponding to the loss of rent.

In cases where the landlord does not intend to carry out the works (see 14.3.2(a)(ii) above), a loss of rent claim may be the principal head of loss. However, in a refurbishment case a tenant should be alive to the possibility that any consequential losses should not be recovered (or should be reduced) because a hypothetical purchaser would not have been able to relet during the refurbishment period in any event. It would not reduce its bid to account for two void periods (i.e. the refurbishment period and a repair period). It would only reduce its bid for one period, the refurbishment period, which would have arisen irrespective of the state of repair. This argument is equally applicable to void rates, service charges and security costs.

(b) The second limb

The second limb of section 18(1) of the 1927 Act involves a consideration of what the actual landlord intends to do with the premises, and not the rather more hypothetical analysis of what a notional purchaser would do. If the tenant is able to show that shortly after the termination date the premises will be demolished, or structural alteration made, the disrepair claim may be extinguished or substantially reduced. This is purely a supersession point.The tenant cannot be required to pay damages to the extent that any repairs it is required to carry out would be rendered valueless by any demolition or structural works a landlord intends to carry out.

Here are the key points of the second limb.

- The second limb requires a consideration of the landlord's actual intentions at the lease termination date.
- This must involve a definite decision by the landlord and an ability to implement that decision. To show a landlord was contemplating demolition or structural alterations on the term date will not be enough.
- To succeed, the tenant must also show that the landlord intended to carry out these works "at or shortly after the date of termination". This is a question of fact and can be a very difficult issue to resolve.
- Events after the termination date are irrelevant and unlikely to be admissible at court. Consequently, a landlord cannot rely upon the fact that the premises are still standing at the date of the court hearing as evidence that it did not have the necessary intention at the termination date. Landlords can and do change their minds. The key issue is what a landlord's intentions were on the termination date.

A corporate occupier is unlikely to come across the opportunity to rely upon the second limb of section 18. This is primarily because the tenant has the burden of proving what the landlord intends. Most institutional landlords will keep their cards close to their chest and may not have decided what to do with the premises by the termination date. Nevertheless, when faced with a dilapidations claim it is always sensible to ask the landlord in open correspondence what its plans are. It is also worth contacting the local planning officer to see if any planning applications have been made and to identify whether the site is ripe for redevelopment. If the landlord misrepresents the position in open correspondence, and this can be shown to be the case later, it may provide the tenant with the basis for setting aside any settlement. If the landlord is evasive, it may be possible to obtain a court order for pre-action disclosure.

(c) Breaches where section 18 does not apply

It is common for a tenant's surveyor or lawyer to blithely rely on section 18 to reduce or extinguish a claim. A landlord and its advisors should remind the tenant that section 18 only applies to repairing obligations, and refer to the other breaches of covenant contained in the schedule of dilapidations such as periodic decoration, reinstatement and statutory requirements.

That is not the end of the matter however. If the tenant can show that it is not reasonable for the landlord to carry out the works, then a court may award diminution in value instead of cost of work (see *Ruxley Electronics and Construction Ltd* v *Forsyth* [1996] AC 344).

14.4 Dilapidations claims during the term

It is unusual for a landlord to pursue a dilapidations claim mid term. This is primarily because a landlord is not usually concerned with the state of the premises mid term provided that the rent is being paid regularly and the state of the premises has not deteriorated to the extent that it affects the value of the freehold reversion. It is also because there are significant legal hurdles to overcome which tend to dissuade a landlord from pursuing a claim in all but cases involving extreme breaches.

However, there are traps for the unwary corporate occupier, and some landlords will attempt to obtain an advantage or financial windfall. For example, if a landlord has recently sold the freehold reversion, the new landlord may instruct a building surveyor to carry out an inspection and then serve the schedule on the tenant and require works to be carried out. The corporate occupier should analyse such requests carefully and robustly resist any request that falls foul of the Leasehold Property (Repairs) Act 1938.

Where a landlord wishes to take action mid term against a tenant who is in breach of its repairing obligation it has 4 potential remedies:

- forfeiture
- damages
- enter and do the works and sue for the cost
- specific performance

14.4.1 Forfeiture/damages — Leasehold Property (Repairs) Act 1938

The 1938 Act was brought in principally to counter speculators who purchased freehold reversions (at a reduced price to reflect the fact that the property was the subject of a long lease) and then served a schedule of dilapidations which the tenant could not comply with for the purpose of obtaining a windfall.

The 1938 Act applies to commercial and residential (but not agricultural) leases granted for seven years or more and with three years or more of the term unexpired.

The 1938 Act prevents a landlord from forfeiting a lease, or suing for damages for breach of repairing covenant unless and until it has gone through the following procedure:

1. the landlord must serve a section 146 notice on the tenant which contains certain prescribed information
2. in particular, the section 146 notice must contain a statement to the effect that the tenant is entitled within a period of 28 days to serve on the landlord a counter notice claiming the benefit of the 1938 Act
3. the section 146 notice must also provide a name and address for the service of the counter notice
4. if the tenant within the 28 day notice period serves a valid counter notice then the landlord will not be permitted to issue proceedings for forfeiture, or peaceably re-enter, or issue proceedings for damages without first obtaining permission from the court.

If the tenant fails to serve a counter notice within the 28 day period the 1938 Act protection ceases to apply (in relation to the breaches identified in the section 146 notice) and the landlord will be free to pursue its remedies. It is therefore of critical importance that a corporate occupier which is served with a section 146 notice mid term instructs its professional team to serve a counter-notice claiming the benefit of the 1938 Act immediately. This is a potential pitfall for corporate occupiers. It is extremely important that it has an efficient procedure for ensuring that any notices received (usually at its registered office) are promptly sent to the appropriate person. Failure to do so could result in a corporate occupier losing its protection, and as a consequence, incurring significant expenditure carrying out repairs it could otherwise have avoided.

If the tenant serves a valid counter notice, the landlord must issue an application in the county court, and permission will only be granted if:

- the landlord proves that one or more of the grounds set out in section 1(5) of the 1938 Act applies and

- the court in exercising its discretion is satisfied that permission ought to be granted to the landlord to allow it to take steps to forfeit the lease or issue proceedings for damages.

The following grounds are contained in section 1(5).

(a) The immediate remedying of the breach is necessary to prevent a substantial diminution in value of the landlord's reversion, or the value of that reversion has already been substantially diminished. Clearly expert valuation evidence will be needed to establish and defend this ground. The reversion must be valued subject to the lease. It is therefore a very difficult ground to prove when there are many years until lease expiry. The hypothetical purchaser will have the benefit of the repairing obligations in the lease, and at rent review, there is likely to be an assumption that the tenant has complied with its repairing obligations. Therefore, it may be difficult to show *any* diminution in value, let alone a substantial diminution.
(b) Immediate repair is necessary to comply with any statute, law, court order or local authority requirement.
(c) Immediate repair is necessary to protect any other occupier of the building.
(d) The cost of immediate repair would be relatively small in comparison with the much greater expense if the work is postponed. A very good example where the landlord might establish this ground is where dry rot has been found in the premises.
(e) Special circumstances exist which make it just and equitable that permission should be granted.

In most cases, the service of a schedule of dilapidations mid term is a "try on" by the landlord and should be resisted by the tenant. It is unlikely that leave will be obtained unless there is very serious disrepair and clear evidence of diminution in the landlord's reversion. It is not unusual for a landlord to serve a schedule, and then abandon its claim on receipt of a valid counter notice. If that is the case, the corporate occupier will be in strong position to resist paying the landlord's statutory costs under section 146(3), which can only be granted if it has made an application to the court. However, the position may be different if the lease contains a contractual costs provision.

The main difficulty which a landlord faces when seeking damages during the term is to show that there has been a diminution in the value of its reversionary interest. In general terms, the longer the unexpired term of the lease, the more difficult it will be. In practice, it is unlikely that a corporate occupier will encounter such a claim. It is much more likely to be faced with a claim for damages at the end of its lease.

14.4.2 Enter and carry out the works

Most modern leases will contain what is commonly referred to in the property world as a *Jervis* v *Harris* clause (named after the case which considered its legal effect). While this clause may take a number of forms, in essence such a clause will provide:

- a right for the landlord to inspect the premises
- that the landlord may serve written notice on the tenant:
  - identifying any breaches of the repairing covenant
  - requiring the tenant to remedy the breaches within a prescribed period (often two months)

- a right for the landlord to enter the premises and carry out works to remedy the items referred to in the notice if the tenant fails to do so within the specified period, and
- a right to recover the cost of works on demand as a contractual debt (as opposed to a claim for damages for breach of repairing covenant).

The importance of such a clause was illustrated in *Jervis* v *Harris* [1996] 1 EGLR 78 where the Court of Appeal held that a landlord was not required to seek the court's permission before bringing a claim based on such a clause. The 1938 Act does not apply to a claim in debt; it only applies to a claim for damages arising out of an alleged breach of a tenant's covenant to repair.

There is no doubt that a *Jervis* v *Harris* clause can provide a landlord with a neat way of avoiding the difficulties of the 1938 Act, and also a way of avoiding the need to establish that there has been a diminution in value of the landlord's reversion. However, there are some significant practical and legal drawbacks from this remedy:

- The courts will construe such clauses restrictively (see *Amsprop Trading Ltd* v *Harris Distribution Ltd* [1997] 2 EGLR 78). The notice must be drafted carefully and the schedule of defects must be detailed and comprehensive. The landlord will only be permitted to recover the cost of those items which it has carried out and are identified in the notice.
- If the landlord carries out repairs which are ultimately found to be unnecessary or which were not identified in the notice, the landlord may be sued for trespass and the cost of those repairs may be irrecoverable.
- The principal disadvantage is that it requires the landlord to expend its own money up front and run the risk it is unable to recover all or some of the expenditure in subsequent court proceedings. There may be arguments as to whether some or all of the items were in fact in disrepair, or whether a cheaper mode of repair would have been sufficient. There may also be arguments about whether the costs are reasonable.
- The major practical drawback is gaining access if the property is occupied by a trading tenant. If the works are extensive, and carrying out those works would disrupt the tenant's use of the property for the duration of the works, a substantial claim for damages (based in trespass) may be brought by the tenant. If a tenant refuses to co-operate, the landlord may be able to obtain an injunction requiring the tenant to allow access. However, this could mean that the procedure requires two separate and costly court actions, first, to obtain access to carry out the works, and second, to recover the cost of the works carried out.
- This remedy will be more attractive to a landlord where the premises are empty (eg where they are surplus). The risks of a counterclaim are significantly reduced and access may be less contentious. However, there will still be practical difficulties for the landlord. There may be problems obtaining an electricity supply and water.

In short, relying on a *Jervis* v *Harris* clause can be a very risky business and there can be no guarantee that a landlord will recover some or indeed any of its outlay. On the contrary, if the tenant is in occupation, and the landlord gets it wrong, it may be unable to recover its outlay and be liable for substantial damages and legal costs.

There is one circumstance where a landlord may be able to use this remedy to great effect. If the premises are empty (eg where the tenant has left before the termination of its lease) a landlord could serve a notice and carry out works before lease expiry. This would have the effect of the landlord being able to avoid any section 18 arguments and have the premises in good condition at lease expiry to facilitate early re-letting. It may still leave arguments about whether certain items were or were not in repair, and the method of repair used, but it would remove one of the more problematic restrictions in terminal dilapidations claims.

14.4.3 Specific performance

The court has a discretionary power to order that a party comply with a contractual obligation. This is known as specific performance and is, in effect, an injunction. Failure to comply with an order for specific performance will be a contempt of court punishable by imprisonment or a fine or both.

For many years it was considered that a landlord, as a matter of law, was not entitled to an order compelling a tenant to comply with its repairing or redecorating covenants (see *Hill* v *Barclay* (1810) 16 Ves Jr 402). However, the decision in *Rainbow Estates Ltd* v *Tokenhold Ltd* [1998] 2 EGLR 34 suggests that an order for specific performance of a tenant's repairing obligation can be made in appropriate cases. The judge in that case did however suggest that it would be a rare case in which such an order would be made. The judge indicated that it would not normally be appropriate where there was a modern commercial lease which contained a forfeiture clause and/or a landlord's right to enter the premises and carry out necessary works. In deciding whether an order for specific performance should be granted, a court will consider, among other things, whether the 1938 Act would have applied if the landlord had sought to forfeit or claim damages as an alternative remedy.

However, it is suggested that there are circumstances where it would be appropriate for an order for specific performance to be granted. Take for example the following facts:

- small scale investor landlord
- substantial tenant occupier (eg a well known retailer) of shop premises
- leaky roof causing damage to both the leased premises and the adjoining premises
- tenant occupier refuses (or fails) to carry out works
- landlord does not want to forfeit (as it will lose a good covenant)
- landlord does not want to exercise *Jervis* v *Harris* clause because it is unwilling (or perhaps unable) to pay for the cost up front and risk not recovering the cost of works (or incurring significant irrecoverable costs) in litigation.

All things being equal, it is considered that the landlord in the above scenario would have a good arguable case for obtaining an order for specific performance of the tenant's repairing obligations.

If a corporate occupier (in its capacity as a reluctant landlord in a legacy situation) is considering applying for an order, it is important that it take prompt action. Delay could result in the court not exercising its discretion in the tenant's favour.

15 Practical Approach to Dilapidations

15.1 Planning

One of the most important aspects of dilapidations claims for a corporate occupier is planning. This will involve an awareness of when operational leases and legacy portfolio subleases are due to expire. This awareness will allow a CREM to develop strategies for properties well in advance of lease expiry dates. The nature of the portfolios dealt with by a CREM will vary and with it the dilapidations risk profile. An office based company that only takes suites in existing office buildings will have a different risk profile to an engineering firm that occupies a large number of stand alone industrial units. However, in the case of most substantial buildings, at lease three years before lease expiry a CREM needs to start the process of considering what is likely to happen on expiry and whether a significant dilapidations claim is likely to follow.

Having reached a decision that a lease is not going to be renewed, a corporate occupier should assemble a professional team to deal with the property. Depending on the size of the premises and any likely claim, that team should probably include a building surveyor, an experienced property litigation lawyer, a valuer and a co-ordinator or project manager. The need to appoint a valuer will be driven by the scale, nature and complexity of the likely dilapidations claim, but if a valuation expert is likely to be required, it is worthwhile identifying who will act at an early stage as there are only a small number who are likely to have sufficient skill and relevant experience to fulfil the role. The need for a co-ordinator or project manager may seem unusual because this is often a role undertaken by a CREM in house. However, in more substantial claims there is considerable merit in employing an external co-ordinator with experience of running litigation work to ensure that focus is maintained on the correct issues and to ensure a measure of objectivity. This role will suit a surveyor with a broad experience of building surveying and valuation issues together with an ability to present the litigation issues to a corporate occupier in terms of a cost benefit analysis at the key decision making points in the process.

15.2 Litigation costs

15.2.1 General principle — winner takes all

A party will not be able to recover the costs of litigating or defending a claim unless there is a court order dealing with costs. Costs orders can either be agreed by the parties (in settling a claim) or awarded by the court at the end of an application or trial.

The costs of a claim are in the discretion of the judge. The starting point is that a court will usually order that the loser should pay the winner's costs. However, that general principle will not always apply. The Civil Procedure Rules (CPR) set out a number of factors which a judge should take into account. These include:

- the conduct of the parties before or during the proceedings:
  - did the parties comply with the protocol (see 15.3.2 (b))?
  - was it reasonable for a party to pursue or contest a particular issue?
  - did the landlord exaggerate its claim?
- whether a party has been successful on some points, even if it has lost overall
- whether there was an open offer, a calderbank offer (ie marked "without prejudice save as to costs" and considered by the judge when judgment has been given) or an offer which complies with CPR Part 36.

Under the CPR the courts can make partial costs orders. A landlord who exaggerates its claim is likely to be penalised by a judge when making a costs order (see *Firle Investments Ltd* v *Datapoint International Ltd* [2001] EWCA Civ 1106). Equally, if a party loses on a part of its case which was very expensive and time consuming, it is likely to be penalised for taking that issue. The court could penalise a winning landlord by either reducing the costs which the landlord is entitled to (eg by reference to a particular issue, by proportionate reduction of the overall costs or by disallowing costs for a particular period of the litigation), or in extreme cases, order the landlord to pay the tenant's costs in relation to a particular issue or period of the litigation.

15.2.2 Part 36 offers

CPR Part 36 provides a means by which both the landlord and the tenant can seek to protect themselves on costs. A Part 36 offer made by one or both of the parties is a matter which the judge cannot take into account when considering whether there is liability, and if so, the amount of damages payable by the tenant, but it is a matter that can be drawn to the judge's attention, and which s/he must take into account, when considering what costs order should be made.

Before 6 April 2007, if proceedings were issued the defendant had to put its money where its mouth was and make a payment of money into court to protect its position. However, the rules have now changed abolishing payments into court and substituting rules enabling a defendant to protect its position on costs by merely making a Part 36 offer.

(a) Part 36 offer by the tenant

A tenant can make a Part 36 offer before or after court proceedings are issued. There are certain technical requirements that must be satisfied:

- the offer must be in writing, state that it is intended to have the consequences of CPR Part 36, and be signed by the tenant or its legal representative
- it must contain an offer to settle for a specified single sum of money, and state whether it relates to the whole claim or part, whether it takes into account any counterclaim or not, and whether it includes interest

- it must specify a period of not less than 21 days within which the tenant offers to pay the landlord's reasonable costs of the proceedings (up to the date of notice of acceptance of the offer)
- the offer must state that if accepted, the tenant will pay the amount within 14 days following the date of acceptance
- the tenant cannot withdraw or change the terms of the offer within the specified period. However, if the offer is not accepted by the landlord within that period, the tenant is then permitted to withdraw or change the terms of the offer.

If the landlord wishes to accept a Part 36 offer, it must serve written notice of acceptance on the tenant. There is no formal notice as such and a letter confirming that the landlord accepts the offer contained in the Part 36 offer will suffice. If the landlord accepts the offer within the period specified in the Part 36 offer, it will be entitled to its reasonable costs to be assessed if not agreed up to the date of the acceptance of the offer. If the landlord accepts a Part 36 offer after the specified period, unless the parties agree the appropriate liability for costs, the court will make a costs order.

If the landlord does not accept a Part 36 offer and fails to recover more than the sum offered at trial, unless a judge considers it is unjust to do so, he/she will order the landlord to pay:

- the tenant's costs from the expiry of the period specified in the Part 36 offer (which must be not less than 21 days) and
- interest on those costs.

However, the landlord will usually be entitled to an order for its costs up to the specified expiry date. It follows that in a case where the tenant accepts some liability, unless and until it makes an open, Part 36 or without prejudice save as to costs offer, it will in effect be paying both parties' costs. This again reinforces the need for the tenant to evaluate the landlord's claim quickly. In short, the sooner that the tenant makes a Part 36 offer, the sooner that it will have some protection on costs. By the same token, the more realistic the offer is, the more pressure that will be placed on the landlord to accept, and if it does not accept the offer, the greater the protection the tenant will have on costs in the subsequent proceedings.

(b) Part 36 offer by landlord

Part 36 also offers a landlord a means of putting the tenant under pressure on costs. A landlord can make a Part 36 offer (both pre and post proceedings). In broad terms the same rules apply (although a landlord's Part 36 offer will indicate a sum which the *landlord* is prepared to settle for and will require the payment of its reasonable costs).

If the tenant does not accept a landlord's Part 36 offer, and at trial the landlord recovers damages which are the same or exceed the figure in its Part 36 offer, then, unless it considers it unjust to do so, the court will penalise the tenant further by awarding:

- interest on the whole or part of the damages at a rate of up to 10% above base rate for some or all of the time after the expiry of the specified period in the Part 36 offer and
- costs on an indemnity basis (ie which results in a higher proportion of actual costs being recovered) and
- interest on costs up to 10%.

Whether a judge will or will not make such dramatic orders will depend on the circumstances of the case. However, it goes without saying that the threat of such an order is a very powerful inducement to settle the case quickly. Every landlord should consider making a Part 36 offer at an early stage. If pitched carefully, this will put the tenant under substantial pressure and is likely, in most cases, to result in early settlement of a case.

15.2.3 Irrecoverable costs

A party which is contemplating resolving a dilapidations dispute (or indeed any dispute) by litigation should be aware of one critical fact. The winner, even if it obtains an order that the loser pay its costs, will not recover *all* of its actual costs. In most cases the victor is only likely to recover between 50% and 60% of its actual legal costs. Occasionally, a court may award 65%–70%.

If a claim is not resolved before trial, the costs of litigating it could be very substantial indeed and sometimes out of all proportion to the amounts at stake. Increasingly, only disputes involving very significant amounts of money can justify the luxury of a full court trial. In the majority of disputes, the irrecoverable costs of litigation (ie the amount by which the winning party's actual costs exceed the amount the loser is ordered to pay) will greatly reduce the amount of damages that are recovered. In some lower value cases the irrecoverable costs may well exceed the damages awarded by the court.

By way of example, if a case involves one building surveyor on each side, the landlord's costs for a three day trial are unlikely to be less than £150,000 (2007 costs) and could be a good deal more. The tenant's costs are likely to be of a similar order. If successful, as indicated above, the winning party is unlikely to be awarded more than 50%–60% of its actual professional costs at an assessment hearing (a hearing after the trial to determine how much the loser should actually pay to the winner). In that example:

- if the landlord succeeds, at best it may only recover £90,000 costs. In a worst case scenario it may only recover £75,000. It will therefore be out of pocket to the tune of £40,000–£75,000 even if it wins. This is often overlooked. It is a strange concept of winning when the victor immediately loses £75,000
- if the landlord loses, in addition to its own costs (which it must meet in full), it is likely to be ordered to pay 50%–60% of the tenant's costs of defending the action. In the example used, the landlord's total costs could be in the region of £240,000.

Clearly the example above assumes that settlement before trial has not been possible, neither party has made a Part 36 offer and a full trial is required. In most cases this will not be the case. However it does illustrate an important aspect of the cost benefit analysis which both parties should consider at all stages of the process, particularly when considering an offer or whether to go to court to see if it can achieve a better result. It is particularly important to carry out a cost benefit analysis in small value cases. The costs involved in litigating a dilapidations dispute will be broadly the same in a low value or high value claims.

15.3 A practical approach to terminal dilapidations claims

A corporate occupier is traditionally a tenant and will often be defending a terminal dilapidations claim. However, in a legacy situation, where it has been left tied into leases with many years to run and which are surplus to requirements, it may sub-let and therefore be both a tenant and landlord (of the subtenant) at the same time. It may find itself in a tripartite dispute, with the superior landlord pursuing a head claim against it, and it in turn pursuing a sub-claim against the subtenant. Alternatively, if the sub-lease ends many years before the headlease, the corporate occupier may have a stand alone claim against the sub-tenant. In the latter case the corporate occupier will want to ensure the repairs are carried out quickly so that the premises can be re-let as soon as possible to provide an income to offset against its head rent liability.

In view of the fact corporate occupiers can be either landlords or tenants (and in some cases both), this part of the chapter suggests a practical approach to a terminal dilapidations claim from both perspectives.

15.3.1 The tenant's decision

Often the first stage of a terminal dilapidations claim is viewed as being the service of a schedule of dilapidations by the landlord. However there are strong commercial reasons why this should not necessarily be the case. A dilapidations claim can represent a significant element of the total accommodation cost (viewed over the whole lease term) and an early evaluation of a potential claim and the implementation of a sensible strategy can result in substantial savings. At any rate, ignoring a potential claim on the basis that it might never happen is rarely a successful tactic in practice. It can mean that the tenant loses control, and then has to react to the landlord's actions and timescales. The business decision about whether to renew or leave is generally made at least 12 months before expiry and the dilapidations should be assessed before that as part of the overall decision making process.

If it appears likely that a corporate occupier will face a substantial claim then it may be cost efficient to evaluate the claim and any potential defences carefully with its professional advisors at that stage, and devise a strategy that will minimise its financial outlay.

The preliminary steps which the tenant should carry out include:

- obtaining an initial assessment by a building surveyor of:
  - the works necessary
  - how long remedial works will take
  - the likely cost of remedy.

- discrete enquires of the local planning department:
  - has a planning application been made?
  - is the site in the local plan for development?
  - is it ripe for redevelopment?

- obtaining an initial view of a valuer as to:
  - what a hypothetical purchaser would be likely to do with the premises
  - how a purchaser/incoming tenant would view, and/or be affected by, the disrepair
  - if the tenant had complied with its covenants, how readily could the landlord have re-let? What would be the usual void period and what if any inducements would the landlord have had to make to secure a tenant?
  - whether there is the potential to run a section 18 defence or whether this would be likely to be a cost of works case.

- obtaining advice on any legal issues that arise.

Once the initial enquires have been completed there are three possible options to consider:

(a) Do the works

Where the preliminary advice and enquiries suggest that there is likely to be liability, the tenant has much to gain by resolving the dispute quickly, and often delay simply increases the amount which the tenant has to pay (which in many cases may include paying a proportion of the landlord's professional fees in addition to its own). The most effective way of resolving the dispute in such circumstances may be for the tenant to carry out the works before the lease expires.

The advantages of this option are set out below.

- If all the works are done, to the required standard, it will avoid:
  - a consequential losses claim (eg loss of rent, service charges, rates; security costs; interest on finance to carry out works). These sums can be significant particularly where the works are likely to take many months to complete and/or where the rent is substantial
  - professional fees involved in a dilapidations dispute (eg building surveyors; valuers; structural and mechanical engineers and lawyers)
  - wasted management time and the hassle of dealing with a dispute.

- The tenant has control of the dispute. It can choose the contractors it uses and has an interest in obtaining the best price. It may be able to arrange for the works to be carried out at a lower price than the landlord would. If there are acceptable alternative ways of repairing the premises, the tenant can chose the cheaper option, where as the landlord is likely to carry out the more expensive repair (see 14.2.1 (b)(iv) above).

- The tenant may be able to recover VAT on works in circumstances where the landlord is not able to do so resulting in an immediate 17.5% reduction in the tenant's liability.

- The tenant can be satisfied that the works are done. In a damages claim where the landlord has not carried out the works by the time of the court hearing, there is no guarantee (or indeed requirement) that the landlord will apply the damages awarded towards carrying out remedial works. If the landlord simply pockets the money, there is no come back for the tenant.

However, carrying out the works before lease expiry is not without potential problems. The most common drawbacks are set out below.

- It may involve the cost and inconvenience of vacating the premises a number of months before lease expiry to allow the works to be carried out. If the tenant is moving its business to alternative premises then it may be problematic to arrange an earlier occupation date, or alternatively it may mean that the tenant is paying rent twice (ie on the new premises which it occupies and on the old premises which it is repairing and which is not generating any income). Clearly the cost of the rent on the new premises needs to be factored into the total cost of doing the works when compared with other options. However, if the premises are surplus and already vacant or if the tenant is downsizing and does not require alternative accommodation, then this may not be a factor.

- It may result in the lost opportunity to raise a substantial defence and/or negotiate a better cash deal.

- There is no guarantee that it will avoid a legal dispute. The landlord may argue that the tenant has failed to carry out works which it should have carried out, or alternatively, it may argue that the works have been carried out to a poor standard or with poor quality materials. That said, it may be able to pass any poor workmanship issues on to the contractors, and if the bulk of the works have been carried out, then the risks that a landlord will want to embark upon an expensive dilapidations claim to recover modest damages may be relatively low. This should not be an issue if the building surveyor has properly specified the works and then supervised the contractor.

If a tenant adopts this strategy, it should make every effort to try and obtain an agreed schedule of works before carrying out those works to avoid a future dispute about the extent and nature of the necessary remedial works.

(b) Quick cash settlement

It may be possible and appropriate to negotiate an early settlement (before the lease expires) based on the payment of a cash sum in lieu of the tenant carrying out any works. This will require the landlord to cooperate and accelerate its decision making process.

The advantages of this approach from the tenant's perspective are:

- it will avoid irrecoverable professional costs of a dispute and possible court proceedings
- it will avoid the possibility of consequential losses such as a substantial loss of rent claim
- if the tenant needs to relocate, it will avoid the need to pay double rent (ie for both the new premises and the old premises) for the duration of the works
- it will achieve certainty
- if the premises are vacant and surplus it can be an aid to an early surrender deal.

The disadvantages of a quick cash settlement may include:

- by taking the initiative, it may encourage the landlord to bring a claim, or at a higher level, in circumstances where it might not have done, or with lower expectations
- the tenant may miss the opportunity to raise a defence which may have substantially reduced (or in some cases extinguished) any damages
- the tenant may lose the opportunity to negotiate a claim down and may pay over the odds (although this has to be balanced against the potential costs of a protracted dispute, and the risks of the landlord bringing a substantial claim for consequential losses).

If a cash settlement is agreed, it is important that the agreement is documented clearly and made in full and final settlement of all claims arising out of the lease. If the premises are empty, the best means of settlement is by way of a formal deed of surrender, with the premises being handed back to the landlord on execution of the deed, and payment of the money.

If the lease still has a number of months to run, the landlord may be concerned that further breaches may occur between the date of any agreement and lease expiry. It may also be concerned not to compromise other potential claims (eg non payment of rent). In those circumstances it may be necessary to expressly provide in the settlement document that the settlement relates solely to dilapidations (and the safest course is to expressly refer to all relevant clauses) and modify the repairing obligation by reference to an attached schedule of condition for the remaining period of the lease. While this is not ideal from the tenant's point of view, unless there is a dramatic deterioration in the condition of the premises in the remaining short period of the term, it is unlikely that a subsequent claim will be forthcoming and the benefits of an earlier settlement will be enjoyed. This approach is most certainly not the best approach in a conditional break clause case (see *Legal & General Assurance Society Ltd* v *Expeditors International (UK) Ltd* [2007] EWCA Civ 7 for an example of where such an agreement resulted in a trip to the Court of Appeal for the parties).

The quick cash settlement option can be used in conjunction with a threat to carry out the works. A useful tactic in some cases is for the tenant to write a without prejudice letter to the landlord indicating that it is prepared to pay cash in lieu of carrying out the works, but that if no agreement is reached on the amount by a given date, the tenant will go ahead and do the works. By giving the landlord a deadline, in most cases, this will cause the landlord to reach a decision quicker than it would otherwise do. For many landlords of older premises who face the need to upgrade their premises, a cash contribution towards the improvement costs is preferable to having to fight a dilapidations claim or see the tenant carry out works that will be wasted.

(c) Wait and see

The final option is to hand the premises back to the landlord in its existing condition and await any formal claim that may be brought. This will allow the tenant to evaluate the landlord's claim and raise any relevant defences at the appropriate time. It may be that by waiting a tenant may avoid a claim altogether. If the landlord decides to redevelop or carry out extensive refurbishment, it may not bring a claim, or alternatively, that claim may be substantially lower than any amount the tenant would have spent on repairs before lease expiry.

A tenant may want to adopt the wait and see option where the building only has a shell or site value, or where it seems likely the premises will require refurbishment, modernisation or modification. Some good examples are set out below.

- An old warehouse or factory building that has reached the end of its useful life and where the site is likely to be redeveloped for residential or modern industrial use.
- A 1960s/1970s office building which requires substantial modernisation works to attract a new tenant (eg new internal configuration; air conditioning; new heating system; IT cabling for computer networks).
- A 1960s industrial warehouse that is of a size and nature which is unlikely to be let without being subdivided into smaller units, and which is only likely to attract a tenant in the secondary warehouse market. Such a tenant will not be concerned with the more cosmetic works which are likely to be contained in a thorough landlord's schedule of dilapidations.
- A retail unit where the incoming tenant is likely to carry out its own substantial refit.

In the above examples, this policy may result in the landlord's claim being significantly reduced (in a supercession case) or extinguished entirely (in a demolition and redevelopment case).

15.3.2 The landlord's schedule

From the landlord's perspective, the first step in a dilapidations claim is usually taken by the landlord serving a terminal schedule of dilapidations on the tenant either in the last few months of the term, or after the lease has expired.

(a) When should the schedule be served?

Whether the schedule is served before or after lease expiry will be determined by the circumstances, and in particular, whether the landlord has immediate plans for the property or not.

The downside from serving a schedule a number of months before termination is that a further inspection will be necessary (at additional expense) at or shortly after the termination date to document any change in condition since the date of service of the first schedule. If the landlord wants to push the claim forward quickly and/or carry out the works to enable it to be relet, there may be merit in serving an early schedule. If it is clear that the tenant is not going to do the works, there is duty on the landlord to mitigate its loss and this may extend to commencing preparations for carrying out the works before the lease expiry (see *Drummond* v *S&U Stores Ltd* [1981] 1 EGLR 42). Failure to do so may extinguish or reduce a consequential losses claim (eg for loss of rent).

In most cases however, the schedule will be served a short time after the lease has expired, and unless the tenant has initiated the process as suggested above, this will commence the dispute process.

(b) Dilapidations pre-action protocol/RICS Dilapidations Guidance Note

Terminal dilapidation claims can be extremely expensive disputes to litigate (see the example given at 15.2.3 above). At their worst, they can involve two building surveyors laboriously giving evidence in court on hundreds of individual items of disrepair, followed by two valuers testifying whether the disrepair has or has not caused a diminution in value to the landlord's reversion. The position can be complicated further if there is a need to call other experts (eg mechanical or structural engineers) or still worse, the dispute involves a landlord, tenant and sub-tenant in circumstances where the tenant is seeking to pass on the landlord's claim (in whole or in part) to the subtenant.

The complex nature of these disputes and the expense of litigating them led the Property Litigation Association to introduce a Pre-action Protocol for Dilapidations (the protocol). The most recent edition was released on 14 September 2006. The PLA worked very closely with the RICS which issued the Dilapidations Guidance Note (guidance note) to assist building surveyors advising in dilapidations cases. The Guidance Note is an extremely helpful and concise checklist of how a dilapidations claim should be approached by a building surveyor and is essential reading before advising in a dilapidations claim.

While the protocol has not yet been formally adopted under the Civil Procedure Rules, it should be followed by all parties so far as possible, and failure to do so could result in the court making an adverse costs order against the offending party. By the same token, the guidance note is likely to be considered as best practice by a court. If a building surveyor departs from the good practice recommended by the guidance note, he should only do so for good reasons and may later be required to explain to the court why he did so.

The principal purpose of the protocol and the guidance note is to promote an exchange of relevant information between the parties before proceedings are issued but within the framework of a realistic timetable. If both sides comply with the spirit of the protocol, and their surveyors apply the best practice contained in the guidance note, the majority of claims will settle without the need for litigation. There will always be cases where certain items give rise to reasonable difference of opinion, and if there is enough at stake, those cases may have to be determined by a judge. However, if the above approach has been followed, the live issues ought to have been cut down to a small number of items, thereby reducing the length of the trial and the overall litigation costs.

(c) The building surveyor's role

A surveyor is likely to be instructed by a landlord to carry out a number of roles. He will inspect the premises and prepare a schedule of dilapidations. He will then probably be required to negotiate a settlement, and if he fails, he will be required to prepare an expert report and give oral evidence at trial. A surveyor clearly has a duty to his client. However, if the matter becomes litigious, his primary duty is to assist the court in its quest to reach the correct conclusion however damaging his evidence may be to his client's case.

(i) *Pre-inspection*

The building surveyor will need to carry out a number of pre-inspection preparations. These are likely to include considering:

- the lease and identifying the appropriate covenants (it is surprising how many schedules are served where the surveyor has not seen a copy of the lease or the relevant covenants)
- any plans verifying the extent of the premises and identifying any common parts or parts retained by the landlord
- any licences for alterations together with plans and specifications
- letters of consent that might have been granted for alterations which have not translated into licences
- any schedules of condition
- any notices served by the landlord (eg section 25 notice — if the notice opposes the grant of a new tenancy on the grounds of redevelopment this may be highly relevant to its claim and in particular section 18 of the 1927 Act)
- any current planning consents or statutory notices
- the landlord's intentions for the premises (if these have been formulated)
- the guidance note (which acts like a checklist).

(ii) *Inspection*

It is very important that an inspection is carried out as soon after the termination of the lease as possible to avoid any allegation by the outgoing tenant that the premises deteriorated after lease expiry and therefore fall outside the tenant's liability. For example, if the landlord claims for missing roof tiles or broken glass and the inspection took place many months after lease expiry, the landlord may not be able to show that these disrepairs occurred during the lease (particularly if there have been recent storms). The onus of proof is on the landlord. The only exception to this is in a case where a tenant is seeking to establish that it has complied with conditions precedent to the operation of break clause. In those circumstances it would be for the tenant to prove that it had handed back the premises in repair.

The surveyor may make notes during the course of inspection. These must be kept and may be needed in subsequent proceedings. In addition to looking at the premises subject to the lease, the surveyor should also consider the nature, age and general standard of repair of the premises in the locality. As discussed above, this will impact upon the standard of repair required for the premises.

On the basis that a picture paints a thousand words, it is essential that the surveyor photographs the items that appear in his schedule. This is especially important if the landlord intends to carry out the works because it will not be possible for the surveyor (or indeed the judge at a site visit) to check the original condition of the premises once the works have been completed. It may be necessary to involve other experts (e.g. electrical or mechanical engineers).

(iii) *Schedule of dilapidations*

There is now a universal format for a schedule of dilapidations and that format should be used. An example is attached to the guidance note and a further example can be found at precedent 15.4.1 at the end of this chapter.

A surveyor when preparing a schedule should be realistic and in relation to each item needs to be prepared to justify its inclusion (in court if needs be). There is a practice among some surveyors of serving a schedule for the purpose of negotiation which includes items which are likely to be negotiated out in the usual horse trade. However, this ignores the surveyor's duty to the court and often, contrary to what the surveyor may think, it is not in his client's interest. Exaggeration or understatement on the part of either surveyor is likely to result in the dispute taking longer to resolve than if both parties took a realistic approach. The longer the dispute goes on, the more money it costs the parties.

There are two other very real risks with the landlord serving an exaggerated schedule. If the matter goes to trial, the surveyor will have to give evidence. He is likely to be cross-examined by the tenant's barrister about the items contained in the original schedule and if he is not able to give a good reason for any significant changes (such as abandoning a large number of items or a number of large items), his credibility will be damaged and the judge may attach little or no weight to his evidence on the items that remain. The second risk is that the court may disapprove of a landlord which has over egged the pudding by penalising it when awarding costs. This is dealt with in more detail at 15.1.1 above.

The correct approach therefore is for a surveyor to ask himself the question: "can I justify this item in the witness-box?" If the answer to that question is no, the item should not be included. Paragraph 3.2.1 of the guidance note summarises the position as follows:

> he must give objective unbiased evidence. It follows that his evidence would be the same whether he acts for the tenant or the landlord, or appears as a single joint expert.

The schedule should be priced. Initially this is usually done by reference to current building cost information data and recognised price books, and by the surveyor using their experience in that locality including speaking to contractors.

(iv) *Scott schedule*

A Scott Schedule is the landlord's schedule of dilapidations in tabular form with a section that enables the tenant to respond, on an item-by-item basis, to the landlord's case. A copy of a Scott Schedule can be found at precedent 15.4.2.

The Scott Schedule allows the tenant, in respect of each item, to set out its case with comments. There are four possible responses:

1. the item and costing are both agreed
2. the item is agreed, but not the costing
3. the costing is agreed, but not the item and
4. the item and the costing are both disputed.

Strictly speaking it is not necessary for a Scott Schedule to be prepared until there is a court order to that effect. Furthermore it is unlikely that the parties will be able to recover the cost of preparing one unless ordered to do so by the court. However, the guidance note observes that it is highly desirable that each party's case be contained in the one document and suggests that the landlord's surveyor should send both a schedule of dilapidations and a Scott Schedule to encourage the tenant's surveyor to respond with its comments in the Scott Schedule. While this may seem excessive, it is likely to save considerable time (and money) as the dispute progresses.

(v) *Claim summary*

It is important that a separate claim summary is prepared. This will usually be prepared by the surveyor, occasionally with input and advice from a solicitor. If possible this should be contained on one side of A4 paper, and should break down the landlord's claim in a logical way. In due course the tenant's solicitor should provide a claim summary in response, using the same headings.

The protocol now requires the claim summary to contain a written endorsement by the landlord's surveyor(s) stating that the overall figure claimed is a fair assessment of the landlord's loss having regard to the principles laid down in the Guidance Note, the common law principles of how that loss should be calculated and section 18(1) of the 1927 Act. Clearly the purpose of this is to cause building surveyors (and valuers) to think carefully before signing a claim off with this endorsement. If they sign off an exaggerated claim summary, they are likely to have very little credibility at trial.

A precedent claim summary with the necessary endorsement can be found at precedent 15.4.3.

(d) The solicitor's role — protocol letter

In most cases it will be prudent for the schedule of dilapidations, Scott Schedule and claim summary to be served by a solicitor. In the past a lawyer's involvement may have been minimal involving little more than serving a schedule with a one line covering letter. That is no longer acceptable. The solicitor must ensure that all the information and documentation required by the protocol is provided to the tenant.

An example of a protocol letter can be found at precedent 15.4.4. Generally it should include the following:

- a summary of the facts on which the claim is based
- copies of the document containing the tenant's obligations eg lease; schedule of condition and licence for alterations
- a summary of the tenant obligations
- schedule of dilapidations (and possibly Scott Schedule)
- an indication whether it is the landlord's present intention to carry out some or all of the works
- a section 18 valuation if it would be reasonable in all the circumstances to do so (see below)
- a summary of claim with the appropriate endorsement by the landlord's surveyors
- any other documentary evidence relied upon by the landlord (eg invoices if some or all of the works have been carried out)

- confirmation of the landlord's VAT status in respect of the premises
- a date by which the tenant must respond (which will depend on what is reasonable but will usually be 56 days).

The landlord's solicitor will have to advise whether it is appropriate or necessary to serve a section 18 valuation or not. Paragraph 10 of the protocol addresses this issue. There are four possibilities.

1. If the landlord has carried out all the necessary remedial works, it is not usually necessary to provide a formal diminution valuation. It can base its claim on the cost of works as evidenced by paid invoices. It will only be required to provide one if "in all the circumstances it would be reasonable to do so".
2. If the landlord has carried out some, but not all of the works, it will not usually be required to provide a section 18 valuation in respect of the works which it has actually carried out (unless it is reasonable to do so).
3. If the landlord does not intend to carry out some or all of the works, then it should provide a formal diminution valuation unless, in all the circumstances, it would be reasonable not to do so.
4. If the landlord has not carried out the works, but intends to do so, the Protocol letter must state precisely what works it intends to carry out, when the works will be done and what steps it has taken towards carrying out those works. The protocol states that a diminution valuation should be provided unless it would be reasonable not to do so.

This aspect of the protocol is perhaps the most contentious. Landlords have traditionally proceeded on the basis of cost of works unless and until a tenant raised a section 18 issue. This is now more difficult in view of the surveyor endorsement requirement. However, there is a world of difference between a valuer's initial view and a fully detailed diminution valuation (supported with comparables) and in our view it is unlikely that many landlords will go to the expense of serving a full report (as opposed to setting out its position) at the protocol letter stage, particularly in low value cases. A landlord is only likely to deal with this issue in any detail once the point has been seriously put in issue by the tenant.

(e) Loss of rent

If the landlord is seeking to recover loss of rent, it should anticipate the tenant's likely response, namely, that even if the premises had been handed back in repair, the state of the market is such that no tenant would have been found within the period of the works. In some circumstances, the answer to this may be to market the premises. If there are potential tenants who are interested, but put off by the dilapidations (or who seek a substantial rent reduction or rent free period because of the condition of the premises), this may assist the landlord in addressing the tenant's argument.

15.3.3 The tenant's response

In most cases the tenant's first involvement will be when it receives a protocol letter and/or a schedule of dilapidations. As indicated in 15.2.1, there is a strong argument in some cases for a tenant evaluation at an earlier stage. If the schedule is served before the termination date, the tenant should consider the points made at 15.2.1 and decide whether to do some or all of the works, make a cash offer or defend the claim (in whole or part).

If the schedule is served at a time when there is no possibility of the tenant carrying out the works (a tenant has no legal right to carry out repairs to the property once the lease expires), the tenant should evaluate the claim quickly and then determine its strategy.

Once served with a schedule of dilapidations, the tenant should appoint a good building surveyor (preferably with court experience) familiar with handling dilapidations disputes. The points made in 15.3.2 (c)(i)–(ii) apply equally to the tenant's surveyor as they do to the landlord's surveyor.

The tenant's surveyor should:

- Meet the landlord's building surveyor at the premises (the protocol suggests within 28 days for a typical case) to review the landlord's schedule. This gives the tenant's surveyor an opportunity to understand the landlord's claim and avoid unnecessary misunderstandings. The meeting is without prejudice and therefore what is said at that meeting cannot be referred to in court unless both sides agree.
- Consider any defences that may be available to the tenant:
  - Is the item capable of falling within the covenant relied upon?
  - If the item alleged is disrepair, has there been any deterioration in the state of the premises (see 14.2.1(b)(i))?
  - Do the works required by the landlord go beyond the scope of the covenant or the standard of repair envisaged by the parties (14.2.1 (b)(ii) and (iii))?
  - Section 18 — what does the landlord intend to do with the premises? What do the planners say about future use? What would a hypothetical purchaser do with the premises? Would any of the items be rendered valueless by works which the landlord (and/or the hypothetical purchaser) would carry out in any event?
- Obtain any additional expert advice that may be necessary eg from M&E consultants.
- Prepare a (draft) counter schedule. The protocol suggests that in a typical case the tenant's response should be provided within 56 days of receipt of the protocol letter. In the first instance a draft should be prepared. This should ideally be contained in a Scott Schedule so that it is possible to see the tenant's response to each and every allegation quickly and easily.
- Take legal advice on any legal issues that arise.

When preparing the counter schedule, it is important that the tenant's building surveyor takes a realistic approach. He/she should consider in relation to smaller items whether it is worth arguing about. Quite often, the parties are better served in the long run by surveyors who are prepared to make sensible concessions than by those who are considered tough negotiators and are reluctant to make such concessions. An unrealistic approach may lead to increased costs, a slower settlement or indeed no settlement and the risk of the court's criticism.

Once a draft counter schedule has been prepared (but before service) it is often sensible to have a meeting with solicitors (or if there is a lot at stake with counsel) to consider the overall position, any legal or surveying issues, and determine what strategy is likely to lead to the most cost effective disposal of the dispute.

The possibilities are:

- defend the claim
- make a realistic offer of settlement (open or Part 36)
- a combination of the two.

In most cases there will be some parts of the claim which are unanswerable, and for which the tenant accepts liability. It may not, however, accept the remedy suggested by the landlord or the cost of carrying out the remedial works. It may also dispute the remainder of the claim. In those circumstances it is very important that the tenant makes a quick and realistic offer of settlement. Unless and until it settles the claim, or makes an open or Part 36 offer to protect its position on costs, it will be at risk of being ordered to pay a significant proportion of the landlord's professional costs of the dispute as well as its own (see 15.3 above).

There is no right or wrong approach to negotiating a settlement. The appropriate strategy will largely depend on the circumstances, and in particular, the nature of the landlord and its approach to such disputes. In some cases an informal discussion of settlement figures between surveyors can be helpful to get a feel for what the landlord is expecting (or at least its opening position). In other cases landlords will keep their cards close to their chest. If, however, the informal approach is not appropriate or does not succeed, in most cases, the most effective tactic is to analyse the claim and the tenant's position very carefully, and put forward a realistic Part 36 offer. A copy of a precedent Part 36 offer by a tenant can be found at 15.4.5. This can be bolstered by an assertion that the tenant has considered the position very carefully, has made one offer which won't be increased, and that if it is not accepted, will be defended in court leaving the landlord to run the risk of paying the tenant's costs, from 21 days after the Part 36 offer. This can be a very effective tactic.

If the tenant is considering running a section 18 defence, it is perfectly sensible to serve the counter schedule without prejudice to the tenant's position on section 18. Indeed, it is highly desirable that the building surveyors try and reach agreement on the items contained in the schedule. The more items they agree, the easier the task of the valuation expert. Whether the tenant serves a full section 18 report or not is a key issue. If the landlord has not, the tenant may feel that to serve a full report at this stage would be premature and give the landlord a tactical edge. If the matter were to be litigated there would, in all likelihood, be a court direction requiring simultaneous exchange of section 18 reports. If the tenant serves a full report at this stage it will allow the landlord's report to be crafted to respond to the points raised in tenant's valuation report. The protocol, suggests that the tenant should disclose its diminution valuation. The answer is probably to set out the diminution valuation in correspondence (or by attaching a separate document) rather than serving a full report (including comparables) which can be both expensive and a very time consuming exercise.

It is not unusual for a tenant's valuation expert to arrive at a diminution in value figure which while not extinguishing the claim, is lower than both the landlord and the tenant's figure for cost of works and consequential costs. There may be a good case in these circumstances for making an open offer. For example, in a case where a landlord intends to carry out the works and it values the cost of works (plus consequential costs) at £3m, and where the tenant's figure (subject to section 18) is £1.5m, but its valuer considers the diminution in value figure to be between £500,000 and £750,000, an open offer at £500,000 coupled with an offer to make an immediate payment on account, could be a powerful tactic. A copy of a precedent paragraph containing an open offer can be found at 15.4.6.

The benefits from this approach are two fold. If the case goes to court, the landlord may seek to recover the interest charges on the money it had to borrow to finance the works. A payment on account would at least reduce that potential liability. Second, this will be a letter which the judge reads. It demonstrates that the tenant has approached the dispute in a sensible way. It also leaves open the possibility of making a Part 36 offer at a higher level if it is thought that the section 18 case is not strong, or is likely to result in damages at a higher level.

15.3.4 The landlord's riposte

If the landlord receives a detailed response to its claim, clearly this will need to be considered carefully by its building surveyor, valuer and lawyer. The landlord should address and respond to the points raised by the tenant.

If the tenant has raised a section 18 defence, one very effective tactic can be to put the works out to tender with a view to the landlord carrying out the works (see 14.3.2 (a) (i) above). In some cases this threat alone will result in a prompt settlement, particularly if the landlord indicates that it will have to borrow money to implement the works and will seek to recover the interest charges as a head of damages. If the landlord does implement its threat, it can then sue on the invoices and seek to recover the actual cost of works. This can put the landlord in a very strong position. It is difficult for a tenant because to succeed it will have to show either the landlord paid over the odds for the works, or alternatively, the landlord acted unreasonably in carrying out the works.

If the landlord intends to base its claim on invoiced works, it is very important that it ensures that the paperwork is both detailed and carefully preserved. In particular:

- all tender documents and responses should be kept
- all contracts with contractors should detail the works which are covered
- all invoices should provide a narrative identifying precisely and with sufficient detail the works which they cover
- all professional invoices should be detailed and preserved
- all bank statements showing interest charges should be kept.

If the landlord does not want to carry out the works, or go through the tender process, it should, at the very least, give a detailed and reasoned response to the tenant's section 18 calculation and set out its own calculation. The landlord should also ensure that the tenant has not attempted to rely upon section 18 for failure to carry out cyclical redecoration or reinstatement (see 14.3.2 (a) and (c)).

The tactic of tendering the works can also be effective if the cost of works is challenged. However, if the landlord has no intentions of carrying out the works it may not want to go the expense of this process. An alternative approach may be to instruct a qualified quantity surveyor to prepare detailed costings. However, while this will be more authoritative than the usual building surveyor's approach (ie using recognised price books), it does not carry as much weight as going to the market, and it will involve the landlord incurring extra fees.

Following the service of a protocol letter, a claim can stall (usually) because of delay on the part of the tenant, or become bogged down as the parties conduct trial by correspondence. The most effective strategy which a landlord can deploy in most cases is to:

- send an open letter before action and/or
- make a realistic Part 36 offer.

Copies of a precedent Part 36 letter can be found at 15.4.6. From a tenant's point of view, if there is likely to be liability, the issue of proceedings will merely result in a significant increase in costs for which the tenant may be liable. If the tenant fails to beat the Part 36 offer, there are significant cost penalties that may be imposed by the court (see 15.2.2 (b)). Often the double barrelled approach (ie a letter before action accompanied by a Part 36 letter) can be very effective. The Part 36 letter is very formal in its appearance and imposes a 21 day deadline. It refers to the consequences of failure to accept the offer. It tends to get the recipient's attention in a way that simple chasing letters do not. It requires the tenant to consider its position quickly and often involves the tenant taking legal advice.

15.3.5 Resolution of the dispute

The spectre of both parties slugging out a dispute in a court room should, in all but exceptional cases, be avoided at all costs. In most cases it will be in neither party's interest to go to court (except perhaps if there are a small number of discrete legal points which can be disposed of without the need for large amounts of oral evidence).

There are a number of alternatives to court proceedings which both parties would be wise to consider. The parties can seek to resolve their dilapidations dispute in a number of ways.

(a) Without prejudice meeting

Some disputes drag on because the parties fail to consider their case or the other side's case with sufficient care. The dispute may also stall because of an ignorance or misunderstanding of the other side's case. In some disputes a without prejudice meeting between the parties, their surveyors and/or in certain circumstances their lawyers may result in an early settlement, or at least in the resolution of some parts of the claim, or clarification of the extent of the dispute.

(b) Mediation

If the parties are not able to settle the dispute between themselves, they can seek assistance in brokering a deal from a qualified mediator. Mediations are becoming a very popular method of dispute resolution. They are flexible and vastly cheaper than going through a formal dispute resolution procedure such as court or arbitration.

The procedure can be determined by the parties. It usually takes place on neutral ground and involves an initial round table meeting with both parties given the opportunity to say a few words about their case. The parties then adjourn to separate rooms and the mediator carries out shuttle diplomacy. He/she will encourage the parties to consider various solutions. It is not the mediator's task to comment on the strength of each party's case, although he/she will do if invited to do so by the parties. If a solution is acceptable to both parties, the mediator will draw up a settlement agreement and the parties will sign up to the agreement there and then. It is therefore of vital importance that the representatives of the landlord and the tenant present at the mediation have authority to settle the dispute.

Importantly, all discussions that take place at the mediation are without prejudice and cannot be referred to in subsequent proceedings unless both parties agree. The courts are increasingly putting pressure on the parties to mediate their disputes, and dilapidations claims (particularly if there are a number of parties) do lend themselves to mediation.

(c) Independent expert determination

One of the cheapest and quickest solutions to a dilapidations dispute is for the parties to jointly instruct a building surveyor (or if appropriate a valuation expert) to determine the dispute. Expert determination is frequently used in rent review disputes. Indeed most commercial leases provide for any dispute about the new rent to be determined by a surveyor acting as an expert. Rarely does a lease contain a similar provision in respect of a dilapidations dispute. However, it can be no less effective.

Often parties are keen to have their disputes adjudicated by a judge. This is probably driven by the misconception that a judge will arrive at the 'right answer'. This however, places too much faith in the court system. Dilapidations claims rarely go to trial. Therefore, there is no guarantee that a judge who hears the case will have any previous knowledge or experience of the legal or technical issues involved. This is particularly so in the county court where the judges deal with a very broad diet of civil disputes. This may be less of an issue in the specialist divisions of the High Court (eg the Technology and Construction Court).

It is entirely possible, therefore, that the parties will have a better quality of decision if the matter is dealt with by an experienced expert who practices in this field. The RICS runs a scheme to facilitate the adjudication of dilapidations disputes by expert determination, and has a panel of expert building surveyors and valuers.

The nature of the procedure will be determined by the contract which requires an expert determination of the dispute. This could be contained in the lease, or more likely in relation to dilapidations, be a separate agreement by the parties entered into after the dispute arises. The decision will be based on the expert's own knowledge and experience, and he/she is not bound to consider evidence or representations put forward by the parties (unless the agreement requires it).

It is unlikely that there will be a trial in the usual way. In most cases the expert determines the case from his own investigations and written submissions from the parties. This can make the process both quicker and very substantially cheaper than court proceedings or arbitration. The downside is that there is no appeal from a decision of an expert, although it may be possible for the parties to sue the expert for negligence in certain circumstances.

Expert determination is in all likelihood only going to be used in relatively small claims. In such claims its speed and low cost make it an ideal way of resolving a dispute.

(d) Arbitration

Some leases require dilapidation disputes to be resolved by arbitration. In many ways arbitration is a lot like court proceedings. The arbitrator decides disputes after hearing evidence and oral representations from the parties. To that extent it can be just as expensive and time consuming as proceedings. However, the Arbitration Act 1996 has made the process more flexible, and the parties have the control to pick an arbitrator with the relevant expertise. There are only very limited grounds for appeal from an arbitrator's award.

(e) Court proceedings — poker

If the parties are unable to agree an alternative to court proceedings, they should make every effort to cut down the areas of dispute as much as possible. This will help to minimise the time and cost involved. In some larger cases where the schedule of dilapidations is challenged and a section 18 defence is being relied upon, there may be an argument for having a split trial with the liability/cost of works issues being tried first, and with the diminution in value aspect being dealt with at a later date.

Some of the most difficult cases to deal with are the low value claims. They often involve the same complexity as higher value claims, but they don't easily lend themselves to litigation. Often the irrecoverable costs of going to full trial will outweigh the amounts at stake. As a rule of thumb, it is doubtful that a cost benefit analysis will support taking a claim to full trial that is likely to result in an award of less than £100,000. Often, in the legacy situation, a corporate occupier is trying to recover much lower sums. In such cases, there is an element of poker to be played, with some landlords threatening (but probably not being prepared to actually issue) court proceedings, and some tenants taking the view that if they do nothing it is unlikely that the landlord will issue proceedings and therefore they can avoid paying anything. In such cases, the bluff and counter bluff of poker will play a part.

15.4 Precedents

15.4.1 Schedule of Dilapidations

Item No	Clause No	Breach complained of	Remedial Works Required	Landlord's Costing
		EXTERNAL		
1	2.3	Water ingress identified in Room 19	Strip off and replace defective sections of felt	£3,000

15.4.2 Scott Schedule

Item No	Clause No	Breach	Remedial Works	Tenant's Comments	Landlord's Response	Landlord's Item		Tenant's Item	
						Landlord's Costing	Tenant's Costing	Landlords' Costing	Tenant's Costing
					EXTERNAL				
1	2.3	Water ingress identified in Room 19	Strip off and replace defective sections of felt	Agreed but patch repair to felt will suffice	Disputed This would not constitute an adequate repair	£3,000	£1,500	£600	£50

15.4.3 Claim summary

ASSESSMENT OF LOSS

Unit 1, Bonkers Hill Industrial Estate, Middlesex

Costs of Works	£1,500,000
Contract, administration and CDM fees @ 6.5%	£97,500
Total cost of building work (excluding VAT)	**£1,597,500**
Loss of rent — 8 months @ £50,000 per month	£400,000
Loss of insurance rent — 8 months @ £3,750	£30,000
Security costs — 2 months @ £4,000	£8,000
Surveyor's fees for preparing the schedule	£3,000
Solicitor's fees for serving the schedule	£500
Landlord's fees for negotiating claim @ 5%	£101,950
TOTAL COSTS	**£2,140,950**

This assessment of loss has been prepared after having regard to the principles laid down in the Royal Institution of Chartered Surveyors Guidance Note on Dilapidations (4th Edition), the common law principles of how loss should be calculated, and in relation to repairing covenants, Section 18(1) of the Landlord and Tenant Act 1927.

We certify that the overall figure claimed is a fair assessment of the landlord's loss

Signed	Signed ....
Pullin Yorchain MRICS...........	Dodge. E. Valuer BSc (Hons) MRICS
(Landlord's building surveyor)	(Landlord's valuer)

Dated: 1 April 2007

15.4.4 Protocol letter

Dear Sirs

Unit 1, Bonkers Hill Industrial Estate, Middlesex ("the premises")

We act for Grab Hall Ucan Limited your former landlord of the above premises under an underlease ("the Underlease") dated 24 April 1996 originally between (1) Grab Hall Ucan Limited and (2) Diddley Squat Limited.

Diddley Squat Limited's underlease expired on 1 March 2007. We enclose a copy of the Underlease for ease of reference.

Undertenant's Repairing Obligations
Diddley Squat Limited covenanted with Grab Hall Ucan Limited in the following terms:

By clause 2.3	"... at all times during the term well and substantially to repair cleanse maintain amend and keep the inside and outside of the buildings on the demised premises ..."
By clause 2.4	"... in every third year (calculated from the 1st March 1998) and in the last year of the term howsoever determined to paint or otherwise treat as the same may require all the outside wood metal and cement work of the buildings on the demised premises (including all new buildings which may at any time during the said term be erected on and all additions made to the demised premises) usually requiring to be painted or otherwise treated with three coats of good paint or other suitable materials of the best quality ..."
By clause 2.5	"... in every fifth year and in the last year of the said term howsoever determined to paint or otherwise treat as the case may require all the inside wood and iron work usually or requiring to be painted or otherwise treated of the buildings then on the demised premises and all additions and fixtures thereto with three coats of good paint or other suitable materials of the best quality ..."
By clause 2.6	"At the expiration or sooner determination of the said term quietly to yield up the demised premises (the building then thereon being so painted treated washed repaired cleansed and kept and if necessary rebuilt as aforesaid) together with all additions and improvements thereto and all fixtures which during the said term may be affixed or fastened to or upon the demised premises (Tenant's or trade fixtures belonging to the Tenant only excepted)"

By clause 2.7 "At all times during the said term to observe and comply in all respects with the provisions and requirements of any and every enactment ..."

Breach of Covenant
Diddley Squat Limited in breach of covenant, failed to yield up the premises in accordance with the repairing and redecoration obligations contained in the Underlease as evidenced by the terminal schedule of dilapidations dated 1 April 2007 prepared by Pullin Yorchain MRICS.

We enclose a copy of Mr Yorchain's schedule of dilapidations by way of service.

Cost of Works
In the circumstances, our client is entitled to claim damages for breach of covenant which we estimate, at this stage, to be £2,140,950. We enclose an assessment of loss in accordance with paragraph 4.8 of the Property Litigation Association Pre-Action Protocol (14 September 2006 edition).

In addition, our client is also entitled to claim interest on these amounts.

Other Information Required by the Protocol
It is Grab Hall Ucan Limited's intention to carry out all the necessary works as soon as possible with a view to re-letting the premises, and the works have already been put out to tender. We enclose a copy of the specification of works for your information.

Our client is able to claim back VAT on the works and therefore is not seeking to recover VAT.

We believe that this letter contains all of the required information for the purposes of the Property Litigation Association Pre-Action Protocol.

In the circumstances we look forward to receiving an acknowledgment of this claim letter within 7 days and look forward to receiving a substantive response within 56 days from receipt of this letter.

You may wish to take legal advice on this letter.

We look forward to hearing from you.

Yours faithfully

15.4.5 Part 36 offer by tenant

Dear Sirs

PART 36 OFFER
Unit 1, Bonkers Hill Industrial Estate, Middlesex ("the premises")

We write further to your claim for damages for breach of covenant in respect of the premises.

This letter constitutes a defendant's offer for the purpose of CPR Part 36.4 and is intended to have the consequences of Part 36.

Diddley Squat Limited is prepared to settle this claim in accordance with CPR Part 36 for the sum of £1,500,000 (inclusive of interest) up to the last date for acceptance. This offer is in respect of the whole of your claim.

It is a requirement of CPR Part 36.2(c) that a period of not less than 21 days be specified within which the defendant will be liable for the claimant's reasonable costs in accordance with CPR Part 36.10 if the offer is accepted. The specified period for the purposes of this offer is 21 days from the date on which you receive this letter.

If you wish to accept the offer, written notice of acceptance must be served. If you serve a notice of acceptance within the 21 day specified period our client will pay your reasonable costs, to be assessed if not agreed, up to the date of the service of the notice of acceptance.

If, after expiry of 21 days, you seek to accept the offer, the parties must either agree the liability for costs or alternatively, the court will make an order as to costs.

If you do not accept this offer and the sum awarded to you at trial is less than the amount offered, we will seek to recover our costs together with interest on those costs from the expiry of the 21 day period specified.

Yours faithfully,

15.4.6 Open offer by tenant

[*set out detailed section 18 defence*]

... Notwithstanding the points which we have made above, Didley Squat Limited recognises that Grab Hall Ucan Limited has suffered a diminution in value of £500,000. To that end, subject to Grab Hall Ucan Limited first disclosing a copy of the relevant building contract (together with any side agreements) which evidence its intention to carry out the works, Diddley Squat Limited is prepared to pay to Grab Hall Ucan Limited £500,000 within 14 days of receiving copies of that documentation, to be disbursed towards the building works which Grab Hall Ucan Limited contends have been occasioned by breaches of repairing covenant.

Whilst it is Diddley Squat Limited's case that Grab Hall Ucan Limited's claim is capped at £500,000 under Section 18(1) of the 1927 Act, it is content for both parties to preserve their respective positions and for Grab Hall Ucan Limited to accept this payment in part settlement of its claim. In other words, it is not being suggested that this sum is offered in full and final settlement of all claims.

15.4.7 Part 36 offer by landlord

Dear Sirs

PART 36 OFFER
Crumbling Mill, Seenbetter Days Industrial Estate, Herts

We write further to our open letter of today's date.

Unless and until a sensible offer is made, you are potentially liable for all our client's legal and professional costs involved with this dilapidations claim, in addition to your own. If this matter is not resolved quickly, our client's legal costs are likely to escalate rapidly when proceedings are prepared and issued.

In an attempt to avoid costly legal proceedings, we are instructed to offer to settle this claim in accordance with CPR Part 36 for the sum of [insert settlement figure] (inclusive of interest up to the last date for acceptance) plus legal costs. This offer is in respect of the whole of our client's claim.

It is a requirement of CPR Part 36.2(c) that a period of not less than 21 days be specified within which the defendant will be liable for the claimant's reasonable costs in accordance with CPR Part 36.10 if the offer is accepted. The specified period for the purposes of this offer is 21 days from the date on which you receive this letter.

If you wish to accept this offer, written notice of acceptance must be served. If you serve notice of acceptance within the 21 day specified period, in addition to the sum of [insert settlement figure], you are liable to pay our client's reasonable costs, to be assessed if not agreed, up to the date of service of the notice of acceptance.

If, after expiry of 21 days, you seek to accept the offer, the parties must either agree the liability for costs or alternatively, the court will make an order as to costs.

If you do not accept this offer and the sum awarded to our client at trial is the same or exceeds this offer, our client will seek to recover its costs on an indemnity basis from the expiry of the 21 day specified period, together with interest on those costs and on any damages awarded at 10% above base rate in accordance with CPR Part 36.14(3).

This letter constitutes a claimants offer for the purpose of CPR Part 36 and is intended to have the consequences of Part 36.

We look forward to hearing from you within 21 days.

Yours faithfully

Wright Bolshy LLP

Appendix
Statutory Material

Law of Distress (Amendment) Act 1908

6. To avoid distress.

In cases where the rent of the immediate tenant of the superior landlord is in arrear it shall be lawful for such superior landlord to serve upon any under tenant or lodger a notice (by registered post addressed to such under tenant or lodger upon the premises) stating the amount of such arrears of rent, and requiring all future payments of rent, whether the same has already accrued due or not, by such under tenant or lodger to be made direct to the superior landlord giving such notice until such arrears shall have been duly paid, and such notice shall operate to transfer to the superior landlord the right to recover, receive, and give a discharge for such rent.

***

Law of Property Act 1925

146 Restrictions on and relief against forfeiture of leases and underleases.

(1) A right of re-entry or forfeiture under any proviso or stipulation in a lease for a breach of any covenant or condition in the lease shall not be enforceable, by action or otherwise, unless and until the lessor serves on the lessee a notice—

(a) specifying the particular breach complained of; and
(b) if the breach is capable of remedy, requiring the lessee to remedy the breach;
and
(c) in any case, requiring the lessee to make compensation in money for the breach;

and the lessee fails, within a reasonable time thereafter, to remedy the breach, if it is capable of remedy, and to make reasonable compensation in money, to the satisfaction of the lessor, for the breach.

(2) Where a lessor is proceeding, by action or otherwise, to enforce such a right of re-entry or forfeiture, the lessee may, in the lessor's action, if any, or in any action brought by himself, apply to the court for relief; and the court may grant or refuse relief, as the court, having regard to the

proceedings and conduct of the parties under the foregoing provisions of this section, and to all the other circumstances, thinks fit; and in case of relief may grant it on such terms, if any, as to costs, expenses, damages, compensation, penalty, or otherwise, including the granting of an injunction to restrain any like breach in the future, as the court, in the circumstances of each case, thinks fit.

(3) A lessor shall be entitled to recover as a debt due to him from a lessee, and in addition to damages (if any), all reasonable costs and expenses properly incurred by the lessor in the employment of a solicitor and surveyor or valuer, or otherwise, in reference to any breach giving rise to a right of re-entry or forfeiture which, at the request of the lessee, is waived by the lessor, or from which the lessee is relieved, under the provisions of this Act.

(4) Where a lessor is proceeding by action or otherwise to enforce a right of re-entry or forfeiture under any covenant, proviso, or stipulation in a lease, or for non-payment of rent, the court may, on application by any person claiming as under-lessee any estate or interest in the property comprised in the lease or any part thereof, either in the lessor's action (if any) or in any action brought by such person for that purpose, make an order vesting, for the whole term of the lease or any less term, the property comprised in the lease or any part thereof in any person entitled as under-lessee to any estate or interest in such property upon such conditions as to execution of any deed or other document, payment of rent, costs, expenses, damages, compensation, giving security, or otherwise, as the court in the circumstances of each case may think fit, but in no case shall any such under-lessee be entitled to require a lease to be granted to him for any longer term than he had under his original sub-lease.

(5) For the purposes of this section—

(a) "Lease" includes an original or derivative under-lease; also an agreement for a lease where the lessee has become entitled to have his lease granted; also a grant at a fee farm rent, or securing a rent by condition;
(b) "Lessee" includes an original or derivative under-lessee, and the persons deriving title under a lessee; also a grantee under any such grant as aforesaid and the persons deriving title under him;
(c) "Lessor" includes an original or derivative under-lessor, and the persons deriving title under a lessor; also a person making such grant as aforesaid and the persons deriving title under him;
(d) "Under-lease" includes an agreement for an under-lease where the under-lessee has become entitled to have his underlease granted;
(e) "Under-lessee" includes any person deriving title under an under-lessee.

(6) This section applies although the proviso or stipulation under which the right of re-entry or forfeiture accrues is inserted in the lease in pursuance of the directions of any Act of Parliament.

(7) For the purposes of this section a lease limited to continue as long only as the lessee abstains from committing a breach of covenant shall be and take effect as a lease to continue for any longer term for which it could subsist, but determinable by a proviso for re-entry on such a breach.

(8) This section does not extend—

(i) To a covenant or condition against assigning, underletting, parting with the possession, or disposing of the land leased where the breach occurred before the commencement of this Act; or
(ii) In the case of a mining lease, to a covenant or condition for allowing the lessor to have access to or inspect books, accounts, records, weighing machines or other things, or to enter or inspect the mine or the workings thereof.

(9) This section does not apply to a condition for forfeiture on the bankruptcy of the lessee or on taking in execution of the lessee's interest if contained in a lease of-

(a) Agricultural or pastoral land;
(b) Mines or minerals
(c) A house used or intended to be used as a public-house or beershop;
(d) A house let as a dwelling-house, with the use of any furniture, books, works of art, or other chattels not being in the nature of fixtures;
(e) Any property with respect to which the personal qualifications of the tenant are of importance for the preservation of the value or character of the property, or on the ground of neighbourhood to the lessor, or to any person holding under him.

(10) Where a condition of forfeiture on the bankruptcy of the lessee or on taking in execution of the lessee's interest is contained in any lease, other than a lease of any of the classes mentioned in the last sub-section, then—

(a) if the lessee's interest is sold within one year from the bankruptcy or taking in execution, this section applies to the forfeiture condition aforesaid;
(b) if the lessee's interest is not sold before the expiration of that year, this section only applies to the forfeiture condition aforesaid during the first year from the date of the bankruptcy or taking in execution.

(11) This section does not, save as otherwise mentioned, affect the law relating to re-entry or forfeiture or relief in case of non-payment of rent.

(12) This section has effect notwithstanding any stipulation to the contrary.

(13) The county court has jurisdiction under this section—

(a) . . .
(b) . . .

***

Landlord and Tenant Act 1927

1. Tenant's right to compensation for improvements.

(1) Subject to the provisions of this Part of this Act, a tenant of a holding to which this Part of this Act applies shall, if a claim for the purpose is made in the prescribed manner [and within the time limited by section forty-seven of the Landlord and Tenant Act, 1954] be entitled, at the termination of the tenancy, on quitting his holding, to be paid by his landlord compensation in respect of any improvement (including the erection of any building) on his holding made by him or his predecessors in title, not being a trade or other fixture which the tenant is by law entitled to remove, which at the termination of the tenancy adds to the letting value of the holding:

Provided that the sum to be paid as compensation for any improvement shall not exceed—

(a) the net addition to the value of the holding as a whole which may be determined to be the direct result of the improvement; or
(b) the reasonable cost of carrying out the improvement at the termination of the tenancy, subject to a deduction of an amount equal to the cost (if any) of putting the works constituting the improvement into a reasonable state of repair, except so far as such cost is covered by the liability of the tenant under any covenant or agreement as to the repair of the premises.

(2) In determining the amount of such net addition as aforesaid, regard shall be had to the purposes for which it is intended that the premises shall be used after the termination of the tenancy, and if it is shown that it is intended to demolish or to make structural alterations in the premises or any part thereof or to use the premises for a different purpose, regard shall be had to the effect of such demolition, alteration or change of user on the additional value attributable to the improvement, and to the length of time likely to elapse between the termination of the tenancy and the demolition, alteration or change of user.

(3) In the absence of agreement between the parties, all questions as to the right to compensation under this section, or as to the amount thereof, shall be determined by the tribunal hereinafter mentioned, and if the tribunal determines that, on account of the intention to demolish or alter or to change the user of the premises, no compensation or a reduced amount of compensation shall be paid, the tribunal may authorise a further application for compensation to be made by the tenant if effect is not given to the intention within such time as may be fixed by the tribunal.

2. Limitation on tenant's right to compensation in certain cases.

(1) A tenant shall not be entitled to compensation under this Part of this Act-

(a) in respect of any improvement made before the commencement of this Act; or
(b) in respect of any improvement made in pursuance of a statutory obligation, or of any improvement which the tenant or his predecessors in title were under an obligation to make in pursuance of a contract entered into, whether before or after the passing of this Act, for valuable consideration, including a building lease; or

(c) in respect of any improvement made less than three years before the termination of the tenancy; or

(d) if within two months after the making of the claim under section one, subsection (1), of this Act the landlord serves on the tenant notice that he is willing and able to grant to the tenant, or obtain the grant to him of, a renewal of the tenancy at such rent and for such term as, failing agreement, the tribunal may consider reasonable; and, where such a notice is so served and the tenant does not within one month from the service of the notice send to the landlord an acceptance in writing of the offer, the tenant shall be deemed to have declined the offer.

(2) Where an offer of the renewal of a tenancy by the landlord under this section is accepted by the tenant, the rent fixed by the tribunal shall be the rent which in the opinion of the tribunal a willing lessee other than the tenant would agree to give and a willing lessor would agree to accept for the premises, having regard to the terms of the lease, but irrespective of the value attributable to the improvement in respect of which compensation would have been payable.

(3) The tribunal in determining the compensation for an improvement shall in reduction of the tenant's claim take into consideration any benefits which the tenant or his predecessors in title may have received from the landlord or his predecessors in title in consideration expressly or impliedly of the improvement.

3. Landlord's right to object.

(1) Where a tenant of a holding to which this Part of this Act applies proposes to make an improvement on his holding, he shall serve on his landlord notice of his intention to make such improvement, together with a specification and plan showing the proposed improvement and the part of the existing premises affected thereby, and if the landlord, within three months after the service of the notice, serves on the tenant notice of objection, the tenant may, in the prescribed manner, apply to the tribunal, and the tribunal may, after ascertaining that notice of such intention has been served upon any superior landlords interested and after giving such persons an opportunity of being heard, if satisfied that the improvement—

(a) is of such a nature as to be calculated to add to the letting value of the holding at the termination of the tenancy; and

(b) is reasonable and suitable to the character thereof; and

(c) will not diminish the value of any other property belonging to the same landlord, or to any superior landlord from whom the immediate landlord of the tenant directly or indirectly holds;

and after making such modifications (if any) in the specification or plan as the tribunal thinks fit, or imposing such other conditions as the tribunal may think reasonable, certify in the prescribed manner that the improvement is a proper improvement:

Provided that, if the landlord proves that he has offered to execute the improvement himself in consideration of a reasonable increase of rent, or of such increase of rent as the tribunal may determine, the tribunal shall not give a certificate under this section unless it is subsequently shown to the satisfaction of the tribunal that the landlord has failed to carry out his undertaking.

(2) In considering whether the improvement is reasonable and suitable to the character of the holding, the tribunal shall have regard to any evidence brought before it by the landlord or any superior landlord (but not any other person) that the improvement is calculated to injure the amenity or convenience of the neighbourhood.

(3) The tenant shall, at the request of any superior landlord or at the request of the tribunal, supply such copies of the plans and specifications of the proposed improvement as may be required.

(4) Where no such notice of objection as aforesaid to a proposed improvement has been served within the time allowed by this section, or where the tribunal has certified an improvement to be a proper improvement, it shall be lawful for the tenant as against the immediate and any superior landlord to execute the improvement according to the plan and specification served on the landlord, or according to such plan and specification as modified by the tribunal or by agreement between the tenant and the landlord or landlords affected, anything in any lease of the premises to the contrary notwithstanding:

Provided that nothing in this subsection shall authorise a tenant to execute an improvement in contravention of any restriction created or imposed-

(a) for naval, military or air force purposes;
(b) for civil aviation purposes under the powers of the Air Navigation Act, 1920;
(c) for securing any rights of the public over the foreshore or bed of the sea.

(5) A tenant shall not be entitled to claim compensation under this Part of this Act in respect of any improvement unless he has, or his predecessors in title have, served notice of the proposal to make the improvement under this section, and (in case the landlord has served notice of objection thereto) the improvement has been certified by the tribunal to be a proper improvement and the tenant has complied with the conditions, if any, imposed by the tribunal, nor unless the improvement is completed within such time after the service on the landlord of the notice of the proposed improvement as may be agreed between the tenant and the landlord or may be fixed by the tribunal, and where proceedings have been taken before the tribunal, the tribunal may defer making any order as to costs until the expiration of the time so fixed for the completion of the improvement.

(6) Where a tenant has executed an improvement of which he has served notice in accordance with this section and with respect to which either no notice of objection has been served by the landlord or a certificate that it is a proper improvement has been obtained from the tribunal, the tenant may require the landlord to furnish to him a certificate that the improvement has been duly executed; and if the landlord refuses or fails within one month after the service of the requisition to do so, the tenant may apply to the tribunal who, if satisfied that the improvement has been duly executed, shall give a certificate to that effect.

Where the landlord furnishes such a certificate, the tenant shall be liable to pay any reasonable expenses incurred for the purpose by the landlord, and if any question arises as to the reasonableness of such expenses, it shall be determined by the tribunal.

18. Provisions as to covenants to repair.

(1) Damages for a breach of a covenant or agreement to keep or put premises in repair during the currency of a lease, or to leave or put premises in repair at the termination of a lease, whether such covenant or agreement is expressed or implied, and whether general or specific, shall in no case exceed the amount (if any) by which the value of the reversion (whether immediate or not) in the premises is diminished owing to the breach of such covenant or agreement as aforesaid; and in particular no damage shall be recovered for a breach of any such covenant or agreement to leave or put premises in repair at the termination of a lease, if it is shown that the premises, in whatever state of repair they might be, would at or shortly after the termination of the tenancy have been or be pulled down, or such structural alterations made therein as would render valueless the repairs covered by the covenant or agreement.

(2) A right of re-entry or forfeiture for a breach of any such covenant or agreement as aforesaid shall not be enforceable, by action or otherwise, unless the lessor proves that the fact that such a notice as is required by section one hundred and forty-six of the Law of Property Act 1925, had been served on the lessee was known either-

(a) to the lessee; or
(b) to an under-lessee holding under an under-lease which reserved a nominal reversion only to the lessee; or
(c) to the person who last paid the rent due under the lease either on his own behalf or as agent for the lessee or under-lessee;

and that a time reasonably sufficient to enable the repairs to be executed had elapsed since the time when the fact of the service of the notice came to the knowledge of any such person.

Where a notice has been sent by registered post addressed to a person at his last known place of abode in the United Kingdom, then, for the purposes of this subsection, that person shall be deemed, unless the contrary is proved, to have had knowledge of the fact that the notice had been served as from the time at which the letter would have been delivered in the ordinary course of post.

This subsection shall be construed as one with section one hundred and forty-six of the Law of Property Act 1925.

(3) This section applies whether the lease was created before or after the commencement of this Act.

19. Provisions as to covenants not to assign, etc, without licence or consent.

(1) In all leases whether made before or after the commencement of this Act containing a covenant condition or agreement against assigning, underletting, charging or parting with the possession of demised premises or any part thereof without licence or consent, such covenant condition or agreement shall, notwithstanding any express provision to the contrary, be deemed to be subject—

(a) to a proviso to the effect that such licence or consent is not to be unreasonably withheld, but this proviso does not preclude the right of the landlord to require payment of a reasonable sum in respect of any legal or other expenses incurred in connection with such licence or consent; and
(b) (if the lease is for more than forty years, and is made in consideration wholly or partially of the erection, or the substantial improvement, addition or alteration of buildings, and the lessor is not a Government department or local or public authority, or a statutory or public utility company) to a proviso to the effect that in the case of any assignment, under-letting, charging or parting with the possession (whether by the holders of the lease or any under-tenant whether immediate or not) effected more than seven years before the end of the term no consent or licence shall be required, if notice in writing of the transaction is given to the lessor within six months after the transaction is effected.

[(1A) Where the landlord and the tenant under a qualifying lease have entered into an agreement specifying for the purposes of this subsection—

(a) any circumstances in which the landlord may withhold his licence or consent to an assignment of the demised premises or any part of them, or
(b) any conditions subject to which any such licence or consent may be granted,

then the landlord—

(i) shall not be regarded as unreasonably withholding his licence or consent to any such assignment if he withholds it on the ground (and it is the case) that any such circumstances exist, and
(ii) if he gives any such licence or consent subject to any such conditions, shall not be regarded as giving it subject to unreasonable conditions;

and section 1 of the Landlord and Tenant Act 1988 (qualified duty to consent to assignment etc.) shall have effect subject to the provisions of this subsection.

(1B) Subsection (1A) of this section applies to such an agreement as is mentioned in that subsection—

(a) whether it is contained in the lease or not, and
(b) whether it is made at the time when the lease is granted or at any other time falling before the application for the landlord's licence or consent is made.

(1C) Subsection (1A) shall not, however, apply to any such agreement to the extent that any circumstances or conditions specified in it are framed by reference to any matter falling to be determined by the landlord or by any other person for the purposes of the agreement, unless under the terms of the agreement—

(a) that person's power to determine that matter is required to be exercised reasonably, or
(b) the tenant is given an unrestricted right to have any such determination reviewed by a person independent of both landlord and tenant whose identity is ascertainable by reference to the agreement,

and in the latter case the agreement provides for the determination made by any such independent person on the review to be conclusive as to the matter in question.

(1D) In its application to a qualifying lease, subsection (1)(b) of this section shall not have effect in relation to any assignment of the lease.

(1E) In subsections (1A) and (1D) of this section-

(a) "qualifying lease" means any lease which is a new tenancy for the purposes of section 1 of the Landlord and Tenant (Covenants) Act 1995 other than a residential lease, namely a lease by which a building or part of a building is let wholly or mainly as a single private residence; and
(b) references to assignment include parting with possession on assignment.]

(2) In all leases whether made before or after the commencement of this Act containing a covenant condition or agreement against the making of improvements without a licence or consent, such covenant condition or agreement shall be deemed, notwithstanding any express provision to the contrary, to be subject to a proviso that such licence or consent is not to be unreasonably withheld; but this proviso does not preclude the right to require as a condition of such licence or consent the payment of a reasonable sum in respect of any damage to or diminution in the value of the premises or any neighbouring premises belonging to the landlord, and of any legal or other expenses properly incurred in connection with such licence or consent nor, in the case of an improvement which does not add to the letting value of the holding, does it preclude the right to require as a condition of such licence or consent, where such a requirement would be reasonable, an undertaking on the part of the tenant to reinstate the premises in the condition in which they were before the improvement was executed.

(3) In all leases whether made before or after the commencement of this Act containing a covenant condition or agreement against the alteration of the user of the demised premises, without licence or consent, such covenant condition or agreement shall, if the alteration does not involve any structural alteration of the premises, be deemed, notwithstanding any express provision to the contrary, to be subject to a proviso that no fine or sum of money in the nature of a fine, whether by way of increase of rent or otherwise, shall be payable for or in respect of such licence or consent; but this proviso does not preclude the right of the landlord to require payment of a reasonable sum in respect of any damage to or diminution in the value of the premises or any neighbouring premises belonging to him and of any legal or other expenses incurred in connection with such licence or consent.

Where a dispute as to the reasonableness of any such sum has been determined by a court of competent jurisdiction, the landlord shall be bound to grant the licence or consent on payment of the sum so determined to be reasonable.

(4) This section shall not apply to leases of agricultural holdings within the meaning of the [Agricultural Holdings Act 1986] [which are leases in relation to which that Act applies, or to farm business tenancies within the meaning of the Agricultural Tenancies Act 1995], and paragraph (b) of subsection (1), subsection (2) and subsection (3) of this section shall not apply to mining leases.

23. Service of notices.

(1) Any notice, request, demand or other instrument under this Act shall be in writing and may be served on the person on whom it is to be served either personally, or by leaving it for him at his last known place of abode in England or Wales, or by sending it through the post in a registered letter addressed to him there, or, in the case of a local or public authority or a statutory or a public utility company, to the secretary or other proper officer at the principal office of such authority or company, and in the case of a notice to a landlord, the person on whom it is to be served shall include any agent of the landlord duly authorised in that behalf.

(2) Unless or until a tenant of a holding shall have received notice that the person theretofore entitled to the rents and profits of the holding (hereinafter referred to as "the original landlord") has ceased to be so entitled, and also notice of the name and address of the person who has become entitled to such rents and profits, any claim, notice, request, demand, or other instrument which the tenant shall serve upon or deliver to the original landlord shall be deemed to have been served upon or delivered to the landlord of such holding.

***

Leasehold Property (Repairs) Act 1938

1. Restriction on enforcement of repairing covenants in long leases of small houses.

(1) Where a lessor serves on a lessee under subsection (1) of section one hundred and forty-six of the Law of Property Act, 1925, a notice that relates to a breach of a covenant or agreement to keep or put in repair during the currency of the lease [all or any of the property comprised in the lease], and at the date of the service of the notice [three] years or more of the term of the lease remain unexpired, the lessee may within twenty-eight days from that date serve on the lessor a counter-notice to the effect that he claims the benefit of this Act.

(2) A right to damages for a breach of such a covenant as aforesaid shall not be enforceable by action commenced at any time at which [three] years or more of the term of the lease remain unexpired unless the lessor has served on the lessee not less than one month before the commencement of the action such a notice as is specified in subsection (1) of section one hundred and forty-six of the Law of Property Act, 1925, and where a notice is served under this subsection, the lessee may, within twenty-eight days from the date of the service thereof, serve on the lessor a counter-notice to the effect that he claims the benefit of this Act.

(3) Where a counter-notice is served by a lessee under this section, then, notwithstanding anything in any enactment or rule of law, no proceedings, by action or otherwise, shall be taken by the lessor for the enforcement of any right of re-entry or forfeiture under any proviso or stipulation in the lease for breach of the covenant or agreement in question, or for damages for breach thereof, otherwise than with the leave of the court.

(4) A notice served under subsection (1) of section one hundred and forty-six of the Law of Property Act, 1925, in the circumstances specified in subsection (1) of this section, and a notice served

under subsection (2) of this section shall not be valid unless it contains a statement, in characters not less conspicuous than those used in any other part of the notice, to the effect that the lessee is entitled under this Act to serve on the lessor a counter-notice claiming the benefit of this Act, and a statement in the like characters specifying the time within which, and the manner in which, under this Act a counter-notice may be served and specifying the name and address for service of the lessor.

(5) Leave for the purposes of this section shall not be given unless the lessor proves—

(a) that the immediate remedying of the breach in question is requisite for preventing substantial diminution in the value of his reversion, or that the value thereof has been substantially diminished by the breach;
(b) that the immediate remedying of the breach is required for giving effect in relation to the [premises] to the purposes of any enactment, or of any byelaw or other provision having effect under an enactment, [or for giving effect to any order of a court or requirement of any authority under any enactment or any such byelaw or other provision as aforesaid];
(c) in a case in which the lessee is not in occupation of the whole of the [premises as respects which the covenant or agreement is proposed to be enforced], that the immediate remedying of the breach is required in the interests of the occupier of [those premises] or of part thereof;
(d) that the breach can be immediately remedied at an expense that is relatively small in comparison with the much greater expense that would probably be occasioned by postponement of the necessary work; or
(e) special circumstances which in the opinion of the court, render it just and equitable that leave should be given.

(6) The court may, in granting or in refusing leave for the purposes of this section, impose such terms and conditions on the lessor or on the lessee as it may think fit.

2. Restriction on right to recover expenses of survey, etc.

A lessor on whom a counter-notice is served under the preceding section shall not be entitled to the benefit of subsection (3) of section one hundred and forty-six of the Law of Property Act, 1925, (which relates to costs and expenses incurred by a lessor in reference to breaches of covenant), so far as regards any costs or expenses incurred in reference to the breach in question, unless he makes an application for leave for the purposes of the preceding section, and on such an application the court shall have power to direct whether and to what extent the lessor is to be entitled to the benefit thereof.

3. Saving for obligation to repair on taking possession.

This Act shall not apply to a breach of a covenant or agreement in so far as it imposes on the lessee an obligation to put [premises] in repair that is to be performed upon the lessee taking possession of the premises or within a reasonable time thereafter.

***

Landlord and Tenant Act 1954

Part II

Security of Tenure for Business, Professional and other Tenants

Tenancies to which Part II applies

23 Tenancies to which Part II applies

(1) Subject to the provisions of this Act, this Part of this Act applies to any tenancy where the property comprised in the tenancy is or includes premises which are occupied by the tenant and are so occupied for the purposes of a business carried on by him or for those and other purposes.

(1A) Occupation or the carrying on of a business—

(a) by a company in which the tenant has a controlling interest; or
(b) where the tenant is a company, by a person with a controlling interest in the company,

shall be treated for the purposes of this section as equivalent to occupation or, as the case may be, the carrying on of a business by the tenant.

(1B) Accordingly references (however expressed) in this Part of this Act to the business of, or to use, occupation or enjoyment by, the tenant shall be construed as including references to the business of, or to use, occupation or enjoyment by, a company falling within subsection (1A)(a) above or a person falling within subsection (1A)(b) above.

(2) In this Part of this Act the expression "business" includes a trade, profession or employment and includes any activity carried on by a body of persons, whether corporate or unincorporate.

(3) In the following provisions of this Part of this Act the expression "the holding", in relation to a tenancy to which this Part of this Act applies, means the property comprised in the tenancy, there being excluded any part thereof which is occupied neither by the tenant nor by a person employed by the tenant and so employed for the purposes of a business by reason of which the tenancy is one to which this Part of this Act applies.

(4) Where the tenant is carrying on a business, in all or any part of the property comprised in a tenancy, in breach of a prohibition (however expressed) of use for business purposes which subsists under the terms of the tenancy and extends to the whole of that property, this Part of this Act shall not apply to the tenancy unless the immediate landlord or his predecessor in title has consented to the breach or the immediate landlord has acquiesced therein.

In this subsection the reference to a prohibition of use for business purposes does not include a prohibition of use for the purposes of a specified business, or of use for purposes of any but a specified business, but save as aforesaid includes a prohibition of use for the purposes of some one or more only of the classes of business specified in the definition of that expression in subsection (2) of this section.

Continuation and renewal of tenancies

24. Continuation of tenancies to which Part II applies and grant of new tenancies

(1) A tenancy to which this Part of this Act applies shall not come to an end unless terminated in accordance with the provisions of this Part of this Act; and, subject to the following provisions of this Act either the tenant or the landlord under such a tenancy may apply to the court for an order for the grant of a new tenancy—

(a) if the landlord has given notice under section 25 of this Act to terminate the tenancy, or
(b) if the tenant has made a request for a new tenancy in accordance with section twenty-six of this Act.

(2) The last foregoing subsection shall not prevent the coming to an end of a tenancy by notice to quit given by the tenant, by surrender or forfeiture, or by the forfeiture of a superior tenancy unless—

(a) in the case of a notice to quit, the notice was given before the tenant had been in occupation in right of the tenancy for one month; . . .
(b) . . .

(2A) Neither the tenant nor the landlord may make an application under subsection (1) above if the other has made such an application and the application has been served.

(2B) Neither the tenant nor the landlord may make such an application if the landlord has made an application under section 29(2) of this Act and the application has been served.

(2C) The landlord may not withdraw an application under subsection (1) above unless the tenant consents to its withdrawal.

(3) Notwithstanding anything in subsection (1) of this section,—

(a) where a tenancy to which this Part of this Act applies ceases to be such a tenancy, it shall not come to an end by reason only of the cesser, but if it was granted for a term of years certain and has been continued by subsection (1) of this section then (without prejudice to the termination thereof in accordance with any terms of the tenancy) it may be terminated by not less than three nor more than six months' notice in writing given by the landlord to the tenant;
(b) where, at a time when a tenancy is not one to which this Part of this Act applies, the landlord gives notice to quit, the operation of the notice shall not be affected by reason that the tenancy becomes one to which this Part of this Act applies after the giving of the notice.

24A Applications for determination of interim rent while tenancy continues

(1) Subject to subsection (2) below, if—

(a) the landlord of a tenancy to which this Part of this Act applies has given notice under section 25 of this Act to terminate the tenancy; or
(b) the tenant of such a tenancy has made a request for a new tenancy in accordance with section 26 of this Act,

either of them may make an application to the court to determine a rent (an "interim rent") which the tenant is to pay while the tenancy ("the relevant tenancy") continues by virtue of section 24 of this Act and the court may order payment of an interim rent in accordance with section 24C or 24D of this Act.

(2) Neither the tenant nor the landlord may make an application under subsection (1) above if the other has made such an application and has not withdrawn it.

(3) No application shall be entertained under subsection (1) above if it is made more than six months after the termination of the relevant tenancy.

24B Date from which interim rent is payable

(1) The interim rent determined on an application under section 24A(1) of this Act shall be payable from the appropriate date.

(2) If an application under section 24A(1) of this Act is made in a case where the landlord has given a notice under section 25 of this Act, the appropriate date is the earliest date of termination that could have been specified in the landlord's notice.

(3) If an application under section 24A(1) of this Act is made in a case where the tenant has made a request for a new tenancy under section 26 of this Act, the appropriate date is the earliest date that could have been specified in the tenant's request as the date from which the new tenancy is to begin.

24C Amount of interim rent where new tenancy of whole premises granted and landlord not opposed

(1) This section applies where—

(a) the landlord gave a notice under section 25 of this Act at a time when the tenant was in occupation of the whole of the property comprised in the relevant tenancy for purposes such as are mentioned in section 23(1) of this Act and stated in the notice that he was not opposed to the grant of a new tenancy; or
(b) the tenant made a request for a new tenancy under section 26 of this Act at a time when he was in occupation of the whole of that property for such purposes and the landlord did not give notice under subsection (6) of that section,

and the landlord grants a new tenancy of the whole of the property comprised in the relevant tenancy to the tenant (whether as a result of an order for the grant of a new tenancy or otherwise).

(2) Subject to the following provisions of this section, the rent payable under and at the commencement of the new tenancy shall also be the interim rent.

(3) Subsection (2) above does not apply where—

(a) the landlord or the tenant shows to the satisfaction of the court that the interim rent under that subsection differs substantially from the relevant rent; or
(b) the landlord or the tenant shows to the satisfaction of the court that the terms of the new tenancy differ from the terms of the relevant tenancy to such an extent that the interim rent under that subsection is substantially different from the rent which (in default of such agreement) the court would have determined under section 34 of this Act to be payable under a tenancy which commenced on the same day as the new tenancy and whose other terms were the same as the relevant tenancy.

(4) In this section "the relevant rent" means the rent which (in default of agreement between the landlord and the tenant) the court would have determined under section 34 of this Act to be payable under the new tenancy if the new tenancy had commenced on the appropriate date (within the meaning of section 24B of this Act).

(5) The interim rent in a case where subsection (2) above does not apply by virtue only of subsection (3)(a) above is the relevant rent.

(6) The interim rent in a case where subsection (2) above does not apply by virtue only of subsection (3)(b) above, or by virtue of subsection (3)(a) and (b) above, is the rent which it is reasonable for the tenant to pay while the relevant tenancy continues by virtue of section 24 of this Act.

(7) In determining the interim rent under subsection (6) above the court shall have regard—

(a) to the rent payable under the terms of the relevant tenancy; and
(b) to the rent payable under any sub-tenancy of part of the property comprised in the relevant tenancy,

but otherwise subsections (1) and (2) of section 34 of this Act shall apply to the determination as they would apply to the determination of a rent under that section if a new tenancy of the whole of the property comprised in the relevant tenancy were granted to the tenant by order of the court and the duration of that new tenancy were the same as the duration of the new tenancy which is actually granted to the tenant.

(8) In this section and section 24D of this Act "the relevant tenancy" has the same meaning as in section 24A of this Act.

24D Amount of interim rent in any other case

(1) The interim rent in a case where section 24C of this Act does not apply is the rent which it is reasonable for the tenant to pay while the relevant tenancy continues by virtue of section 24 of this Act.

(2) In determining the interim rent under subsection (1) above the court shall have regard-

(a) to the rent payable under the terms of the relevant tenancy; and
(b) to the rent payable under any sub-tenancy of part of the property comprised in the relevant tenancy,

but otherwise subsections (1) and (2) of section 34 of this Act shall apply to the determination as they would apply to the determination of a rent under that section if a new tenancy from year to year of the whole of the property comprised in the relevant tenancy were granted to the tenant by order of the court.

(3) If the court—

(a) has made an order for the grant of a new tenancy and has ordered payment of interim rent in accordance with section 24C of this Act, but
(b) either—
(i) it subsequently revokes under section 36(2) of this Act the order for the grant of a new tenancy; or
(ii) the landlord and tenant agree not to act on the order,

the court on the application of the landlord or the tenant shall determine a new interim rent in accordance with subsections (1) and (2) above without a further application under section 24A(1) of this Act.

25 Termination of tenancy by the landlord

(1) The landlord may terminate a tenancy to which this Part of this Act applies by a notice given to the tenant in the prescribed form specifying the date at which the tenancy is to come to an end (hereinafter referred to as "the date of termination"):

Provided that this subsection has effect subject to the provisions of section 29B(4) of this Act and the provisions of Part IV of this Act as to the interim continuation of tenancies pending the disposal of applications to the court.

(2) Subject to the provisions of the next following subsection, a notice under this section shall not have effect unless it is given not more than twelve nor less than six months before the date of termination specified therein.

(3) In the case of a tenancy which apart from this Act could have been brought to an end by notice to quit given by the landlord—

(a) the date of termination specified in a notice under this section shall not be earlier than the earliest date on which apart from this Part of this Act the tenancy could have been brought to an end by notice to quit given by the landlord on the date of the giving of the notice under this section; and
(b) where apart from this Part of this Act more than six months' notice to quit would have been required to bring the tenancy to an end, the last foregoing subsection shall have effect with the substitution for twelve months of a period six months longer than the length of notice to quit which would have been required as aforesaid.

(4) In the case of any other tenancy, a notice under this section shall not specify a date of termination earlier than the date on which apart from this Part of this Act the tenancy would have come to an end by effluxion of time.
(5) . . .
(6) A notice under this section shall not have effect unless it states whether the landlord is opposed to the grant of a new tenancy to the tenant.

(7) A notice under this section which states that the landlord is opposed to the grant of a new tenancy to the tenant shall not have effect unless it also specifies one or more of the grounds specified in section 30(1) of this Act as the ground or grounds for his opposition.

(8) A notice under this section which states that the landlord is not opposed to the grant of a new tenancy to the tenant shall not have effect unless it sets out the landlord's proposals as to—

(a) the property to be comprised in the new tenancy (being either the whole or part of the property comprised in the current tenancy);
(b) the rent to be payable under the new tenancy; and
(c) the other terms of the new tenancy.

26 Tenant's request for a new tenancy

(1) A tenant's request for a new tenancy may be made where the current tenancy is a tenancy granted for a term of years certain exceeding one year, whether or not continued by section twenty-four of this Act, or granted for a term of years certain and thereafter from year to year.

(2) A tenant's request for a new tenancy shall be for a tenancy beginning with such date, not more than twelve nor less than six months after the making of the request, as may be specified therein:

Provided that the said date shall not be earlier than the date on which apart from this Act the current tenancy would come to an end by effluxion of time or could be brought to an end by notice to quit given by the tenant.

(3) A tenant's request for a new tenancy shall not have effect unless it is made by notice in the prescribed form given to the landlord and sets out the tenant's proposals as to the property to be comprised in the new tenancy (being either the whole or part of the property comprised in the current tenancy), as to the rent to be payable under the new tenancy and as to the other terms of the new tenancy.

(4) A tenant's request for a new tenancy shall not be made if the landlord has already given notice under the last foregoing section to terminate the current tenancy, or if the tenant has already given notice to quit or notice under the next following section; and no such notice shall be given by the landlord or the tenant after the making by the tenant of a request for a new tenancy.

(5) Where the tenant makes a request for a new tenancy in accordance with the foregoing provisions of this section, the current tenancy shall, subject to the provisions of sections 29B(4) and 36(2) of this Act and the provisions of Part IV of this Act as to the interim continuation of tenancies, terminate immediately before the date specified in the request for the beginning of the new tenancy.

(6) Within two months of the making of a tenant's request for a new tenancy the landlord may give notice to the tenant that he will oppose an application to the court for the grant of a new tenancy, and any such notice shall state on which of the grounds mentioned in section thirty of this Act the landlord will oppose the application.

27 Termination by tenant of tenancy for fixed term

(1) Where the tenant under a tenancy to which this Part of this Act applies, being a tenancy granted for a term of years certain, gives to the immediate landlord, not later than three months before the date on which apart from this Act the tenancy would come to an end by effluxion of time, a notice in writing that the tenant does not desire the tenancy to be continued, section twenty-four of this Act shall not have effect in relation to the tenancy unless the notice is given before the tenant has been in occupation in right of the tenancy for one month.

(1A) Section 24 of this Act shall not have effect in relation to a tenancy for a term of years certain where the tenant is not in occupation of the property comprised in the tenancy at the time when, apart from this Act, the tenancy would come to an end by effluxion of time.

(2) A tenancy granted for a term of years certain which is continuing by virtue of section twenty-four of this Act shall not come to an end by reason only of the tenant ceasing to occupy the property comprised in the tenancy but may be brought to an end on any ... day by not less than three months' notice in writing given by the tenant to the immediate landlord, whether the notice is given ... after the date on which apart from this Act the tenancy would have come to an end or before that date, but not before the tenant has been in occupation in right of the tenancy for one month.

(3) Where a tenancy is terminated under subsection (2) above, any rent payable in respect of a period which begins before, and ends after, the tenancy is terminated shall be apportioned, and any rent paid by the tenant in excess of the amount apportioned to the period before termination shall be recoverable by him.

28 Renewal of tenancies by agreement

Where the landlord and tenant agree for the grant to the tenant of a future tenancy of the holding, or of the holding with other land, on terms and from a date specified in the agreement, the current tenancy shall continue until that date but no longer, and shall not be a tenancy to which this Part of this Act applies.

29 Order by court for grant of new tenancy or termination of current tenancy

(1) Subject to the provisions of this Act, on an application under section 24(1) of this Act, the court shall make an order for the grant of a new tenancy and accordingly for the termination of the current tenancy immediately before the commencement of the new tenancy.

(2) Subject to the following provisions of this Act, a landlord may apply to the court for an order for the termination of a tenancy to which this Part of this Act applies without the grant of a new tenancy—

(a) if he has given notice under section 25 of this Act that he is opposed to the grant of a new tenancy to the tenant; or
(b) if the tenant has made a request for a new tenancy in accordance with section 26 of this Act and the landlord has given notice under subsection (6) of that section.

(3) The landlord may not make an application under subsection (2) above if either the tenant or the landlord has made an application under section 24(1) of this Act.

(4) Subject to the provisions of this Act, where the landlord makes an application under subsection (2) above—

(a) if he establishes, to the satisfaction of the court, any of the grounds on which he is entitled to make the application in accordance with section 30 of this Act, the court shall make an order for the termination of the current tenancy in accordance with section 64 of this Act without the grant of a new tenancy; and
(b) if not, it shall make an order for the grant of a new tenancy and accordingly for the termination of the current tenancy immediately before the commencement of the new tenancy.

(5) The court shall dismiss an application by the landlord under section 24(1) of this Act if the tenant informs the court that he does not want a new tenancy.

(6) The landlord may not withdraw an application under subsection (2) above unless the tenant consents to its withdrawal.

29A Time limits for applications to court

(1) Subject to section 29B of this Act, the court shall not entertain an application—

(a) by the tenant or the landlord under section 24(1) of this Act; or
(b) by the landlord under section 29(2) of this Act,

if it is made after the end of the statutory period.

(2) In this section and section 29B of this Act "the statutory period" means a period ending—

(a) where the landlord gave a notice under section 25 of this Act, on the date specified in his notice; and

(b) where the tenant made a request for a new tenancy under section 26 of this Act, immediately before the date specified in his request.

(3) Where the tenant has made a request for a new tenancy under section 26 of this Act, the court shall not entertain an application under section 24(1) of this Act which is made before the end of the period of two months beginning with the date of the making of the request, unless the application is made after the landlord has given a notice under section 26(6) of this Act.

29B Agreements extending time limits

(1) After the landlord has given a notice under section 25 of this Act, or the tenant has made a request under section 26 of this Act, but before the end of the statutory period, the landlord and tenant may agree that an application such as is mentioned in section 29A(1) of this Act, may be made before the end of a period specified in the agreement which will expire after the end of the statutory period.

(2) The landlord and tenant may from time to time by agreement further extend the period for making such an application, but any such agreement must be made before the end of the period specified in the current agreement.

(3) Where an agreement is made under this section, the court may entertain an application such as is mentioned in section 29A(1) of this Act if it is made before the end of the period specified in the agreement.

(4) Where an agreement is made under this section, or two or more agreements are made under this section, the landlord's notice under section 25 of this Act or tenant's request under section 26 of this Act shall be treated as terminating the tenancy at the end of the period specified in the agreement or, as the case may be, at the end of the period specified in the last of those agreements.

30 Opposition by landlord to application for a new tenancy

(1) The grounds on which a landlord may oppose an application under section 24(1) of this Act, or make an application under section 29(2) of this Act, are such of the following grounds as may be stated in the landlord's notice under section twenty-five of this Act or, as the case may be, under subsection (6) of section twenty-six thereof, that is to say:—

(a) where under the current tenancy the tenant has any obligations as respects the repair and maintenance of the holding, that the tenant ought not to be granted a new tenancy in view of the state of repair of the holding, being a state resulting from the tenant's failure to comply with the said obligations;
(b) that the tenant ought not to be granted a new tenancy in view of his persistent delay in paying rent which has become due;
(c) hat the tenant ought not to be granted a new tenancy in view of other substantial breaches by him of his obligations under the current tenancy, or for any other reason connected with the tenant's use or management of the holding;

(d) that the landlord has offered and is willing to provide or secure the provision of alternative accommodation for the tenant, that the terms on which the alternative accommodation is available are reasonable having regard to the terms of the current tenancy and to all other relevant circumstances, and that the accommodation and the time at which it will be available are suitable for the tenant's requirements (including the requirement to preserve goodwill) having regard to the nature and class of his business and to the situation and extent of, and facilities afforded by, the holding;

(e) where the current tenancy was created by the sub-letting of part only of the property comprised in a superior tenancy and the landlord is the owner of an interest in reversion expectant on the termination of that superior tenancy, that the aggregate of the rents reasonably obtainable on separate lettings of the holding and the remainder of that property would be substantially less than the rent reasonably obtainable on a letting of that property as a whole, that on the termination of the current tenancy the landlord requires possession of the holding for the purpose of letting or otherwise disposing of the said property as a whole, and that in view thereof the tenant ought not to be granted a new tenancy;

(f) that on the termination of the current tenancy the landlord intends to demolish or reconstruct the premises comprised in the holding or a substantial part of those premises or to carry out substantial work of construction on the holding or part thereof and that he could not reasonably do so without obtaining possession of the holding;

(g) subject as hereinafter provided, that on the termination of the current tenancy the landlord intends to occupy the holding for the purposes, or partly for the purposes, of a business to be carried on by him therein, or as his residence.

(1A) Where the landlord has a controlling interest in a company, the reference in subsection (1)(g) above to the landlord shall be construed as a reference to the landlord or that company.

(1B) Subject to subsection (2A) below, where the landlord is a company and a person has a controlling interest in the company, the reference in subsection (1)(g) above to the landlord shall be construed as a reference to the landlord or that person.

(2) The landlord shall not be entitled to oppose an application under section 24(1) of this Act, or make an application under section 29(2) of this Act, on the ground specified in paragraph (g) of the last foregoing subsection if the interest of the landlord, or an interest which has merged in that interest and but for the merger would be the interest of the landlord, was purchased or created after the beginning of the period of five years which ends with the termination of the current tenancy, and at all times since the purchase or creation thereof the holding has been comprised in a tenancy or successive tenancies of the description specified in subsection (1) of section twenty-three of this Act.

(2A) Subsection (1B) above shall not apply if the controlling interest was acquired after the beginning of the period of five years which ends with the termination of the current tenancy, and at all times since the acquisition of the controlling interest the holding has been comprised in a tenancy or successive tenancies of the description specified in section 23(1) of this Act.

(3) . . .

31 Dismissal of application for new tenancy where landlord successfully opposes

(1) If the landlord opposes an application under subsection (1) of section twenty-four of this Act on grounds on which he is entitled to oppose it in accordance with the last foregoing section and establishes any of those grounds to the satisfaction of the court, the court shall not make an order for the grant of a new tenancy.

(2) Where the landlord opposes an application under section 24(1) of this Act, or makes an application under section 29(2) of this Act, on one or more of the grounds specified in section 30(1)(d) to (f) of this Act but establishes none of those grounds, and none of the other grounds specified in section 30(1) of this Act, to the satisfaction of the court, then if the court would have been satisfied on any of the grounds specified in section 30(1)(d) to (f) of this Act if the date of termination specified in the landlord's notice or, as the case may be, the date specified in the tenant's request for a new tenancy as the date from which the new tenancy is to begin, had been such later date as the court may determine, being a date not more than one year later than the date so specified,—

(a) the court shall make a declaration to that effect, stating of which of the said grounds the court would have been satisfied as aforesaid and specifying the date determined by the court as aforesaid, but shall not make an order for the grant of a new tenancy;
(b) if, within fourteen days after the making of the declaration, the tenant so requires the court shall make an order substituting the said date for the date specified in the said landlord's notice or tenant's request, and thereupon that notice or request shall have effect accordingly.

31A Grant of new tenancy in some cases where section 30(1)(f) applies

(1) Where the landlord opposes an application under section 24(1) of this Act on the ground specified in paragraph (f) of section 30(1) of this Act, or makes an application under section 29(2) of this Act on that ground, the court shall not hold that the landlord could not reasonably carry out the demolition, reconstruction or work of construction intended without obtaining possession of the holding if—

(a) the tenant agrees to the inclusion in the terms of the new tenancy of terms giving the landlord access and other facilities for carrying out the work intended and, given that access and those facilities, the landlord could reasonably carry out the work without obtaining possession of the holding and without interfering to a substantial extent or for a substantial time with the use of the holding for the purposes of the business carried on by the tenant; or
(b) the tenant is willing to accept a tenancy of an economically separable part of the holding and either paragraph (a) of this section is satisfied with respect to that part or possession of the remainder of the holding would be reasonably sufficient to enable the landlord to carry out the intended work.

(2) For the purposes of subsection (1)(b) of this section a part of a holding shall be deemed to be an economically separable part if, and only if, the aggregate of the rents which, after the completion

of the intended work, would be reasonably obtainable on separate lettings of that part and the remainder of the premises affected by or resulting from the work would not be substantially less than the rent which would then be reasonably obtainable on a letting of those premises as a whole.

32 Property to be comprised in new tenancy

(1) Subject to the following provisions of this section, an order under section twenty-nine of this Act for the grant of a new tenancy shall be an order for the grant of a new tenancy of the holding; and in the absence of agreement between the landlord and the tenant as to the property which constitutes the holding the court shall in the order designate that property by reference to the circumstances existing at the date of the order.

(1A) Where the court, by virtue of paragraph (b) of section 31A(1) of this Act, makes an order under section 29 of this Act for the grant of a new tenancy in a case where the tenant is willing to accept a tenancy of part of the holding, the order shall be an order for the grant of a new tenancy of that part only.

(2) The foregoing provisions of this section shall not apply in a case where the property comprised in the current tenancy includes other property besides the holding and the landlord requires any new tenancy ordered to be granted under section twenty-nine of this Act to be a tenancy of the whole of the property comprised in the current tenancy; but in any such case—

(a) any order under the said section twenty-nine for the grant of a new tenancy shall be an order for the grant of a new tenancy of the whole of the property comprised in the current tenancy, and
(b) references in the following provisions of this Part of this Act to the holding shall be construed as references to the whole of that property.

(3) Where the current tenancy includes rights enjoyed by the tenant in connection with the holding, those rights shall be included in a tenancy ordered to be granted under section twenty-nine of this Act except as otherwise agreed between the landlord and the tenant or, in default of such agreement, determined by the court.

33 Duration of new tenancy

Where on an application under this Part of this Act the court makes an order for the grant of a new tenancy, the new tenancy shall be such tenancy as may be agreed between the landlord and the tenant, or, in default of such an agreement, shall be such a tenancy as may be determined by the court to be reasonable in all the circumstances, being, if it is a tenancy for a term of years certain, a tenancy for a term not exceeding fifteen years, and shall begin on the coming to an end of the current tenancy.

34 Rent under new tenancy

(1) The rent payable under a tenancy granted by order of the court under this Part of this Act shall be such as may be agreed between the landlord and the tenant or as, in default of such agreement, may be determined by the court to be that at which, having regard to the terms of

the tenancy (other than those relating to rent), the holding might reasonably be expected to be let in the open market by a willing lessor, there being disregarded—

(a) any effect on rent of the fact that the tenant has or his predecessors in title have been in occupation of the holding,
(b) any goodwill attached to the holding by reason of the carrying on thereat of the business of the tenant (whether by him or by a predecessor of his in that business),
(c) any effect on rent of an improvement to which this paragraph applies,
(d) in the case of a holding comprising licensed premises, any addition to its value attributable to the licence, if it appears to the court that having regard to the terms of the current tenancy and any other relevant circumstances the benefit of the licence belongs to the tenant.

(2) Paragraph (c) of the foregoing subsection applies to any improvement carried out by a person who at the time it was carried out was the tenant, but only if it was carried out otherwise than in pursuance of an obligation to his immediate landlord, and either it was carried out during the current tenancy or the following conditions are satisfied, that is to say,—

(a) that it was completed not more than twenty-one years before the application to the court was made; and
(b) that the holding or any part of it affected by the improvement has at all times since the completion of the improvement been comprised in tenancies of the description specified in section 23(1) of this Act; and
(c) that at the termination of each of those tenancies the tenant did not quit.

(2A) If this Part of this Act applies by virtue of section 23(1A) of this Act, the reference in subsection (1)(d) above to the tenant shall be construed as including—

(a) a company in which the tenant has a controlling interest, or
(b) where the tenant is a company, a person with a controlling interest in the company.

(3) Where the rent is determined by the court the court may, if it thinks fit, further determine that the terms of the tenancy shall include such provision for varying the rent as may be specified in the determination.

(4) It is hereby declared that the matters which are to be taken into account by the court in determining the rent include any effect on rent of the operation of the provisions of the Landlord and Tenant (Covenants) Act 1995.

35 Other terms of new tenancy

(1) The terms of a tenancy granted by order of the court under this Part of this Act (other than terms as to the duration thereof and as to the rent payable thereunder), including, where different persons own interests which fulfil the conditions specified in section 44(1) of this Act in different parts of it, terms as to the apportionment of the rent, shall be such as may be agreed between the landlord and the tenant or as, in default of such agreement, may be determined by the court; and in determining those terms the court shall have regard to the terms of the current tenancy and to all relevant circumstances.

(2) In subsection (1) of this section the reference to all relevant circumstances includes (without prejudice to the generality of that reference) a reference to the operation of the provisions of the Landlord and Tenant (Covenants) Act 1995.

36 Carrying out of order for new tenancy

(1) Where under this Part of this Act the court makes an order for the grant of a new tenancy, then, unless the order is revoked under the next following subsection or the landlord and the tenant agree not to act upon the order, the landlord shall be bound to execute or make in favour of the tenant, and the tenant shall be bound to accept, a lease or agreement for a tenancy of the holding embodying the terms agreed between the landlord and the tenant or determined by the court in accordance with the foregoing provisions of this Part of this Act; and where the landlord executes or makes such a lease or agreement the tenant shall be bound, if so required by the landlord, to execute a counterpart or duplicate thereof.

(2) If the tenant, within fourteen days after the making of an order under this Part of this Act for the grant of a new tenancy, applies to the court for the revocation of the order the court shall revoke the order; and where the order is so revoked, then, if it is so agreed between the landlord and the tenant or determined by the court, the current tenancy shall continue, beyond the date at which it would have come to an end apart from this subsection, for such period as may be so agreed or determined to be necessary to afford to the landlord a reasonable opportunity for reletting or otherwise disposing of the premises which would have been comprised in the new tenancy; and while the current tenancy continues by virtue of this subsection it shall not be a tenancy to which this Part of this Act applies.

(3) Where an order is revoked under the last foregoing subsection any provision thereof as to payment of costs shall not cease to have effect by reason only of the revocation; but the court may, if it thinks fit, revoke or vary any such provision or, where no costs have been awarded in the proceedings for the revoked order, award such costs.

(4) A lease executed or agreement made under this section, in a case where the interest of the lessor is subject to a mortgage, shall be deemed to be one authorised by section ninety-nine of the Law of Property Act 1925 (which confers certain powers of leasing on mortgagors in possession), and subsection (13) of that section (which allows those powers to be restricted or excluded by agreement) shall not have effect in relation to such a lease or agreement.

37 Compensation where order for new tenancy precluded on certain grounds

(1) Subject to the provisions of this Act, in a case specified in subsection (1A), (1B) or (1C) below (a "compensation case") the tenant shall be entitled on quitting the holding to recover from the landlord by way of compensation an amount determined in accordance with this section.

(1A) The first compensation case is where on the making of an application by the tenant under section 24(1) of this Act the court is precluded (whether by subsection (1) or subsection (2) of section 31 of this Act) from making an order for the grant of a new tenancy by reason of any of the grounds specified in paragraphs (e), (f) and (g) of section 30(1) of this Act (the "compensation grounds") and not of any grounds specified in any other paragraph of section 30(1).

(1B) The second compensation case is where on the making of an application under section 29(2) of this Act the court is precluded (whether by section 29(4)(a) or section 31(2) of this Act) from making an order for the grant of a new tenancy by reason of any of the compensation grounds and not of any other grounds specified in section 30(1) of this Act.

(1C) The third compensation case is where—

(a) the landlord's notice under section 25 of this Act or, as the case may be, under section 26(6) of this Act, states his opposition to the grant of a new tenancy on any of the compensation grounds and not on any other grounds specified in section 30(1) of this Act; and

(b) either—

(i) no application is made by the tenant under section 24(1) of this Act or by the landlord under section 29(2) of this Act; or

(ii) such an application is made but is subsequently withdrawn.

(2) Subject to the following provisions of this section, compensation under this section shall be as follows, that is to say,—

(a) where the conditions specified in the next following subsection are satisfied in relation to the whole of the holding it shall be the product of the appropriate multiplier and twice the rateable value of the holding,

(b) in any other case it shall be the product of the appropriate multiplier and the rateable value of the holding.

(3) The said conditions are—

(a) that, during the whole of the fourteen years immediately preceding the termination of the current tenancy, premises being or comprised in the holding have been occupied for the purposes of a business carried on by the occupier or for those and other purposes;

(b) that, if during those fourteen years there was a change in the occupier of the premises, the person who was the occupier immediately after the change was the successor to the business carried on by the person who was the occupier immediately before the change.

(3A) If the conditions specified in subsection (3) above are satisfied in relation to part of the holding but not in relation to the other part, the amount of compensation shall be the aggregate of sums calculated separately as compensation in respect of each part, and accordingly, for the purpose of calculating compensation in respect of a part any reference in this section to the holding shall be construed as a reference to that part.

(3B) Where section 44(1A) of this Act applies, the compensation shall be determined separately for each part and compensation determined for any part shall be recoverable only from the person who is the owner of an interest in that part which fulfils the conditions specified in section 44(1) of this Act.

(4) Where the court is precluded from making an order for the grant of a new tenancy under this Part of this Act in a compensation case, the court shall on the application of the tenant certify that fact.

(5) For the purposes of subsection (2) of this section the rateable value of the holding shall be determined as follows:—

(a) where in the valuation list in force at the date on which the landlord's notice under section twenty-five or, as the case may be, subsection (6) of section twenty-six of this Act is given a value is then shown as the annual value (as hereinafter defined) of the holding, the rateable value of the holding shall be taken to be that value;
(b) where no such value is so shown with respect to the holding but such a value or such values is or are so shown with respect to premises comprised in or comprising the holding or part of it, the rateable value of the holding shall be taken to be such value as is found by a proper apportionment or aggregation of the value or values so shown;
(c) where the rateable value of the holding cannot be ascertained in accordance with the foregoing paragraphs of this subsection, it shall be taken to be the value which, apart from any exemption from assessment to rates, would on a proper assessment be the value to be entered in the said valuation list as the annual value of the holding;

and any dispute arising, whether in proceedings before the court or otherwise, as to the determination for those purposes of the rateable value of the holding shall be referred to the Commissioners of Inland Revenue for decision by a valuation officer.

An appeal shall lie to the Lands Tribunal from any decision of a valuation officer under this subsection, but subject thereto any such decision shall be final.

(5A) If part of the holding is domestic property, as defined in section 66 of the Local Government Finance Act 1988,—

(a) the domestic property shall be disregarded in determining the rateable value of the holding under subsection (5) of this section; and
(b) if, on the date specified in subsection (5)(a) of this section, the tenant occupied the whole or any part of the domestic property, the amount of compensation to which he is entitled under subsection (1) of this section shall be increased by the addition of a sum equal to his reasonable expenses in removing from the domestic property.

(5B) Any question as to the amount of the sum referred to in paragraph (b) of subsection (5A) of this section shall be determined by agreement between the landlord and the tenant or, in default of agreement, by the court.

(5C) If the whole of the holding is domestic property, as defined in section 66 of the Local Government Finance Act 1988, for the purposes of subsection (2) of this section the rateable value of the holding shall be taken to be an amount equal to the rent at which it is estimated the holding might reasonably be expected to let from year to year if the tenant undertook to pay all usual tenant's rates and taxes and to bear the cost of the repairs and insurance and the other expenses (if any) necessary to maintain the holding in a state to command that rent.

(5D) The following provisions shall have effect as regards a determination of an amount mentioned in subsection (5C) of this section—

(a) the date by reference to which such a determination is to be made is the date on which the landlord's notice under section 25 or, as the case may be, subsection (6) of section 26 of this Act is given;
(b) any dispute arising, whether in proceedings before the court or otherwise, as to such a determination shall be referred to the Commissioners of Inland Revenue for decision by a valuation officer;
(c) an appeal shall lie to the Lands Tribunal from such a decision but subject to that, such a decision shall be final.

(5E) Any deduction made under paragraph 2A of Schedule 6 to the Local Government Finance Act 1988 (deduction from valuation of hereditaments used for breeding horses etc) shall be disregarded, to the extent that it relates to the holding, in determining the rateable value of the holding under subsection (5) of this section.

(6) The Commissioners of Inland Revenue may by statutory instrument make rules prescribing the procedure in connection with references under this section.

(7) In this section—

the reference to the termination of the current tenancy is a reference to the date of termination specified in the landlord's notice under section twenty-five of this Act or, as the case may be, the date specified in the tenant's request for a new tenancy as the date from which the new tenancy is to begin;
the expression "annual value" means rateable value except that where the rateable value differs from the net annual value the said expression means net annual value;
the expression "valuation officer" means any officer of the Commissioners of Inland Revenue for the time being authorised by a certificate of the Commissioners to act in relation to a valuation list.

(8) In subsection (2) of this section "the appropriate multiplier" means such multiplier as the Secretary of State may by order made by statutory instrument prescribe and different multipliers may be so prescribed in relation to different cases.
(9) A statutory instrument containing an order under subsection (8) of this section shall be subject to annulment in pursuance of a resolution of either House of Parliament.

37A Compensation for possession obtained by misrepresentation

(1) Where the court—

(a) makes an order for the termination of the current tenancy but does not make an order for the grant of a new tenancy, or
(b) refuses an order for the grant of a new tenancy,

and it subsequently made to appear to the court that the order was obtained, or the court was induced to refuse the grant, by misrepresentation or the concealment of material facts, the court may order the landlord to pay to the tenant such sum as appears sufficient as compensation for damage or loss sustained by the tenant as the result of the order or refusal.

(2) Where—
  (a) the tenant has quit the holding—
    (i) after making but withdrawing an application under section 24(1) of this Act; or
    (ii) without making such an application; and
  (b) it is made to appear to the court that he did so by reason of misrepresentation or the concealment of material facts,

the court may order the landlord to pay to the tenant such sum as appears sufficient as compensation for damage or loss sustained by the tenant as the result of quitting the holding.

38 Restriction on agreements excluding provisions of Part II

(1) Any agreement relating to a tenancy to which this Part of this Act applies (whether contained in the instrument creating the tenancy or not) shall be void (except as provided by section 38A of this Act) in so far as it purports to preclude the tenant from making an application or request under this Part of this Act or provides for the termination or the surrender of the tenancy in the event of his making such an application or request or for the imposition of any penalty or disability on the tenant in that event.

(2) Where—

  (a) during the whole of the five years immediately preceding the date on which the tenant under a tenancy to which this Part of this Act applies is to quit the holding, premises being or comprised in the holding have been occupied for the purposes of a business carried on by the occupier or for those and other purposes, and
  (b) if during those five years there was a change in the occupier of the premises, the person who was the occupier immediately after the change was the successor to the business carried on by the person who was the occupier immediately before the change,

any agreement (whether contained in the instrument creating the tenancy or not and whether made before or after the termination of that tenancy) which purports to exclude or reduce compensation under section 37 of this Act shall to that extent be void, so however that this subsection shall not affect any agreement as to the amount of any such compensation which is made after the right to compensation has accrued.

(3) In a case not falling within the last foregoing subsection the right to compensation conferred by section 37 of this Act may be excluded or modified by agreement.

(4) . . .

38A Agreements to exclude provisions of Part 2

(1) The persons who will be the landlord and the tenant in relation to a tenancy to be granted for a term of years certain which will be a tenancy to which this Part of this Act applies may agree that the provisions of sections 24 to 28 of this Act shall be excluded in relation to that tenancy.

(2) The persons who are the landlord and the tenant in relation to a tenancy to which this Part of this Act applies may agree that the tenancy shall be surrendered on such date or in such circumstances as may be specified in the agreement and on such terms (if any) as may be so specified.

(3) An agreement under subsection (1) above shall be void unless—

(a) the landlord has served on the tenant a notice in the form, or substantially in the form, set out in Schedule 1 to the Regulatory Reform (Business Tenancies) (England and Wales) Order 2003 ("the 2003 Order"); and
(b) the requirements specified in Schedule 2 to that Order are met.

(4) An agreement under subsection (2) above shall be void unless—

(a) the landlord has served on the tenant a notice in the form, or substantially in the form, set out in Schedule 3 to the 2003 Order; and
(b) the requirements specified in Schedule 4 to that Order are met.

General and supplementary provisions

39 Saving for compulsory acquisitions

(1) . . .
(2) If the amount of the compensation which would have been payable under section thirty-seven of this Act if the tenancy had come to an end in circumstances giving rise to compensation under that section and the date at which the acquiring authority obtained possession had been the termination of the current tenancy exceeds the amount of the compensation payable under section 121 of the Lands Clauses Consolidation Act 1845 or section 20 of the Compulsory Purchase Act 1965 in the case of a tenancy to which this Part of this Act applies, that compensation shall be increased by the amount of the excess.
(3) Nothing in section twenty-four of this Act shall affect the operation of the said section one hundred and twenty-one.

40 Duties of tenants and landlords of business premises to give information to each other

(1) Where a person who is an owner of an interest in reversion expectant (whether immediately or not) on a tenancy of any business premises has served on the tenant a notice in the prescribed form requiring him to do so, it shall be the duty of the tenant to give the appropriate person in writing the information specified in subsection (2) below.

(2) That information is—

(a) whether the tenant occupies the premises or any part of them wholly or partly for the purposes of a business carried on by him;
(b) whether his tenancy has effect subject to any sub-tenancy on which his tenancy is immediately expectant and, if so—
(i) what premises are comprised in the sub-tenancy;

(ii) for what term it has effect (or, if it is terminable by notice, by what notice it can be terminated);
(iii) what is the rent payable under it;
(iv) who is the sub-tenant;
(v) (to the best of his knowledge and belief) whether the sub-tenant is in occupation of the premises or of part of the premises comprised in the sub-tenancy and, if not, what is the sub-tenant's address;
(vi) whether an agreement is in force excluding in relation to the sub-tenancy the provisions of sections 24 to 28 of this Act; and
(vii) whether a notice has been given under section 25 or 26(6) of this Act, or a request has been made under section 26 of this Act, in relation to the sub-tenancy and, if so, details of the notice or request; and

(c) (to the best of his knowledge and belief) the name and address of any other person who owns an interest in reversion in any part of the premises.

(3) Where the tenant of any business premises who is a tenant under such a tenancy as is mentioned in section 26(1) of this Act has served on a reversioner or a reversioner's mortgagee in possession a notice in the prescribed form requiring him to do so, it shall be the duty of the person on whom the notice is served to give the appropriate person in writing the information specified in subsection (4) below.

(4) That information is—

(a) whether he is the owner of the fee simple in respect of the premises or any part of them or the mortgagee in possession of such an owner,

(b) if he is not, then (to the best of his knowledge and belief)—
(i) the name and address of the person who is his or, as the case may be, his mortgagor's immediate landlord in respect of those premises or of the part in respect of which he or his mortgagor is not the owner in fee simple;
(ii) for what term his or his mortgagor's tenancy has effect and what is the earliest date (if any) at which that tenancy is terminable by notice to quit given by the landlord; and
(iii) whether a notice has been given under section 25 or 26(6) of this Act, or a request has been made under section 26 of this Act, in relation to the tenancy and, if so, details of the notice or request;

(c) (to the best of his knowledge and belief) the name and address of any other person who owns an interest in reversion in any part of the premises; and

(d) if he is a reversioner, whether there is a mortgagee in possession of his interest in the premises and, if so, (to the best of his knowledge and belief) what is the name and address of the mortgagee.

(5) A duty imposed on a person by this section is a duty—

(a) to give the information concerned within the period of one month beginning with the date of service of the notice; and

(b) if within the period of six months beginning with the date of service of the notice that person becomes aware that any information which has been given in pursuance of the notice is not, or is no longer, correct, to give the appropriate person correct information within the period of one month beginning with the date on which he becomes aware.

(6) This section shall not apply to a notice served by or on the tenant more than two years before the date on which apart from this Act his tenancy would come to an end by effluxion of time or could be brought to an end by notice to quit given by the landlord.

(7) Except as provided by section 40A of this Act, the appropriate person for the purposes of this section and section 40A(1) of this Act is the person who served the notice under subsection (1) or (3) above.

(8) In this section—

"business premises" means premises used wholly or partly for the purposes of a business;
"mortgagee in possession" includes a receiver appointed by the mortgagee or by the court who is in receipt of the rents and profits, and "his mortgagor" shall be construed accordingly;
"reversioner" means any person having an interest in the premises, being an interest in reversion expectant (whether immediately or not) on the tenancy;
"reversioner's mortgagee in possession" means any person being a mortgagee in possession in respect of such an interest; and
"sub-tenant" includes a person retaining possession of any premises by virtue of the Rent (Agriculture) Act 1976 or the Rent Act 1977 after the coming to an end of a sub-tenancy, and "sub-tenancy" includes a right so to retain possession.

40A Duties in transfer cases

(1) If a person on whom a notice under section 40(1) or (3) of this Act has been served has transferred his interest in the premises or any part of them to some other person and gives the appropriate person notice in writing—

(a) of the transfer of his interest; and
(b) of the name and address of the person to whom he transferred it,

on giving the notice he ceases in relation to the premises or (as the case may be) to that part to be under any duty imposed by section 40 of this Act.

(2) If—

(a) the person who served the notice under section 40(1) or (3) of this Act ("the transferor") has transferred his interest in the premises to some other person ("the transferee"); and
(b) the transferor or the transferee has given the person required to give the information notice in writing—
(i) of the transfer; and
(ii) of the transferee's name and address,
the appropriate person for the purposes of section 40 of this Act and subsection (1) above is the

transferee.

(3) If—

(a) a transfer such as is mentioned in paragraph (a) of subsection (2) above has taken place; but

(b) neither the transferor nor the transferee has given a notice such as is mentioned in paragraph (b) of that subsection,

any duty imposed by section 40 of this Act may be performed by giving the information either to the transferor or to the transferee.

40B Proceedings for breach of duties to give information

A claim that a person has broken any duty imposed by section 40 of this Act may be made the subject of civil proceedings for breach of statutory duty; and in any such proceedings a court may order that person to comply with that duty and may make an award of damages.

41 Trusts

(1) Where a tenancy is held on trust, occupation by all or any of the beneficiaries under the trust, and the carrying on of a business by all or any of the beneficiaries, shall be treated for the purposes of section twenty-three of this Act as equivalent to occupation or the carrying on of a business by the tenant; and in relation to a tenancy to which this Part of this Act applies by virtue of the foregoing provisions of this subsection—

(a) references (however expressed) in this Part of this Act and in the Ninth Schedule to this Act to the business of, or to carrying on of business, use, occupation or enjoyment by, the tenant shall be construed as including references to the business of, or to carrying on of business, use, occupation or enjoyment by, the beneficiaries or beneficiary;

(b) the reference in paragraph (d) of subsection (1) of section thirty-four of this Act to the tenant shall be construed as including the beneficiaries or beneficiary; and

(c) a change in the persons of the trustees shall not be treated as a change in the person of the tenant.

(2) Where the landlord's interest is held on trust the references in paragraph (g) of subsection (1) of section thirty of this Act to the landlord shall be construed as including references to the beneficiaries under the trust or any of them; but, except in the case of a trust arising under a will or on the intestacy of any person, the reference in subsection (2) of that section to the creation of the interest therein mentioned shall be construed as including the creation of the trust.

41A Partnerships

(1) The following provisions of this section shall apply where—

(a) a tenancy is held jointly by two or more persons (in this section referred to as the joint tenants); and

(b) the property comprised in the tenancy is or includes premises occupied for the purposes of a business; and

(c) the business (or some other business) was at some time during the existence of the tenancy carried on in partnership by all the persons who were then the joint tenants or by those and other persons and the joint tenants' interest in the premises was then partnership property; and

(d) the business is carried on (whether alone or in partnership with other persons) by one or some only of the joint tenants and no part of the property comprised in the tenancy is occupied, in right of the tenancy, for the purposes of a business carried on (whether alone or in partnership with other persons) by the other or others.

(2) In the following provisions of this section those of the joint tenants who for the time being carry on the business are referred to as the business tenants and the others as the other joint tenants.

(3) Any notice given by the business tenants which, had it been given by all the joint tenants, would have been—

(a) a tenant's request for a new tenancy made in accordance with section 26 of this Act; or

(b) a notice under subsection (1) or subsection (2) of section 27 of this Act;

shall be treated as such if it states that it is given by virtue of this section and sets out the facts by virtue of which the persons giving it are the business tenants; and references in those sections and in section 24A of this Act to the tenant shall be construed accordingly.

(4) A notice given by the landlord to the business tenants which, had it been given to all the joint tenants, would have been a notice under section 25 of this Act shall be treated as such a notice, and references in that section to the tenant shall be construed accordingly.

(5) An application under section 24(1) of this Act for a new tenancy may, instead of being made by all the joint tenants, be made by the business tenants alone; and where it is so made—

(a) this Part of this Act shall have effect, in relation to it, as if the references therein to the tenant included references to the business tenants alone; and

(b) the business tenants shall be liable, to the exclusion of the other joint tenants, for the payment of rent and the discharge of any other obligation under the current tenancy for any rental period beginning after the date specified in the landlord's notice under section 25 of this Act or, as the case may be, beginning on or after the date specified in their request for a new tenancy.

(6) Where the court makes an order under section 29 of this Act for the grant of a new tenancy it may order the grant to be made to the business tenants or to them jointly with the persons carrying on the business in partnership with them, and may order the grant to be made subject to the satisfaction, within a time specified by the order, of such conditions as to guarantors, sureties or otherwise as appear to the court equitable, having regard to the omission of the other joint tenants from the persons who will be the tenant under the new tenancy.

(7) The business tenants shall be entitled to recover any amount payable by way of compensation

under section 37 or section 59 of this Act.

42 Groups of companies

(1) For the purposes of this section two bodies corporate shall be taken to be members of a group if and only if one is a subsidiary of the other or both are subsidiaries of a third body corporate or the same person has a controlling interest in both.

. . .

(2) Where a tenancy is held by a member of a group, occupation by another member of the group, and the carrying on of a business by another member of the group, shall be treated for the purposes of section twenty-three of this Act as equivalent to occupation or the carrying on of a business by the member of the group holding the tenancy; and in relation to a tenancy to which this Part of this Act applies by virtue of the foregoing provisions of this subsection—

(a) references (however expressed) in this Part of this Act and in the Ninth Schedule to this Act to the business of or to use occupation or enjoyment by the tenant shall be construed as including references to the business of or to use occupation or enjoyment by the said other member;
(b) the reference in paragraph (d) of subsection (1) of section thirty-four of this Act to the tenant shall be construed as including the said other member; and
(c) an assignment of the tenancy from one member of the group to another shall not be treated as a change in the person of the tenant.

(3) Where the landlord's interest is held by a member of a group—

(a) the reference in paragraph (g) of subsection (1) of section 30 of this Act to intended occupation by the landlord for the purposes of a business to be carried on by him shall be construed as including intended occupation by any member of the group for the purposes of a business to be carried on by that member; and
(b) the reference in subsection (2) of that section to the purchase or creation of any interest shall be construed as a reference to a purchase from or creation by a person other than a member of the group.

43 Tenancies excluded from Part II

(1) This Part of this Act does not apply—

(a) to a tenancy of an agricultural holding which is a tenancy in relation to which the Agricultural Holdings Act 1986 applies or a tenancy which would be a tenancy of an agricultural holding in relation to which that Act applied if subsection (3) of section 2 of that Act did not have effect or, in a case where approval was given under subsection (1) of that section, if that approval had not been given;
(aa) to a farm business tenancy;
(b) to a tenancy created by a mining lease;
(c), (d) . . .

(2) This Part of this Act does not apply to a tenancy granted by reason that the tenant was the holder of an office, appointment or employment from the grantor thereof and continuing only so long as the tenant holds the office, appointment or employment, or terminable by the grantor on the tenant's ceasing to hold it, or coming to an end at a time fixed by reference to the time at which the tenant ceases to hold it:
Provided that this subsection shall not have effect in relation to a tenancy granted after the commencement of this Act unless the tenancy was granted by an instrument in writing which expressed the purpose for which the tenancy was granted.

(3) This Part of this Act does not apply to a tenancy granted for a term certain not exceeding six months unless—

(a) the tenancy contains provision for renewing the term or for extending it beyond six months from its beginning; or
(b) the tenant has been in occupation for a period which, together with any period during which any predecessor in the carrying on of the business carried on by the tenant was in occupation, exceeds twelve months.

43A Jurisdiction of county court to make declaration

Where the rateable value of the holding is such that the jurisdiction conferred on the court by any other provision of this Part of this Act is, by virtue of section 63 of this Act, exercisable by the county court, the county court shall have jurisdiction (but without prejudice to the jurisdiction of the High Court) to make any declaration as to any matter arising under this Part of this Act, whether or not any other relief is sought in the proceedings.

44 Meaning of "the landlord" in Part II, and provisions as to mesne landlords, etc

(1) Subject to subsections (1A) and (2) below, in this Part of this Act the expression "the landlord", in relation to a tenancy (in this section referred to as "the relevant tenancy"), means the person (whether or not he is the immediate landlord) who is the owner of that interest in the property comprised in the relevant tenancy which for the time being fulfils the following conditions, that is to say—

(a) that it is an interest in reversion expectant (whether immediately or not) on the termination of the relevant tenancy, and
(b) that it is either the fee simple or a tenancy which will not come to an end within fourteen months by effluxion of time and, if it is such a tenancy, that no notice has been given by virtue of which it will come to an end within fourteen months or any further time by which it may be continued under section 36(2) or section 64 of this Act,

and is not itself in reversion expectant (whether immediately or not) on an interest which fulfils those conditions.

(1A) The reference in subsection (1) above to a person who is the owner of an interest such as is mentioned in that subsection is to be construed, where different persons own such interests in different parts of the property, as a reference to all those persons collectively.

(2) References in this Part of this Act to a notice to quit given by the landlord are references to a notice to quit given by the immediate landlord.

(3) The provisions of the Sixth Schedule to this Act shall have effect for the application of this Part of this Act to cases where the immediate landlord of the tenant is not the owner of the fee simple in respect of the holding.

45 . . .

. . .

46 Interpretation of Part II

(1) In this Part of this Act:—

"business" has the meaning assigned to it by subsection (2) of section twenty-three of this Act;
"current tenancy" means the tenancy under which the tenant holds for the time being;
"date of termination" has the meaning assigned to it by subsection (1) of section twenty-five of this Act;
subject to the provisions of section thirty-two of this Act, "the holding" has the meaning assigned to it by subsection (3) of section twenty-three of this Act;
"interim rent" has the meaning given by section 24A(1) of this Act;
"mining lease" has the same meaning as in the Landlord and Tenant Act 1927.

(2) For the purposes of this Part of this Act, a person has a controlling interest in a company, if, had he been a company, the other company would have been its subsidiary; and in this Part —
"company" has the meaning given by section 735 of the Companies Act 1985; and
"subsidiary" has the meaning given by section 736 of that Act

64 Interim continuation of tenancies pending determination by court

(1) In any case where—

(a) a notice to terminate a tenancy has been given under Part I or Part II of this Act or a request for a new tenancy has been made under Part II thereof, and
(b) an application to the court has been made under the said Part I or [under section 24(1) or 29(2) of this Act], as the case may be, and
(c) apart from this section the effect of the notice or request would be to terminate the tenancy before the expiration of the period of three months beginning with the date on which the application is finally disposed of,

the effect of the notice or request shall be to terminate the tenancy at the expiration of the said period of three months and not at any other time.

(2) The reference in paragraph (c) of subsection (1) of this section to the date on which an

application is finally disposed of shall be construed as a reference to the earliest date by which the proceedings on the application (including any proceedings on or in consequence of an appeal) have been determined and any time for appealing or further appealing has expired, except that if the application is withdrawn or any appeal is abandoned the reference shall be construed as a reference to the date of the withdrawal or abandonment.

SCHEDULE 6

PROVISIONS FOR PURPOSES OF PART II WHERE IMMEDIATE LANDLORD IS NOT THE FREEHOLDER

Section 44

Definitions

1

In this Schedule the following expressions have the meanings hereby assigned to them in relation to a tenancy (in this Schedule referred to as "the relevant tenancy"), that is to say:—

"the competent landlord" means the person who in relation to the tenancy is for the time being the landlord (as defined by section forty-four of this Act) for the purposes of Part II of this Act;
"mesne landlord" means a tenant whose interest is intermediate between the relevant tenancy and the interest of the competent landlord; and
"superior landlord" means a person (whether the owner of the fee simple or a tenant) whose interest is superior to the interest of the competent landlord.

Power of court to order reversionary tenancies

2

Where the period for which in accordance with the provisions of Part II of this Act it is agreed or determined by the court that a new tenancy should be granted thereunder will extend beyond the date on which the interest of the immediate landlord will come to an end, the power of the court under Part II of this Act to order such a grant shall include power to order the grant of a new tenancy until the expiration of that interest and also to order the grant of such a reversionary tenancy or reversionary tenancies as may be required to secure that the combined effects of those grants will be equivalent to the grant of a tenancy for that period; and the provisions of Part II of this Act shall, subject to the necessary modifications, apply in relation to the grant of a tenancy together with one or more reversionary tenancies as they apply in relation to the grant of one new tenancy.

Acts of competent landlord binding on other landlords

3

(1) Any notice given by the competent landlord under Part II of this Act to terminate the relevant tenancy, and any agreement made between that landlord and the tenant as to the granting, duration, or terms of a future tenancy, being an agreement made for the purposes of the said Part II, shall bind the interest of any mesne landlord notwithstanding that he has not consented to the giving of the notice or was not a party to the agreement.

(2) The competent landlord shall have power for the purposes of Part II of this Act to give effect to

any agreement with the tenant for the grant of a new tenancy beginning with the coming to an end of the relevant tenancy notwithstanding that the competent landlord will not be the immediate landlord at the commencement of the new tenancy, and any instrument made in the exercise of the power conferred by this sub-paragraph shall have effect as if the mesne landlord had been a party thereto.

(3) Nothing in the foregoing provisions of this paragraph shall prejudice the provisions of the next following paragraph.

Provisions as to consent of mesne landlord to acts of competent landlord

4

(1) If the competent landlord, not being the immediate landlord, gives any such notice or makes any such agreement as is mentioned in sub-paragraph (1) of the last foregoing paragraph without the consent of every mesne landlord, any mesne landlord whose consent has not been given thereto shall be entitled to compensation from the competent landlord for any loss arising in consequence of the giving of the notice or the making of the agreement.

(2) If the competent landlord applies to any mesne landlord for his consent to such a notice or agreement, that consent shall not be unreasonably withheld, but may be given subject to any conditions which may be reasonable (including conditions as to the modification of the proposed notice or agreement or as to the payment of compensation by the competent landlord).

(3) Any question arising under this paragraph whether consent has been unreasonably withheld or whether any conditions imposed on the giving of consent are unreasonable shall be determined by the court.

Consent of superior landlord required for agreements affecting his interest

5

An agreement between the competent landlord and the tenant made for the purposes of Part II of this Act in a case where—

(a) the competent landlord is himself a tenant, and

(b) the agreement would apart from this paragraph operate as respects any period after the coming to an end of the interest of the competent landlord,

shall not have effect unless every superior landlord who will be the immediate landlord of the tenant during any part of that period is a party to the agreement.

Withdrawal by competent landlord of notice given by mesne landlord

6

Where the competent landlord has given a notice under section 25 of this Act to terminate the relevant tenancy and, within two months after the giving of the notice, a superior landlord—

(a) becomes the competent landlord; and

(b) gives to the tenant notice in the prescribed form that he withdraws the notice previously given,

the notice under section 25 of this Act shall cease to have effect, but without prejudice to the giving of a further notice under that section by the competent landlord.

Duty to inform superior landlords

7

If the competent landlord's interest in the property comprised in the relevant tenancy is a tenancy which will come or can be brought to an end within sixteen months (or any further time by which it may be continued under section 36(2) or section 64 of this Act) and he gives to the tenant under the relevant tenancy a notice under section 25 of this Act to terminate the tenancy or is given by him a notice under section 26(3) of this Act:—

(a) the competent landlord shall forthwith send a copy of the notice to his immediate landlord; and
(b) any superior landlord whose interest in the property is a tenancy shall forthwith send to his immediate landlord any copy which has been sent to him in pursuance of the preceding sub-paragraph or this sub-paragraph.

SCHEDULE 8

APPLICATION OF PART II TO LAND BELONGING TO CROWN AND DUCHIES OF LANCASTER AND CORNWALL

Section 56

1

Where an interest in any property comprised in a tenancy belongs to Her Majesty in right of the Duchy of Lancaster, then for the purposes of Part II of this Act the Chancellor of the Duchy shall represent Her Majesty and shall be deemed to be the owner of the interest.

2

Where an interest in any property comprised in a tenancy belongs to the Duchy of Cornwall, then for the purposes of Part II of this Act such person as the Duke of Cornwall, or other the possessor for the time being of the Duchy of Cornwall, appoints shall represent the Duke of Cornwall or other the possessor aforesaid, and shall be deemed to be the owner of the interest and may do any act or thing under the said Part II which the owner of that interest is authorised or required to do thereunder.

3

. . .

4

The amount of any compensation payable under section thirty-seven of this Act by the Chancellor of the Duchy of Lancaster shall be raised and paid as an expense incurred in improvement of land belonging to Her Majesty in right of the Duchy within section twenty-five of the Act of the fifty-seventh year of King George the Third, Chapter ninety-seven.

5

Any compensation payable under section thirty-seven of this Act by the person representing the Duke of Cornwall or other the possessor for the time being of the Duchy of Cornwall shall be paid, and advances therefor made, in the manner and subject to the provisions of section eight of the Duchy of Cornwall Management Act 1863 with respect to improvements of land mentioned in that section.

***

County Courts Act 1984

138. Provisions as to forfeiture for non-payment of rent

(1) This section has effect where a lessor is proceeding by action in a county court (being an action in which the county court has jurisdiction) to enforce against a lessee a right of re-entry or forfeiture in respect of any land for non-payment of rent.

(2) If the lessee pays into court [or to the lessor] not less than 5 clear days before the return day all the rent in arrear and the costs of the action, the action shall cease, and the lessee shall hold the land according to the lease without any new lease.

(3) If—

(a) the action does not cease under subsection (2); and
(b) the court at the trial is satisfied that the lessor is entitled to enforce the right of re-entry or forfeiture,

the court shall order possession of the land to be given to the lessor at the expiration of such period, not being less than 4 weeks from the date of the order, as the court thinks fit, unless within that period the lessee pays into court [or to the lessor]all the rent in arrear and the costs of the action.

(4) The court may extend the period specified under subsection (3) at any time before possession of the land is recovered in pursuance of the order under that subsection.

(5) ... if—

(a) within the period specified in the order; or
(b) within that period as extended under subsection (4), the lessee pays into court [or to the lessor]—
(i) all the rent in arrear; and
(ii) the costs of the action,

He shall hold the land according to the lease without any new lease.

(6) Subsection (2) shall not apply where the lessor is proceeding in the same action to enforce a right of re-entry or forfeiture on any other ground as well as for non-payment of rent, or to enforce any other claim as well as the right of re-entry or forfeiture and the claim for arrears of rent.

(7) If the lessee does not—

(a) within the period specified in the order; or
(b) within that period as extended under subsection (4), pay into court [or to the lessor]—
(i) all the rent in arrear; and
(ii) the costs of the action,

the order shall be [enforceable] in the prescribed manner and so long as the order remains unreversed the lessee shall [subject to subsection (8) and (9A),] be barred from all relief.

(8) The extension under subsection (4) of a period fixed by a court shall not be treated as relief from which the lessee is barred by subsection (7) if he fails to pay into court [or to the lessor] all the rent in arrear and the costs of the action within that period.

(9) Where the court extends a period under subsection (4) at a time when-

(a) that period has expired; and
(b) a warrant has been issued for the possession of the land,

the court shall suspend the warrant for the extended period; and, if, before the expiration of the extended period, the lessee pays into court [F1 or to the lessor]all the rent in arrear and all the costs of the action, the court shall cancel the warrant.

(9A) Where the lessor recovers possession of the land at any time after the making of the order under subsection (3) (whether as a result of the enforcement of the order or otherwise) the lessee may, at any time within six months from the date on which the lessor recovers possession, apply to the court for relief; and on any such application the court may, if it thinks fit, grant to the lessee such relief, subject to such terms and conditions, as it thinks fit.

(9B) Where the lessee is granted relief on an application under subsection (9A) he shall hold the land according to the lease without any new lease.

(9C) An application under subsection (9A) may be made by a person with an interest under a lease of the land derived (whether immediately or otherwise) from the lessee's interest therein in like manner as if he were the lessee; and on any such application the court may make an order which (subject to such terms and conditions as the court thinks fit) vests the land in such a person, as lessee of the lessor, for the remainder of the term of the lease under which he has any such interest as aforesaid, or for any lesser term.

In this subsection any reference to the land includes a reference to a part of the land.]

(10) Nothing in this section or section 139 shall be taken to affect-

(a) the power of the court to make any order which it would otherwise have power to make as respects a right of re-entry of forfeiture on any ground other than non-payment of rent; or
(b) section 146(4) of the Law of Property Act 1925 (relief against forfeiture).

***

Insolvency Act 1986

1A. Moratorium

(1) Where the directors of an eligible company intend to make a proposal for a voluntary arrangement, they may take steps to obtain a moratorium for the company.

(2) The provisions of Schedule A1 to this Act have effect with respect to—

(a) companies eligible for a moratorium under this section,
(b) the procedure for obtaining such a moratorium,
(c) the effects of such a moratorium, and
(d) the procedure applicable (in place of sections 2 to 6 and 7) in relation to the approval and implementation of a voluntary arrangement where such a moratorium is or has been in force.

5. Effect of approval

(1) This section applies where a decision approving a voluntary arrangement has effect under section 4A.

(2) The . . . voluntary arrangement-

(a) takes effect as if made by the company at the creditors' meeting, and
(b) binds every person who in accordance with the rules—
 (i) was entitled to vote at that meeting (whether or not he was present or represented at it), or
 (ii) would have been so entitled if he had had notice of it, as if he were a party to the voluntary arrangement.

6. Challenge of decisions

(1) Subject to this section, an application to the court may be made, by any of the persons specified below, on one or both of the following grounds, namely—

(a) that a voluntary arrangement which has effect under section 4A unfairly prejudices the interests of a creditor, member or contributory of the company;
(b) that there has been some material irregularity at or in relation to either of the meetings.

(2) The persons who may apply under this section are—

(a) a person entitled, in accordance with the rules, to vote at either of the meetings;
(aa) a person who would have been entitled, in accordance with the rules, to vote at the creditors' meeting if he had had notice of it]
(b) the nominee or any person who has replaced him under section 2(4) or 4(2); and

(c) if the company is being wound up or an administration order is in force, the liquidator or administrator.

(3) An application under this section shall not be made[(a)]after the end of the period of 28 days beginning with the first day on which each of the reports required by section 4(6) has been made to the court[or

(b) in the case of a person who was not given notice of the creditors' meeting, after the end of the period of 28 days beginning with the day on which he became aware that the meeting had taken place,

but (subject to that) an application made by a person within subsection (2)(aa) on the ground that the voluntary arrangement prejudices his interests may be made after the arrangement has ceased to have effect, unless it came to an end prematurely.]

(4) Where on such an application the court is satisfied as to either of the grounds mentioned in subsection (1), it may do one or both of the following, namely—

(a) revoke or suspend[any decision approving the voluntary arrangement which has effect under section 4A] or, in a case falling within subsection (1)(b), any[decision taken by the meeting in question which has effect under that section];
(b) give a direction to any person for the summoning of further meetings to consider any revised proposal the person who made the original proposal may make or, in the case falling within subsection (1)(b), a further company or (as the case may be) creditors' meeting to reconsider the original proposal.

(5) Where at any time after giving a direction under subsection (4)(b) for the summoning of meetings to consider a revised proposal the court is satisfied that the person who made the original proposal does not intend to submit a revised proposal, the court shall revoke the direction and revoke or suspend any[decision approving the voluntary arrangement which has effect under section 4A].

(6) In a case where the court, on an application under this section with respect to any meeting-

(a) gives a direction under subsection (4)(b), or
(b) revokes or suspends an approval under subsection (4)(a) or (5),

the court may give such supplemental directions as it thinks fit and, in particular, directions with respect to things done[under the voluntary arrangement since it took effect.

(7) Except in pursuance of the preceding provisions of this section, a decision taken at a meeting summoned under section 3 is not invalidated by any irregularity at or in relation to the meeting.

123. Definition of inability to pay debts

(1) A company is deemed unable to pay its debts—

(a) if a creditor (by assignment or otherwise) to whom the company is indebted in a sum exceeding £750 then due has served on the company, by leaving it at the company's registered office, a written demand (in the prescribed form) requiring the company to pay the sum so due and the company has for 3 weeks thereafter neglected to pay the sum or to secure or compound for it to the reasonable satisfaction of the creditor, or
(b) if, in England and Wales, execution or other process issued on a judgment, decree or order of any court in favour of a creditor of the company is returned unsatisfied in whole or in part, or
(c) if, in Scotland, the induciae of a charge for payment on an extract decree, or an extract registered bond, or an extract registered protest, have expired without payment being made, or
(d) if, in Northern Ireland, a certificate of unenforceability has been granted in respect of a judgment against the company, or
(e) if it is proved to the satisfaction of the court that the company is unable to pay its debts as they fall due.

(2) A company is also deemed unable to pay its debts if it is proved to the satisfaction of the court that the value of the company's assets is less than the amount of its liabilities, taking into account its contingent and prospective liabilities.

(3) The money sum for the time being specified in subsection (1)(a) is subject to increase or reduction by order under section 416 in Part XV.

126. Power to stay or restrain proceedings against company.

(1) At any time after the presentation of a winding-up petition, and before a winding-up order has been made, the company, or any creditor or contributory, may—

(a) where any action or proceeding against the company is pending in the High Court or Court of Appeal in England and Wales or Northern Ireland, apply to the court in which the action or proceeding is pending for a stay of proceedings therein, and
(b) where any other action or proceeding is pending against the company, apply to the court having jurisdiction to wind up the company to restrain further proceedings in the action or proceeding;

and the court to which the application is so made may (as the case may be) stay, sist or restrain the proceedings accordingly on such terms as it thinks fit.

130. Consequences of winding-up order.

(1) On the making of a winding-up order, a copy of the order must forthwith be forwarded by the company (or otherwise as may be prescribed) to the registrar of companies, who shall enter it in his records relating to the company.

(2) When a winding-up order has been made or a provisional liquidator has been appointed, no action or proceeding shall be proceeded with or commenced against the company or its property, except by leave of the court and subject to such terms as the court may impose.

(3) When an order has been made for winding up a company registered under section 680 of the Companies Act, no action or proceeding shall be commenced or proceeded with against the company or its property or any contributory of the company, in respect of any debt of the company, except by leave of the court, and subject to such terms as the court may impose.

(4) An order for winding up a company operates in favour of all the creditors and of all contributories of the company as if made on the joint petition of a creditor and of a contributory.

178. Power to disclaim onerous property.

(1) This and the next two sections apply to a company that is being wound up in England and Wales.

(2) Subject as follows, the liquidator may, by the giving of the prescribed notice, disclaim any onerous property and may do so notwithstanding that he has taken possession of it, endeavoured to sell it, or otherwise exercised rights of ownership in relation to it.

(3) The following is onerous property for the purposes of this section—

(a) any unprofitable contract, and
(b) any other property of the company which is unsaleable or not readily saleable or is such that it may give rise to a liability to pay money or perform any other onerous act.

(4) A disclaimer under this section—

(a) operates so as to determine, as from the date of the disclaimer, the rights, interests and liabilities of the company in or in respect of the property disclaimed; but
(b) does not, except so far as is necessary for the purpose of releasing the company from any liability, affect the rights or liabilities of any other person.

(5) A notice of disclaimer shall not be given under this section in respect of any property if—

(a) a person interested in the property has applied in writing to the liquidator or one of his predecessors as liquidator requiring the liquidator or that predecessor to decide whether he will disclaim or not, and
(b) the period of 28 days beginning with the day on which that application was made, or such longer period as the court may allow, has expired without a notice of disclaimer having been given under this section in respect of that property.

(6) Any person sustaining loss or damage in consequence of the operation of a disclaimer under this section is deemed a creditor of the company to the extent of the loss or damage and accordingly may prove for the loss or damage in the winding up.

181. Powers of court (general)

(1) This section and the next apply where the liquidator has disclaimed property under section 178.

(2) An application under this section may be made to the court by—

(a) any person who claims an interest in the disclaimed property, or
(b) any person who is under any liability in respect of the disclaimed property, not being a liability discharged by the disclaimer.

(3) Subject as follows, the court may on the application make an order, on such terms as it thinks fit, for the vesting of the disclaimed property in, or for its delivery to—

(a) a person entitled to it or a trustee for such a person, or
(b) a person subject to such a liability as is mentioned in subsection (2)(b) or a trustee for such a person.

(4) The court shall not make an order under subsection (3)(b) except where it appears to the court that it would be just to do so for the purpose of compensating the person subject to the liability in respect of the disclaimer.

(5) The effect of any order under this section shall be taken into account in assessing for the purpose of section 178(6) the extent of any loss or damage sustained by any person in consequence of the disclaimer.

(6) An order under this section vesting property in any person need not be completed by conveyance, assignment or transfer.

252. Interim order of court

(1) In the circumstances specified below, the court may in the case of a debtor (being an individual) make an interim order under this section.

(2) An interim order has the effect that, during the period for which it is in force—

(a) no bankruptcy petition relating to the debtor may be presented or proceeded with,
(aa) no landlord or other person to whom rent is payable may exercise any right of forfeiture by peaceable re-entry in relation to premises let to the debtor in respect of a failure by the debtor to comply with any term or condition of his tenancy of such premises, except with the leave of the court] and
(b) no other proceedings, and no execution or other legal process, may be commenced or continued[and no distress may be levied] against the debtor or his property except with leave of the court.

260. Effect of approval

(1) This section has effect where the meeting summoned under section 257 approves the proposed voluntary arrangement (with or without modifications).

(2) The approved arrangement—

(a) takes effect as if made by the debtor at the meeting, and
(b) binds every person who in accordance with the rules—
(i) was entitled to vote at the meeting (whether or not he was present or represented at it), or
(ii) would have been so entitled if he had had notice of it,

as if he were a party to the arrangement.

(2A) If—

(a) when the arrangement ceases to have effect any amount payable under the arrangement to a person bound by virtue of subsection (2)(b)(ii) has not been paid, and
(b) the arrangement did not come to an end prematurely, the debtor shall at that time become liable to pay to that person the amount payable under the arrangement.]

(3) The Deeds of Arrangement Act 1914 does not apply to the approved voluntary arrangement.

(4) Any interim order in force in relation to the debtor immediately before the end of the period of 28 days beginning with the day on which the report with respect to the creditors' meeting was made to the court under section 259 ceases to have effect at the end of that period.

This subsection applies except to such extent as the court may direct for the purposes of any application under section 262 below.

(5) Where proceedings on a bankruptcy petition have been stayed by an interim order which ceases to have effect under subsection (4), that petition is deemed, unless the court otherwise orders, to have been dismissed.

262. Challenge of meeting's decision

(1) Subject to this section, an application to the court may be made, by any of the persons specified below, on one or both of the following grounds, namely—

(a) that a voluntary arrangement approved by a creditors' meeting summoned under section 257 unfairly prejudices the interests of a creditor of the debtor;
(b) that there has been some material irregularity at or in relation to such a meeting.

(2) The persons who may apply under this section are—

(a) the debtor;

(b) a person who—
(i) was entitled, in accordance with the rules, to vote at the creditors' meeting, or
(ii) would have been so entitled if he had had notice of it]
(c) the nominee (or his replacement under section [256(3), 256A(4)] or 258(3)); and
(d) if the debtor is an undischarged bankrupt, the trustee of his estate or the official receiver.

(3) An application under this section shall not be made

(a) after the end of the period of 28 days beginning with the day on which the report of the creditors' meeting was made to the court under section 259 [or
(b) in the case of a person who was not given notice of the creditors' meeting, after the end of the period of 28 days beginning with the day on which he became aware that the meeting had taken place,

but (subject to that) an application made by a person within subsection (2)(b)(ii) on the ground that the arrangement prejudices his interests may be made after the arrangement has ceased to have effect, unless it has come to an end prematurely.]

(4) Where on an application under this section the court is satisfied as to either of the grounds mentioned in subsection (1), it may do one or both of the following, namely—

(a) revoke or suspend any approval given by the meeting;
(b) give a direction to any person for the summoning of a further meeting of the debtor's creditors to consider any revised proposal he may make or, in a case falling within subsection (1)(b), to reconsider his original proposal.

(5) Where at any time after giving a direction under subsection (4)(b) for the summoning of a meeting to consider a revised proposal the court is satisfied that the debtor does not intend to submit such a proposal, the court shall revoke the direction and revoke or suspend any approval given at the previous meeting.

(6) Where the court gives a direction under subsection (4)(b), it may also give a direction continuing or, as the case may require, renewing, for such period as may be specified in the direction, the effect in relation to the debtor of any interim order.

(7) In any case where the court, on an application made under this section with respect to a creditors' meeting, gives a direction under subsection (4)(b) or revokes or suspends an approval under subsection (4)(a) or (5), the court may give such supplemental directions as it thinks fit and, in particular, directions with respect to—

(a) things done since the meeting under any voluntary arrangement approved by the meeting, and
(b) such things done since the meeting as could not have been done if any interim order had been in force in relation to the debtor when they were done.

(8) Except in pursuance of the preceding provisions of this section, an approval given at a creditors' meeting summoned under section 257 is not invalidated by any irregularity at or in relation to the meeting.

285. Restriction on proceedings and remedies

(1) At any time when proceedings on a bankruptcy petition are pending or an individual has been adjudged bankrupt the court may stay any action, execution or other legal process against the property or person of the debtor or, as the case may be, of the bankrupt.

(2) Any court in which proceedings are pending against any individual may, on proof that a bankruptcy petition has been presented in respect of that individual or that he is an undischarged bankrupt, either stay the proceedings or allow them to continue on such terms as it thinks fit.

(3) After the making of a bankruptcy order no person who is a creditor of the bankrupt in respect of a debt provable in the bankruptcy shall-

(a) have any remedy against the property or person of the bankrupt in respect of that debt, or
(b) before the discharge of the bankrupt, commence any action or other legal proceedings against the bankrupt except with the leave of the court and on such terms as the court may impose.

This is subject to sections 346 (enforcement procedures) and 347 (limited right to distress).

(4) Subject as follows, subsection (3) does not affect the right of a secured creditor of the bankrupt to enforce his security.

(5) Where any goods of an undischarged bankrupt are held by any person by way of pledge, pawn or other security, the official receiver may, after giving notice in writing of his intention to do so, inspect the goods.

Where such a notice has been given to any person, that person is not entitled, without leave of the court, to realise his security unless he has given the trustee of the bankrupt's estate a reasonable opportunity of inspecting the goods and of exercising the bankrupt's right of redemption.

(6) References in this section to the property or goods of the bankrupt are to any of his property or goods, whether or not comprised in his estate.

315. Disclaimer (general power)

(1) Subject as follows, the trustee may, by the giving of the prescribed notice, disclaim any onerous property and may do so notwithstanding that he has taken possession of it, endeavoured to sell it or otherwise exercised rights of ownership in relation to it.

(2) The following is onerous property for the purposes of this section, that is to say—

(a) any unprofitable contract, and
(b) any other property comprised in the bankrupt's estate which is unsaleable or not readily saleable, or is such that it may give rise to a liability to pay money or perform any other onerous act.

(3) A disclaimer under this section—

(a) operates so as to determine, as from the date of the disclaimer, the rights, interests and liabilities of the bankrupt and his estate in or in respect of the property disclaimed, and
(b) discharges the trustee from all personal liability in respect of that property as from the commencement of his trusteeship,

but does not, except so far as is necessary for the purpose of releasing the bankrupt, the bankrupt's estate and the trustee from any liability, affect the rights or liabilities of any other person.

(4) A notice of disclaimer shall not be given under this section in respect of any property that has been claimed for the estate under section 307 (after-acquired property) or 308 (personal property of bankrupt exceeding reasonable replacement value)[or 308A], except with the leave of the court.

(5) Any person sustaining loss or damage in consequence of the operation of a disclaimer under this section is deemed to be a creditor of the bankrupt to the extent of the loss or damage and accordingly may prove for the loss or damage as a bankruptcy debt.

320. Court order vesting disclaimed property

(1) This section and the next apply where the trustee has disclaimed property under section 315.

(2) An application may be made to the court under this section by—
(a) any person who claims an interest in the disclaimed property,
(b) any person who is under any liability in respect of the disclaimed property, not being a liability discharged by the disclaimer, or
(c) where the disclaimed property is property in a dwelling house, any person who at the time when the bankruptcy petition was presented was in occupation of or entitled to occupy the dwelling house.

(3) Subject as follows in this section and the next, the court may, on an application under this section, make an order on such terms as it thinks fit for the vesting of the disclaimed property in, or for its delivery to—

(a) a person entitled to it or a trustee for such a person,
(b) a person subject to such a liability as is mentioned in subsection (2)(b) or a trustee for such a person, or

(c) where the disclaimed property is property in a dwelling house, any person who at the time when the bankruptcy petition was presented was in occupation of or entitled to occupy the dwelling house.

(4) The court shall not make an order by virtue of subsection (3)(b) except where it appears to the court that it would be just to do so for the purpose of compensating the person subject to the liability in respect of the disclaimer.

(5) The effect of any order under this section shall be taken into account in assessing for the purposes of section 315(5) the extent of any loss or damage sustained by any person in consequence of the disclaimer.

(6) An order under this section vesting property in any person need not be completed by any conveyance, assignment or transfer.

376. Time-limits

Where by any provision in this Group of Parts or by the rules the time for doing anything is limited, the court may extend the time, either before or after it has expired, on such terms, if any, as it thinks fit.

Schedule A1

Part II Obtaining a moratorium

Nominee's statement

6. (1) Where the directors of a company wish to obtain a moratorium, they shall submit to the nominee—

(a) a document setting out the terms of the proposed voluntary arrangement,
(b) a statement of the company's affairs containing—
(i) such particulars of its creditors and of its debts and other liabilities and of its assets as may be prescribed, and
(ii) such other information as may be prescribed, and

(c) any other information necessary to enable the nominee to comply with sub-paragraph (2) which he requests from them.

(2) The nominee shall submit to the directors a statement in the prescribed form indicating whether or not, in his opinion—

(a) the proposed voluntary arrangement has a reasonable prospect of being approved and implemented,
(b) the company is likely to have sufficient funds available to it during the proposed moratorium to enable it to carry on its business, and

(c) meetings of the company and its creditors should be summoned to consider the proposed voluntary arrangement.

(3) In forming his opinion on the matters mentioned in sub-paragraph (2), the nominee is entitled to rely on the information submitted to him under sub-paragraph (1) unless he has reason to doubt its accuracy.

(4) The reference in sub-paragraph (2)(b) to the company's business is to that business as the company proposes to carry it on during the moratorium.

Documents to be submitted to court

7.(1) To obtain a moratorium the directors of a company must file (in Scotland, lodge) with the court—

(a) a document setting out the terms of the proposed voluntary arrangement,
(b) a statement of the company's affairs containing—
  (i) such particulars of its creditors and of its debts and other liabilities and of its assets as may be prescribed, and
  (ii) such other information as may be prescribed,
(c) a statement that the company is eligible for a moratorium,
(d) a statement from the nominee that he has given his consent to act, and
(e) a statement from the nominee that, in his opinion—
  (i) the proposed voluntary arrangement has a reasonable prospect of being approved and implemented,
  (ii) the company is likely to have sufficient funds available to it during the proposed moratorium to enable it to carry on its business, and
  (iii) meetings of the company and its creditors should be summoned to consider the proposed voluntary arrangement.

(2) Each of the statements mentioned in sub-paragraph (1)(b) to (e), except so far as it contains the particulars referred to in paragraph (b)(i), must be in the prescribed form.

(3) The reference in sub-paragraph (1)(e)(ii) to the company's business is to that business as the company proposes to carry it on during the moratorium.

(4) The Secretary of State may by regulations modify the requirements of this paragraph as to the documents required to be filed (in Scotland, lodged) with the court in order to obtain a moratorium.

Duration of moratorium

8. (1) A moratorium comes into force when the documents for the time being referred to in paragraph 7(1) are filed or lodged with the court and references in this Schedule to "the beginning of the moratorium" shall be construed accordingly.

(2) A moratorium ends at the end of the day on which the meetings summoned under paragraph 29(1) are first held (or, if the meetings are held on different days, the later of those days), unless it is extended under paragraph 32.

(3) If either of those meetings has not first met before the end of the period of 28 days beginning with the day on which the moratorium comes into force, the moratorium ends at the end of the day on which those meetings were to be held (or, if those meetings were summoned to be held on different days, the later of those days), unless it is extended under paragraph 32.

(4) If the nominee fails to summon either meeting within the period required by paragraph 29(1), the moratorium ends at the end of the last day of that period.

(5) If the moratorium is extended (or further extended) under paragraph 32, it ends at the end of the day to which it is extended (or further extended).

(6) Sub-paragraphs (2) to (5) do not apply if the moratorium comes to an end before the time concerned by virtue of—

(a) paragraph 25(4) (effect of withdrawal by nominee of consent to act),
(b) an order under paragraph 26(3), 27(3) or 40 (challenge of actions of nominee or directors), or
(c) a decision of one or both of the meetings summoned under paragraph 29.

(7) If the moratorium has not previously come to an end in accordance with sub-paragraphs (2) to (6), it ends at the end of the day on which a decision under paragraph 31 to approve a voluntary arrangement takes effect under paragraph 36.

(8) The Secretary of State may by order increase or reduce the period for the time being specified in sub-paragraph (3).

Notification of beginning of moratorium

9.(1) When a moratorium comes into force, the directors shall notify the nominee of that fact forthwith.
(2) If the directors without reasonable excuse fail to comply with sub-paragraph (1), each of them is liable to imprisonment or a fine, or both.

10.(1) When a moratorium comes into force, the nominee shall, in accordance with the rules—

(a) advertise that fact forthwith, and
(b) notify the registrar of companies, the company and any petitioning creditor of the company of whose claim he is aware of that fact.

(2) In sub-paragraph (1)(b), "petitioning creditor" means a creditor by whom a winding-up petition has been presented before the beginning of the moratorium, as long as the petition has not been dismissed or withdrawn.

(3) If the nominee without reasonable excuse fails to comply with sub-paragraph (1)(a) or (b), he is liable to a fine.

Notification of end of moratorium

11.(1) When a moratorium comes to an end, the nominee shall, in accordance with the rules—

(a) advertise that fact forthwith, and
(b) notify the court, the registrar of companies, the company and any creditor of the company of whose claim he is aware of that fact.

(2) If the nominee without reasonable excuse fails to comply with sub-paragraph (1)(a) or (b), he is liable to a fine.

Part III Effects of moratorium

Effect on creditors, etc.

12.(1) During the period for which a moratorium is in force for a company-—

(a) no petition may be presented for the winding up of the company,
(b) no meeting of the company may be called or requisitioned except with the consent of the nominee or the leave of the court and subject (where the court gives leave) to such terms as the court may impose,
(c) no resolution may be passed or order made for the winding up of the company,
(d) no petition for an administration order in relation to the company may be presented,
(e) no administrative receiver of the company may be appointed,
(f) no landlord or other person to whom rent is payable may exercise any right of forfeiture by peaceable re-entry in relation to premises let to the company in respect of a failure by the company to comply with any term or condition of its tenancy of such premises, except with the leave of the court and subject to such terms as the court may impose,
(g) no other steps may be taken to enforce any security over the company's property, or to repossess goods in the company's possession under any hire-purchase agreement, except with the leave of the court and subject to such terms as the court may impose, and
(h) no other proceedings and no execution or other legal process may be commenced or continued, and no distress may be levied, against the company or its property except with the leave of the court and subject to such terms as the court may impose.

(2) Where a petition, other than an excepted petition, for the winding up of the company has been presented before the beginning of the moratorium, section 127 shall not apply in relation to any disposition of property, transfer of shares or alteration in status made during the moratorium or at a time mentioned in paragraph 37(5)(a).

(3) In the application of sub-paragraph (1)(h) to Scotland, the reference to execution being commenced or continued includes a reference to diligence being carried out or continued, and the reference to distress being levied is omitted.

(4) Paragraph (a) of sub-paragraph (1) does not apply to an excepted petition and, where such a petition has been presented before the beginning of the moratorium or is presented during the moratorium, paragraphs (b) and (c) of that sub-paragraph do not apply in relation to proceedings on the petition.

(5) For the purposes of this paragraph, "excepted petition" means a petition under—

(a) section 124A of this Act,
(b) section 72 of the Financial Services Act 1986 on the ground mentioned in subsection (1)(b) of that section, or
(c) section 92 of the Banking Act 1987 on the ground mentioned in subsection (1)(b) of that section.
(d) section 367 of the Financial Services and Markets Act 2000 on the ground mentioned in subsection (3)(b) of that section.]

13.(1) This paragraph applies where there is an uncrystallised floating charge on the property of a company for which a moratorium is in force.

(2) If the conditions for the holder of the charge to give a notice having the effect mentioned in sub-paragraph (4) are met at any time, the notice may not be given at that time but may instead be given as soon as practicable after the moratorium has come to an end.

(3) If any other event occurs at any time which (apart from this sub-paragraph) would have the effect mentioned in sub-paragraph (4), then—

(a) the event shall not have the effect in question at that time, but
(b) if notice of the event is given to the company by the holder of the charge as soon as is practicable after the moratorium has come to an end, the event is to be treated as if it had occurred when the notice was given.

(4) The effect referred to in sub-paragraphs (2) and (3) is—

(a) causing the crystallisation of the floating charge, or
(b) causing the imposition, by virtue of provision in the instrument creating the charge, of any restriction on the disposal of any property of the company.

(5) Application may not be made for leave under paragraph 12(1)(g) or (h) with a view to obtaining—
(a) the crystallisation of the floating charge, or
(b) the imposition, by virtue of provision in the instrument creating the charge, of any restriction on the disposal of any property of the company.

14. Security granted by a company at a time when a moratorium is in force in relation to the company may only be enforced if, at that time, there were reasonable grounds for believing that it would benefit the company.

Schedule B1

42 Moratorium on insolvency proceedings

(1) This paragraph applies to a company in administration.

(2) No resolution may be passed for the winding up of the company.

(3) No order may be made for the winding up of the company.

(4) Sub-paragraph (3) does not apply to an order made on a petition presented under—

(a) section 124A (public interest), or
(b) section 367 of the Financial Services and Markets Act 2000 (c. 8) (petition by Financial Services Authority).

(5) If a petition presented under a provision referred to in sub-paragraph (4) comes to the attention of the administrator, he shall apply to the court for directions under paragraph 63.

43 Moratorium on other legal process

(1) This paragraph applies to a company in administration.

(2) No step may be taken to enforce security over the company's property except—

(a) with the consent of the administrator, or
(b) with the permission of the court.

(3) No step may be taken to repossess goods in the company's possession under a hire-purchase agreement except—

(a) with the consent of the administrator, or
(b) with the permission of the court.

(4) A landlord may not exercise a right of forfeiture by peaceable re-entry in relation to premises let to the company except—

(a) with the consent of the administrator, or
(b) with the permission of the court.

(5) In Scotland, a landlord may not exercise a right of irritancy in relation to premises let to the company except—

(a) with the consent of the administrator, or
(b) with the permission of the court.

(6) No legal process (including legal proceedings, execution, distress and diligence) may be instituted or continued against the company or property of the company except—

(a) with the consent of the administrator, or
(b) with the permission of the court.

(7) Where the court gives permission for a transaction under this paragraph it may impose a condition on or a requirement in connection with the transaction.

(8) In this paragraph "landlord" includes a person to whom rent is payable.

74 Challenge to administrator's conduct of company

(1) A creditor or member of a company in administration may apply to the court claiming that—

(a) the administrator is acting or has acted so as unfairly to harm the interests of the applicant (whether alone or in common with some or all other members or creditors), or
(b) the administrator proposes to act in a way which would unfairly harm the interests of the applicant (whether alone or in common with some or all other members or creditors).

(2) A creditor or member of a company in administration may apply to the court claiming that the administrator is not performing his functions as quickly or as efficiently as is reasonably practicable.

(3) The court may—

(a) grant relief;
(b) dismiss the application;
(c) adjourn the hearing conditionally or unconditionally;
(d) make an interim order;
(e) make any other order it thinks appropriate.

(4) In particular, an order under this paragraph may—

(a) regulate the administrator's exercise of his functions;
(b) require the administrator to do or not do a specified thing;
(c) require a creditors' meeting to be held for a specified purpose;
(d) provide for the appointment of an administrator to cease to have effect;
(e) make consequential provision.

(5) An order may be made on a claim under sub-paragraph (1) whether or not the action complained of—

(a) is within the administrator's powers under this Schedule;
(b) was taken in reliance on an order under paragraph 71 or 72.

(6) An order may not be made under this paragraph if it would impede or prevent the implementation of—

(a) a voluntary arrangement approved under Part I,
(b) a compromise or arrangement sanctioned under section 425 of the Companies Act (compromise with creditors and members), or
(c) proposals or a revision approved under paragraph 53 or 54 more than 28 days before the day on which the application for the order under this paragraph is made.

***

Landlord and Tenant Act 1988

1. Qualified duty to consent to assigning, underletting etc. of premises.

(1) This section applies in any case where—

(a) a tenancy includes a covenant on the part of the tenant not to enter into one or more of the following transactions, that is—
(i) assigning,
(ii) underletting,
(iii) charging, or
(iv) parting with the possession of, the premises comprised in the tenancy or any part of the premises without the consent of the landlord or some other person, but
(b) the covenant is subject to the qualification that the consent is not to be unreasonably withheld (whether or not it is also subject to any other qualification).

(2) In this section and section 2 of this Act—

(a) references to a proposed transaction are to any assignment, underletting, charging or parting with possession to which the covenant relates, and
(b) references to the person who may consent to such a transaction are to the person who under the covenant may consent to the tenant entering into the proposed transaction.

(3) Where there is served on the person who may consent to a proposed transaction a written application by the tenant for consent to the transaction, he owes a duty to the tenant within a reasonable time—

(a) to give consent, except in a case where it is reasonable not to give consent,
(b) to serve on the tenant written notice of his decision whether or not to give consent specifying in addition-
(i) if the consent is given subject to conditions, the conditions,
(ii) if the consent is withheld, the reasons for withholding it.

(4) Giving consent subject to any condition that is not a reasonable condition does not satisfy the duty under subsection (3)(a) above.

(5) For the purposes of this Act it is reasonable for a person not to give consent to a proposed transaction only in a case where, if he withheld consent and the tenant completed the transaction, the tenant would be in breach of a covenant.

(6) It is for the person who owed any duty under subsection (3) above—

(a) if he gave consent and the question arises whether he gave it within a reasonable time, to show that he did,
(b) if he gave consent subject to any condition and the question arises whether the condition was a reasonable condition, to show that it was,

(c) if he did not give consent and the question arises whether it was reasonable for him not to do so, to show that it was reasonable,

and, if the question arises whether he served notice under that subsection within a reasonable time, to show that he did.

2. Duty to pass on applications.

(1) If, in a case where section 1 of this Act applies, any person receives a written application by the tenant for consent to a proposed transaction and that person—

(a) is a person who may consent to the transaction or (though not such a person) is the landlord, and
(b) believes that another person, other than a person who he believes has received the application or a copy of it, is a person who may consent to the transaction,

he owes a duty to the tenant (whether or not he owes him any duty under section 1 of this Act) to take such steps as are reasonable to secure the receipt within a reasonable time by the other person of a copy of the application.

(2) The reference in section 1(3) of this Act to the service of an application on a person who may consent to a proposed transaction includes a reference to the receipt by him of an application or a copy of an application (whether it is for his consent or that of another).

3 Qualified duty to approve consent by another.

(1) This section applies in any case where—

(a) a tenancy includes a covenant on the part of the tenant not without the approval of the landlord to consent to the sub-tenant—
(i) assigning,
(ii) underletting,
(iii) charging, or
(iv) parting with the possession of,
the premises comprised in the sub-tenancy or any part of the premises, but

(b) the covenant is subject to the qualification that the approval is not to be unreasonably withheld (whether or not it is also subject to any other qualification).

(2) Where there is served on the landlord a written application by the tenant for approval or a copy of a written application to the tenant by the sub-tenant for consent to a transaction to which the covenant relates the landlord owes a duty to the sub-tenant within a reasonable time—

(a) to give approval, except in a case where it is reasonable not to give approval,
(b) to serve on the tenant and the sub-tenant written notice of his decision whether or not to give approval specifying in addition—

(i) if approval is given subject to conditions, the conditions,
(ii) if approval is withheld, the reasons for withholding it.

(3) Giving approval subject to any condition that is not a reasonable condition does not satisfy the duty under subsection (2)(a) above.

(4) For the purposes of this section it is reasonable for the landlord not to give approval only in a case where, if he withheld approval and the tenant gave his consent, the tenant would be in breach of covenant.

(5) It is for a landlord who owed any duty under subsection (2) above—

(a) if he gave approval and the question arises whether he gave it within a reasonable time, to show that he did,
(b) if he gave approval subject to any condition and the question arises whether the condition was a reasonable condition, to show that it was,
(c) if he did not give approval and the question arises whether it was reasonable for him not to do so, to show that it was reasonable,

and, if the question arises whether he served notice under that subsection within a reasonable time, to show that he did.

5. Interpretation.

(1) In this Act—

"covenant" includes condition and agreement,
"consent" includes licence,
"landlord" includes any superior landlord from whom the tenant's immediate landlord directly or indirectly holds,
"tenancy", subject to subsection (3) below, means any lease or other tenancy (whether made before or after the coming into force of this Act) and includes—

(a) a sub-tenancy, and
(b) an agreement for a tenancy

and references in this Act to the landlord and to the tenant are to be interpreted accordingly, and

"tenant", where the tenancy is affected by a mortgage (within the meaning of the Law of Property Act 1925) and the mortgagee proposes to exercise his statutory or express power of sale, includes the mortgagee.

(2) An application or notice is to be treated as served for the purposes of this Act if—

(a) served in any manner provided in the tenancy, and

(b) in respect of any matter for which the tenancy makes no provision, served in any manner provided by section 23 of the Landlord and Tenant Act 1927.

(3) This Act does not apply to a secure tenancy (defined in section 79 of the Housing Act 1985)[or to an introductory tenancy (within the meaning of Chapter I of Part V of the Housing Act 1996)].

(4) This Act applies only to applications for consent or approval served after its coming into force.

Index